Sustainable Artificial Intelligence-Powered Applications

IEREK Interdisciplinary Series for Sustainable Development

The interdisciplinary series "Sustainable Artificial intelligence Powered Applications," (SAIPA) aims to cover a wide range of Artificial Intelligence applications with a specific focus on sustainability to further support the sustainable development goals. AI systems are able to perform a wide range of tasks, including high-level image and video recognition, perspective modeling, smart automation, advanced simulation, and complex analytics among many others, and has high learning and improvement potential through machine learning and deep learning. These capabilities have many uses and provide many benefits in all industries applications. SAIPA includes a diverse range of these applications, perspectives, and expertise, where it seeks to advance the understanding and implementation of sustainable AI-powered solutions across various scientific disciplines. Its aim is to become a comprehensive guide to different experimental sciences. The "Sustainable AI-Powered Applications" publishes conference proceedings, monographs, and textbooks that address the intersection of artificial intelligence, sustainability, and advanced technologies. Authors, researchers, contributors, and scientists from all scientific disciplines related to AI and advanced technologies, are all invited to contribute to this series to shape the future of AI in a sustainable and responsible manner. Thus, the series represents an excellent endeavor that promotes a collaborative and interdisciplinary strategy for the development of sustainable AI-driven applications. The scope of this series includes, but is not limited to: Promoting sustainability and human well- being, health and medical use, optimizing smart cities planning, transportation systems management, infrastructure development, creating livable urban environments, renewable energy, agricultural Innovations, precision agriculture, weather agriculture, sustainable food production, crop & animal monitoring, and agricultural robotics, Socio-Economic Impacts, optimizing energy consumption in buildings, factories in energy usage and lower carbon, smart grids applications, smart grids management, climate modelling, environmental monitoring, waste management, Internet of things (IOT) and other domains related to the use of AI applications that improve overall environmental impact.

Series Editor: Mourad Amer **editor@saipa-series.com**

Aniss Moumen · Brahim El Bhiri ·
Mohamed-Mustapha Zarrouk · Charaf Hajjaj ·
Mohammed Zouiten
Editors

Artificial Intelligence and Data Analytics for Innovative Applications in Engineering, Sustainability and Technology

Springer

Editors
Aniss Moumen
Department of Computer Science, Mathematics and Logistics
National School of Applied Science—Kenitra
Kenitra, Morocco

Mohamed-Mustapha Zarrouk
Euromed University of Fez
Fez, Morocco

Mohammed Zouiten
Laboratory Artificial Intelligence, Innovation and Computer Science
Departement of Geography
Faculty of Polydisciplinary of TAZA
Faculty of Sciences Dhar El Mehraz of Fez
Sidi Mohammed Ben Abdellah University
Taza, Morocco

Brahim El Bhiri
Harmony Technology
Rabat, Morocco

Charaf Hajjaj
Department of Physics
Faculty of Science El Jadida
El Jadida, Morocco

ISSN 3005-1762 ISSN 3005-1770 (electronic)
Sustainable Artificial Intelligence-Powered Applications
ISBN 978-3-031-90317-5 ISBN 978-3-031-90318-2 (eBook)
https://doi.org/10.1007/978-3-031-90318-2

This Springer imprint is published by the registered company Springer Nature Switzerland AG
The registered company address is: Gewerbestrasse 11, 6330 Cham, Switzerland

Contents

Understanding Public e-Service Adoption: A Conceptual Framework for Moroccan Urban Agencies

Imane Abouliatim and Karima Tounsi

Abstract

To understand and to clarify the motivations for the acceptability of e-services offered by Moroccan urban agencies. (MUA) and to identify the main reasons for the online adoption rate among citizens, the study will build an electronic government adoption model. By means of the technology adoption model (TAM) base, the central goal of this paper consists of creating a conceptual model of elucidating the many drivers that influence the government online adoption behavior of urban agency users, it is imperative to delve into the variable involved. Our research assesses the essential elements including perceived usefulness, perceived ease of use, social influence, trust, and user satisfaction which determine citizens' adoption behavior.

Keywords

Intention to use · E-government · TAM · Urban agency · User satisfaction · Social influence · Trust

1 Introduction

For the last twenty years, information technology (the Internet, cell phones, and all the other tools used to gather, store, analyze, and share information) has advanced, allowing for easier access to communication tools, an increase in information sources, new genres of entertainment, and a revolution in how citizens and businesses connect with their government, referred to "e-government."

I. Abouliatim (✉) · K. Tounsi
Management of Information Technology, Institut National des Postes et des Telecommunications (INPT), Rabat, Morocco
e-mail: imaneab@gmail.com

Chen et al. [1] describe the electronic government like the strategic approach of governmental entities and establishments in utilizing the capabilities of internet technology to improve their operational efficiency and productivity.

Electronic government has several advantageous economic and social effects. By using it, government operations can be carried out with greater transparency, corruption can be decreased, and citizens, businesses, and even the government itself can save time and money [2].

The correct use of electronic government can reduce bureaucracy and raise the accurateness, efficacy, and fluidity of operations inside the government agencies. It also contributes to facilitating access to public services and reduces the need to physically visit administrations, minimizing travel and helping to save costs effort, space, and time [3].

Moreover, giving citizens access to government e-services has many advantages, including enhancing access to health and education as well as advancing gender equality, and enabling citizen participation in governmental decision-making [4]. The access difficulties to these online services, which can be all sorts of things like financial, technical, cultural or cognitive, for example must be reduced for governments to realize the aforementioned benefits. Reading and understanding are necessary in addition to having a computer and an Internet to gain access to online services.

Al-Yafi et al. [5] argue that e-government in Arab and African nations, for instance, is not being used to its full potential despite the investments made by these governments. So, it's clear that despite efforts to advance e-government, Morocco still struggles to rank highly worldwide.

The country of Morocco has shown notable progress between the years 2008 and 2016, as seen by its performance on the United Nations' E-Government Development Index. During this period, Morocco ascended from the 140th position to the 85th position, indicating a significant advancement in its e-government capabilities. But it experienced significant

A. Moumen et al. (eds.), *Artificial Intelligence and Data Analytics for Innovative Applications in Engineering, Sustainability and Technology*, Sustainable Artificial Intelligence-Powered Applications, https://doi.org/10.1007/978-3-031-90318-2_1

regression in 2018 when it fell to 110th place, and in 2020, it rose to 106th place. However, Morocco advanced five spots in 2022, moving up to place 101.

The execution of electronic services is heavily reliant on technology, nonetheless, the accomplishment or absence of digital governance is not restricted exclusively to technological facets, as it is also affected by an assortment of other elements, including organizational, political, legislative, and economical factors on the part of the supply side of government. The same is true of cultural and social elements affecting the demand side's adoption and acceptance of electronic services [6].

Subsequently, the MUA are pioneering organizations in the e-government sector. They provide 11 e-services for the benefit of businesses and citizens, and by 2022 they have completely dematerialized their administrative processes, but people still go in person to obtain their services.

This paper identifies a set of factors from different domains, such as demographic, social, behavioral, and technological factors. These drivers could potentially impact the implementation of electronic services among MUA and need to be identified and analyzed to facilitate successful adoption in the future. If these administrations do not understand why citizens prefer traditional methods to e-service delivery channels, they cannot take the strategic steps necessary to achieve their goals for citizen adoption of these channels.

Thus, this study aims to elucidate the intricacies of the proposed conceptual model by meticulously analyzing its constituent elements. A comprehensive understanding of these variables is imperative for the effective formulation, execution, and implementation of electronic services by MUA in the imminent future.

2 Research Background

2.1 Presentation of the Urban Agencies

In 2022, Morocco's urbanization rate was close to 64.6%, and by 2050, it was projected to be 73.6% [7]. Constraints arise for the government when attempting to meet citizens' needs for housing, infrastructure, equipment, investment, and employment.

In response to the evolving landscape of urban areas, MUA were instituted in 1984 upon the establishment of the Urban Agency of Casablanca. Subsequently, beginning in 1993, these agencies were extensively integrated across the entirety of the Kingdom. This marked a significant step toward the optimization of urban development and management in Morocco.

There is now 30 MUA spread throughout the country:

- 3 Regional Urban Agencies: AU Laâyoune, AU Goulmime Oued Noun, and AU Dakhla Oued Eddahab.
- 6 Metropolitan Urban Agencies: AU Agadir, Tangier, Meknes, Rabat-Salé, Fez and Marrakech.
- 20 Urban Agencies of Intermediate Cities.
- And the specific case of the Urban Agency of Casablanca which is supervised by the Ministry of Interior. Casablanca is the largest industrial and financial metropolis in the country, with the highest number of factories and banks, and a population of over 4 million. This represents 12.6% of the total population of the kingdom.

The MUA falls under the category of a public institution, which is defined by its possession of legal personality in addition to financial autonomy. Its territorial jurisdiction is determined by decree of creation of each Urban Agency which corresponds to one or more prefectures and/or provinces.

2.2 Missions of the Urban Agencies

The MUA have various missions which have objectives to:

- Promote the development of territories through new territorial engineering dedicating the progressive evolution to project urbanism (territorial projects, city projects, urban projects, the urban renewal, etc.).
- Promote investment through the opening of new urbanization areas.
- Promote a mode of management that values the approach of the citizen/client.
- Support social housing programs.
- Integrate the environmental dimension and take into account the imperatives of sustainable development.

2.3 E-Services in Urban Agencies

As part of the follow-up to the implementation of preventive measures against the proliferation of the coronavirus pandemic (Covid-19), the MUAs have been encouraged to speed up the process of dematerializing the various town planning procedures by setting up platforms. This process has allowed generalizing the e-services for the benefit of citizens, investors, and territorial managers to a minimum of 90% in all 29 MMUA that are under the supervision of the Ministry [8]. These e-services are accessible through the MUAs' websites.

MUA support and provide a variety of e-services linked to their business. E-services, like the E-note of urbanistic information, E-instruction, and E-result, enable the simplification of procedures and the circuits of instruction of the requests for authorization and information to optimize the times of instruction and treatment and to avoid the administrative comings and goings of the files and the citizens.

The MUA also offers the possibility of access to the Geoportal, which is a homogeneous and cartographic interactive platform of all the approved urban planning documents within the territorial jurisdiction of each urban agency, accessible to the public via its website.

In terms of e-support services, public administrations provide support services to businesses and citizens, for example:

- E-requests which deal with citizens' complaints.
- E-regulation which allows the online consultation of the regulations and procedures to be followed in the field of town planning.
- E-payment which aims to facilitate exchanges between service providers and the Urban Agency.
- E-sale, which allows the applicant to obtain the documents produced by the Agency online.
- E-prestation which allows the online payment of the services rendered by the Urban Agency.
- E-appointment which allows contacting the Agency's managers by targeting the needs and directing the applicant; and the Digital Order Desk, which replaces the traditional order desk by allowing documents to be filed online.

3 Literature Review and Hypothesis Development

3.1 Adoption of e-Services

Acceptance technologies have attracted considerable interest from academics, researchers, and practitioners, especially since the establishment of IT systems in companies [9], and citizen acceptance of e-government services represents one of their dimensions, as a consumer's choice of an electronic service delivery channel over traditional channels can be seen as a technology adoption issue.

In addition, we find in the literature terms that are interchangeable with the acceptance of electronic services, such as "adoption" or "intention" or "willingness". Different explanations describe these terms [10]. However, Warkentin et al. [11] offer a broader description of online government adoption, characterizing it as "the intention to engage in e-government", including the intention to obtain information and to ask for administrative services online.

Exploring of the variables that impact human attitudes and convictions regarding the adoption of informational technology, such as electronic services, has become a subject of interest to scholars. Literature has been analyzed in depth and we have discerned a series of theoretical models that have been constructed to elucidate on the technological adoption process. Among them:

- Theory of Reasoned Action (TRA) [12];
- The Theory of Planned Behavior (TPB) [13];
- The Theory of Acceptance Model (TAM) [14].

However, TAM, first introduced by Davis from 1986, has been used extensively by scholars in their studies of user acceptance of technology [15] because it offers a simple model that can be suitable to any situation [16] and allows its extension with external variables that represent individual or organizational peculiarities. With more than 700 quotes from its original model [16], various exploration efforts have employed the TAM by integrating supplementary factors or alternative theoretical frameworks to analyze the acceptance of electronic public services.

Carter and Belanger [17] integrated the DOI framework with the TAM to construct a theoretical model that explores the drivers inducing the acceptability of electronic governance. In several prior studies, the TAM was integrated with the constructs of perceived confidence and risk in order to offer a comprehensive perspective on the adoption of e-government services. Thus, these undertakings have imparted a broader and more varied stance on the approval of digital government amenities [18].

The TAM has experienced considerable evolution to clarify the intricacies of human interaction concerning the acceptance or resistance of technology [15]. In his inspiring research, Davis [14] postulated a focal role for the drivers of perceived utility and perceived ease of use act a fundamental function in shaping the acceptability of technology by users.

Davis posits that computer users' attitudes, intentions, and perceived behavior are influenced by both their perceptions of usefulness (PU) and their perceptions of ease of use (PEOU). PU describes the extent to which a user feels that using a particular system could have a favorable or positive effect on his job.

Conversely, PEOU could be understood as an individual's estimation of the effort needed to utilize a particular technology (see Fig. 1).

Given the aforementioned circumstances, coupled with the absence of a universally applicable framework that accommodates diverse technologies, cultures, civilizations, and

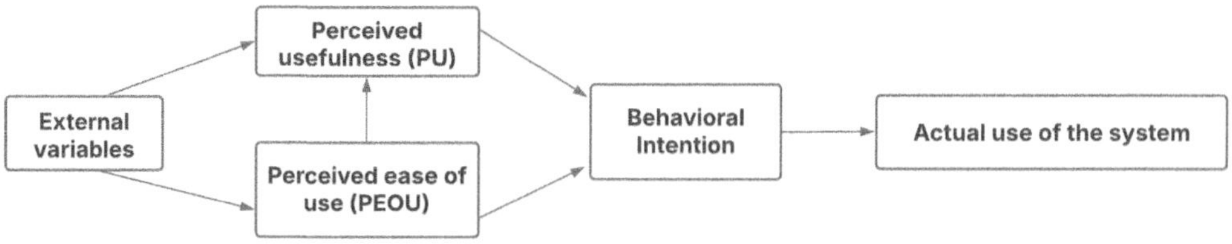

Fig. 1 Lasted version of the TAM (Venkatesh and Davis, 1996)

domains, particularly in less developed nations. The TAM has been deemed a robust and commanding framework for elucidating and presaging the espousal of technology, and our verdict has been bolstered in support of this paradigm [19] thus allowing it to embrace variables specific to our sociocultural context.

The current inquiry utilized three essential constructs of the TAM, including intention to use, perceived ease of use, and perceived usefulness, to illustrate the application of electronic services provided by MUA among their users. Thus, resulting in the following assumptions:

H3: It appears there may be a correlation worth exploring among the perceived ease of use and the perceived use of online services provided by the MUA.
H4: The relationship observed seems to be significant in terms of perceived ease of use and motivation to use MUA online services.
H5: It is interesting to note the confirmed correlation concerning perceived usefulness and the willingness to use MUA technological offerings.

3.2 Other Contributing Factors

The proposed model offered in our research contracts the TAM by introducing three additional concepts, namely user satisfaction, trust, and social influence as well as a range of important demographic variables.

Demographic Factors

Studies have shown that personal characteristics, including but not limited to gender, age, educational level, and internet experience, can affect technology acceptance [10, 20]. Older people tend to be more uncomfortable using modern technologies and may be unwilling to adopt them. Furthermore, women might be more likely not to use specific technologies, especially those traditionally associated with men. People who are more educated are more familiar with technology and adopt it more quickly. Comprehending these variables proves to be advantageous in discerning the elements that impact the conformity or opposition of the system.

The incorporation of individual differences as moderating variables has been employed to assess their potential influence on the connection between determinants and acceptance of electronic services by Urban Agencies. As a result, the following assumptions were made:

H1: The perceived ease of use of electronic services given by the MUA is significantly influenced by demographic variables.
H1a: There is a tendency for women to receive lower evaluations compared to males in terms of their judgment of the ease of use of online government services offered by Urban Agencies.
H1b: The older population tends to provide lower ratings to the user-friendliness of government online services presented by the MUA in comparison with the younger demographic.
H1c: Level of education is probably an encouraging influence on the perceived ease of use of online government services provided by the MUA**.**
H1d: The past experience of utilizing web-based platforms will have a good outcome on the perceived ease of use of MUA e-administration.
H2: Demographic factors exert a strong stimulus on the perceived usefulness of an MUA online service.
H2a: Women have a lower predisposition compared to men in evaluating the perceived use of online administrative services provided by Urban Agencies.
H2b: Older people will be less likely to evaluate the perceived utility of online administration services of Urban Agencies than the young generation.
H2c: Educational level will influence the perceived utility of online administrative services provided by the Urban Agencies.
H2d: Previous experience of using the Internet is likely to develop a more optimistic outcome on the perceived usefulness of online administrative services provided by Urban Agencies.

Social Influence

Implementing eGovernment services in underdeveloped or Arab countries tends to be different from doing so in more developed countries, owing to the influence of cultural norms, historical traditions, religious beliefs, and societal values on the process and also by the social influence that comes from the discipline of psychology the concept denotes an individual's comprehension that the majority of individuals who hold significance in their life hold a certain stance on whether they should or should not embrace a particular conduct [12].

Given the significance of examining the impact of the social environment on the usability of ICT services, the current inquiry examined the consequences of social aspects on the implementation of electronic services provided by Urban Agencies. It is improbable that individuals would encounter a feeling of compulsion to comply with the anticipations of others who consider the implementation of a novel system as vital [21]. The impact might be imposed by the individual's personal or professional milieu, encompassing factors such as family, friends, colleagues, supervisors, and so on.

Therefore, we have made the following assumptions:

H6: The concept of social influence was shown to have a favorable effect on the perceived usefulness of e-services offered by MUA.
H7: The phenomenon of social influence exerts a beneficial impact on individuals' desire to utilize electronic services provided by MUA.

User Satisfaction

A recent approach to the deployment of e-services in public bodies is introduced, where the measurement of quality is determined by the satisfaction of citizens. This approach could be considered innovative.

In contemporary public administration, it has become increasingly indispensable to possess a comprehensive requirement of possible citizens and their cognitions. This is of utmost importance in furnishing the most exceptional services to the citizens. Therefore, scrutinizing the quality of virtual services can be an indispensable element in elucidating the acceptance of e-services. This investigation could aid in determining whether users would persist in availing of the digital system or not. In the online administration environment that may differ from the conventional environment, users can acquire new practices by using e-services and may not be satisfied with the services offered unlike conventional ones [22] and therefore do not use them, so customer-focused e-services and quality of service are sought after.

Assessing service excellence is essential in determining customer satisfaction with the service experience. Assessing the level of customer satisfaction derived from the service experience holds immense significance in this evaluation, and the realm of e-government, it is a rate to which an

Table 1 Servqual measures

Servqual measurement	Definition
Reactivity	The perception of the service provider's attentiveness and helpfulness as perceived by the client
Reliability	The capacity to deliver the committed service reliably and precisely
Empathy	The company's care and attention to its customers

Source Table adapted from (Parasuraman et al. 1988)

online administration portal or website allows effective access to online public services to help citizens and complete their transactions with the government, to evaluate customer satisfaction with service quality.

Parasuraman et al. [23] established the SERVQUAL scale which is composed of five dimensions that include real elements, dependability, promptness, self-confidence, and understanding, which collectively provide a comprehensive framework for measuring service quality; this measure has been extensively tested by researchers, but only three of them apply to e-service quality [24] (see Table 1).

Consequently, the present research endeavors to put forward the subsequent suppositions:

H8: The perceived usefulness of using an electronic service from the Urban Agency has an encouraging result on user satisfaction.
H9: Ease of use when using an Urban Agency electronic service has an encouraging result on user satisfaction.
H10: Enhancing the perceived quality of electronic services offered by Urban Agencies is expected to positively impact consumer satisfaction.
H10.1: There is a clear and direct correlation responsiveness and user satisfaction in respect to the e-services offered by the Urban Agency.
H10.2: The dependability of the Urban Agency's e-services is positively correlated with user satisfaction.
H10.3: A strong and positive relationship is present in empathy and user satisfaction with the e-services delivered by the MUA.
H11: The uptake of MUA e-services is positively influenced by user happiness.

Trust

Most of the studied literature underscores the importance of trust to be a basic driver toward the adoption of electronic services. These studies underscore the essential role that trust plays in the choice-executive process of individuals [6, 25].

Carter and Bélanger [17] has been asserted that a significant obstacle to the broad acceptance of electronic services is the dearth of confidence and trust, specifically concerning privacy and security of personal records and financial data.

Citizens are worried about exposing their sensitive personal data to the authorities via the vastness of the web, as they are wary of the potential abuse of their data and the infringement of their privacy.

Bélanger and Carter [26] propose two forms of trust in the field of online government. First, trust in government is linked to the perceived ability of an institution to provide reliable online services that prioritize privacy and security.

Second, trust in the net is correlated with citizens' awareness of the various elements within the organizational framework, such as organization, regulation, and legislation that make for a sense in security and safety [27].

Thus, the following assumptions are stated:

H12: The presence of trust is expected to create a favorable impression on individuals' intention to utilize electronic services provided by MUA.

H13: The level of trust has a beneficial influence on the perceived utility of electronic services provided by MUA.

4 The Conceptual Model

The factors previously mentioned, and the hypotheses put forward allowed us to build our conceptual framework (see Fig. 2), which considers the concepts debated and incorporates different concepts from different subjects like sociology, information structures, and public management.

5 Conclusion

The emerging field of e-government has resulted in substantial transformations in the dynamics of interaction between governments and citizens, as many nations strive to enhance citizen use of public electronic service. In this way, today we are witnessing the commitment of administrations and authorities to understand and meet the needs of their beneficiaries to satisfy them, as the citizen has moved from the position of a citizen to a customer. Governments want to increase the delivery of public services, foster progressive engagement with business, provide simplified accessibility to information, and optimize government management through the use of efficient public administration practices [22].

This academic article presents a research model that can clarify the various determinants that may be influencing the uptake of electronic services among Urban Agencies in Morocco. The proposed inquiry model is securely established in the extensively recognized and empirically authenticated TAM. Using this hypothetical foundation, the writers intend to present a thorough knowledge of the basic aspects that might either hasten or impede the incorporation of e-services in this framework, thus making a contribution to the present literature in the realm of information systems. The factors borrowed are intention to use, perceived usefulness, and perceived ease of use. Other drivers considered useful were trust, user satisfaction, and social influence. In addition, demographical drivers were incorporated as a set of moderators in the model. Also, empirical work is currently underway through a questionnaire survey targeting users of the Urban Agencies' e-services to test the research model produced through econometric modeling.

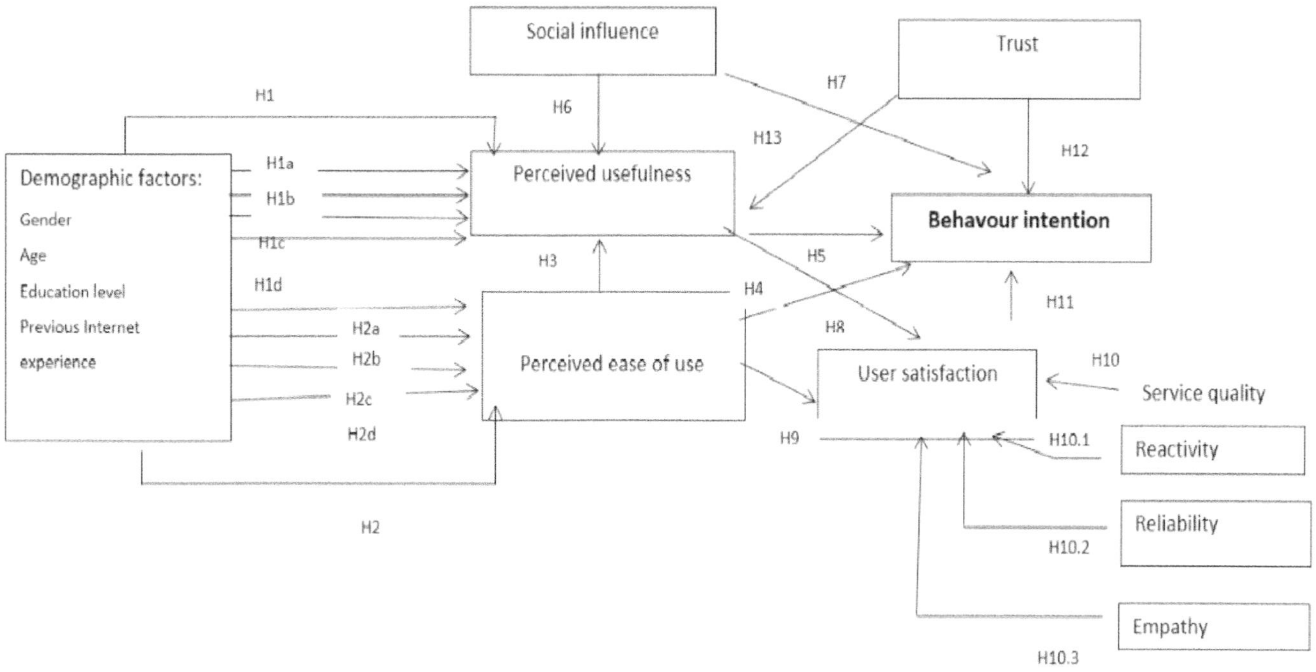

Fig. 2 Research model (self-made)

References

1. Chen Y, Chen HM, Ching RK, Huang WW (2007) Electronic government implementation: a comparison between developed and developing countries. Int J Electron GovMent Res (IJEGR) 3(2):45–61
2. Asogwa BE (2013) Electronic government as a paradigm shift for efficient public services: opportunities and challenges for Nigerian government. Library Hi Tech 31(1):141–159. https://doi.org/10.1108/07378831311303985
3. World bank, "e-Government," World Bank. Accessed 09 Feb 2023. [Online]. Available: https://www.worldbank.org/en/topic/digitaldevelopment/brief/e-government
4. Mensah IK, Jianing M, Durrani DK (2017) Factors influencing citizens' intention to use e-government services: a case study of south Korean students in China. Int J Electron Govt Res (IJEGR) 13(1):14–32
5. Al-Yafi K, Hindi N, Osman I (2014) Exploring user satisfaction of the public e-services in the State of Qatar: case of traffic violations e-service provided by the ministry of interior
6. AlHujran O, Aloudat A, Altarawneh I (2013) Factors influencing citizen adoption of e-government in developing countries: the case of Jordan. Int J Technol Human Interact 9(2):1–19. https://doi.org/10.4018/jthi.2013040101
7. HCP, "Taux d'urbanisation (en %) par année : 1960–2050.," Taux d'urbanisation (en %) par année : 1960 – 2050. Accessed 30 Mar 2021. [Online]. Available: https://www.hcp.ma/Taux-d-urbanisation-en-par-annee-1960-2050_a682.html
8. Fédération des Agences Urbaines au Maroc, "Les Agences Urbaines marocaines, d'une mission de régulation à un organe de développement du territoire," Rabat, May 2022
9. Rogers EM (1995) the innovation decision process. In: Diffusion of innovations. The free Press, New York, NY, USA
10. Colesca SE, Dobrica L (2008) Adoption and use of E-Government services: the case of Romania. JART 6(03). https://doi.org/10.22201/icat.16656423.2008.6.03.526
11. Warkentin M, Gefen D, Pavlou PA, Rose GM (2002) Encouraging citizen adoption of e-government by building trust. Electron Mark 12(3):157–162
12. Ajzen I, Fishbein M (1975) A Bayesian analysis of attribution processes. Psychological Bulletin 82(2):261
13. Ajzen I (1991) The theory of planned behavior. Organ Behav Hum Decis Process 50(2):179–211. https://doi.org/10.1016/0749-5978(91)90020-T
14. Davis FD (1985) A technology acceptance model for empirically testing new end-user information systems: theory and results. Ph.D. Thesis, Massachusetts Institute of Technology
15. Marangunić N, Granić A (2015) Technology acceptance model: a literature review from 1986 to 2013. Univ Access Inf Soc 14(1):81–95. https://doi.org/10.1007/s10209-014-0348-1
16. Chuttur M (2009) Overview of the technology acceptance model: origins, developments and future directions
17. Carter L, Bélanger F (2005) The utilization of e-government services: citizen trust, innovation and acceptance factors. Inf Syst J 15(1):5–25. https://doi.org/10.1111/j.1365-2575.2005.00183.x
18. Al-Adawi Z, Yousafzai S, Pallister J (2005) Conceptual model of citizen adoption of e-government. In: The second international conference on innovations in information technology, IT Innovation Dubai, pp 1–10
19. Davis FD, Venkatesh V (1996) A critical assessment of potential measurement biases in the technology acceptance model: three experiments. Int J Hum Comput Stud 45(1):19–45
20. Venkatesh V, Morris MG (2000) Why don't men ever stop to ask for directions? Gender, social influence, and their role in technology acceptance and usage behaviour. MIS Quarterly 115–139

21. Venkatesh V, Morris MG, Davis GB, Davis FD (2003) User acceptance of information technology: toward a unified view. MIS Q 27(3):425–478. https://doi.org/10.2307/30036540
22. Hussein R, Mohamed N, Ahlan AR, Mahmud M (2011) E-government application: an integrated model on G2C adoption of online tax. Transform Govt People Process Policy 5:225–248. https://doi.org/10.1108/17506161111155388
23. Parasuraman A, Zeithaml VA, Malhotra A (2005) E-S-QUAL: a multiple-item scale for assessing electronic service quality. J Serv Res 7(3):213–233. https://doi.org/10.1177/1094670504271156
24. Alshibly HH (2015) Exploring the key factors for establishing electronic commerce loyalty 22(3):25
25. Rana NP, Williams MD, Dwivedi YK, Williams J (2011) Diversity and Diffusion of theories, models, and theoretical constructs in eGovernment research. In: Janssen M, Scholl HJ, Wimmer MA, Tan Y (eds) Electronic government. Lecture notes in computer science, vol. 6846. Springer, Berlin, Heidelberg, pp 1–12. https://doi.org/10.1007/978-3-642-22878-0_1
26. Bélanger F, Carter L (2008) Trust and risk in e-government adoption. J Strateg Inf Syst 17(2):165–176. https://doi.org/10.1016/j.jsis.2007.12.002
27. Fakhoury R, Aubert B (2015) Citizenship, trust, and behavioural intentions to use public e-services: the case of Lebanon. Int J Inf Manage 35(3):346–351

A Global Review and Empirical Study on Artificial Intelligence in Industry 4.0 with Insights from Moroccan Students

Anass Ben Abdelouahab, Adham Chaibi, Samia Haman, Zineb Bentassil, and Aniss Moumen

Abstract

This study examines the role of artificial intelligence in Industry 4.0 by merging a review of global literature with in-depth interviews conducted with Moroccan engineering students from a state school in Kenitra and Tangier. The literature review points out key trends and challenges from improvements in production efficiency, predictive maintenance, and quality control to difficulties with technical integration, skills shortages, and data security. In parallel the interviews provide insights into the theoretical opinions of students regarding the ongoing industrial changes and reveal a gap between global trends and local educational practices. Finally, the findings suggest that while AI has the potential to significantly enhance industrial operations, many obstacles remain before its full benefits can be realized. Addressing these challenges and research gaps is crucial for ensuring that AI can ensure a sustainable growth and innovation across industries.

Keywords

Intelligence · Industry 4.0 · Manufacturing · Artificial · Machine learning · Supply chain · Deep learning · Predictive maintenance · Quality · Students

A. Ben Abdelouahab (✉) · S. Haman · Z. Bentassil · A. Moumen
Engineering Science Laboratory, National School of Applied Sciences Ibn Tofail University, Kenitra, Morocco
e-mail: Anass.benabdelouahab@uit.ac.ma

A. Chaibi
Advanced Systems Engineering Laboratory, National School of Applied Sciences Ibn Tofail University, Kenitra, Morocco

S. Haman
Smartilab Laboratory, Emsi, Rabat, Morocco

1 Introduction

Artificial intelligence and Industry 4.0 are changing how operations are managed. Integrating AI into the industrial sector offers many possibilities for efficiency and improvement, but it also presents several challenges that need to be addressed. To gain a practical perspective on these issues, we conducted interviews with five students, focusing on their point of views regarding AI and Industry 4.0. Their answers covered topics ranging from definitions to the expected benefits of AI applications. This study summarizes the main points from those interviews, highlighting key application areas, challenges, commonly used methods, and potential gains. It also checks if the students' responses match global trends, providing insights into how Moroccan students perceive these developments. Focusing on these perspectives, this work helps us understand the ongoing transformation of the industrial sector as AI continues to advance.

2 AI and Industry 4.0

2.1 Global Adoption Trends

In literature, we find that the definition of Industry 4.0 is the fourth revolution following automation, marked by the introduction of new technologies such as the Internet of Things with AI, digital twins, augmented and virtual reality, and Big Data [1–3]. This work outlines two distinct phases of Industry 4.0 and highlights emerging research directions, with a particular focus on practical manufacturing applications. It examines the latest trends in Intelligent Autonomous Systems and emphasizes key AI developments along with their benefits, challenges, and uses within Industry 4.0. At the same time, it clarifies the fundamental characteristics of

A. Moumen et al. (eds.), *Artificial Intelligence and Data Analytics for Innovative Applications in Engineering, Sustainability and Technology*, Sustainable Artificial Intelligence-Powered Applications, https://doi.org/10.1007/978-3-031-90318-2_2

Industry 4.0, placing it within the broader context of the fourth industrial revolution while addressing security risks posed by modern digital technologies. An analysis of digitalization experiences in China and the United States reveals how these leading economies have leveraged technological innovation to maintain their competitive edge. Moreover, AI especially machine learning and deep learning is being applied across various regions to improve production efficiency, quality control, and predictive maintenance. In the area of industry 4.0 and sustainability, some companies prioritize operational efficiency, which can indirectly benefit the environment, while others explicitly embrace eco-friendly innovations. In general, companies in the eastern regions tend to prioritize efficiency, Western firms tend toward environmental objectives, and companies in Africa and South America are still in the early stages of integrating Industry 4.0 with sustainable practices. Despite these trends, Industry 4.0 has yet to fully evolve into a catalyst for sustainability-based business models on a global scale.

2.2 Challenges Identified

Several works highlight significant obstacles that must be addressed for AI to fully realize its potential in Industry 4.0. Also, we found that deep learning DL holds promise for Industrial Internet of Things IIoT applications, it also presents hurdles such as algorithmic complexity and data handling requirements that need further exploration [4]. Decision-making approaches, spanning strategic to real-time levels, can be hindered by obstacles related to rapid information processing and organizational readiness [5]. Additionally, during our research multiple studies underscore a range of impediments to AI adoption, including IT infrastructure limitations, shortages of qualified personnel, data quality issues, ambiguous business cases, and regulatory constraints [6]. These issues often leave industries at modest Technology Readiness Levels TRL 3–4 for AI infrastructure and 3–6 for software platforms, necessitating tailored recommendations to advance higher TRLs.

Although industries share similar difficulties, the solutions are frequently sector-specific, complicating efforts to transfer best practices across different domains [7]. Indeed, industrial AI has proven advantageous, but various technical and operational barriers persist [8]. The same holds true at a broader level, where the widespread integration of AI faces notable challenges and open questions [9]. In practice, machine learning methods dominate AI initiatives in Industry 4.0 used in 41% of the cases studied and Python emerges as the primary tool for simulations; however, these approaches still encounter gaps in product quality control and other operational aspects [10]. Vision systems, for instance, show promise when combined with machine learning, yet integrating such solutions into mechanical engineering contexts introduces unique technical and methodological hurdles that demand further investigation [11].

2.3 Research Gaps and Conclusion

In summary, we found that the literature reveals several clear research gaps in Industry 4.0. There remains a need to better tackle issues in predictive quality and to standardize effective solutions [12]. Although examples like the LEGO factory using explainable AI show what is possible, further work is needed to integrate these methods into everyday industrial practice [13]. Predictive maintenance, while critical, still requires refinement [14], and despite a broad understanding of Industry 4.0, practical implementation challenges persist [15]. In the same way we found that for the decision-making frameworks from strategic planning to real-time actions face significant obstacles also, and while innovative approaches such as combining AI with blockchain are promising, they must be validated across sectors [16, 17]. Many comparative digitalization studies indicate that countries like China and the United States are leading, highlighting the need for more global comparisons, example of the study [3]. Moreover, even though hybrid models like CNN–SVM have achieved high accuracy in specific tasks, scaling these solutions remains an open question [18]. Finally, we noticed that the research into the link between Industry 4.0 and sustainability shows two distinct paths with regional differences, yet a unified, sustainability-driven approach is still missing [19].

In conclusion, we found that the global adoption of AI in Industry 4.0 presents many opportunities to improve the productivity and operational excellence. And also, there are many significant challenges and research gaps must be addressed. Based on many studies cross different sector, promoting human–machine collaboration, and ensuring robust data governance, future research can open the way for more effective, sustainable, and secure industrial AI applications. This synthesis serves not only as a reflection of current trends and challenges but also as a roadmap for future inquiry in the field.

In addition to our research which was conducted to capture these global trends, challenges, and research gaps a series of interviews were carried out with Moroccan students to gain local insights and practical perspectives.

3 Methodology

Qualitative studies include all studies using qualitative methods for the collection and description of qualitative data. In this article, we perform a qualitative analysis to analyze the interviews conducted with 5 persons interviewed in the field of industry.

To do our interviews we determined the sample of our interview by specifying its characteristics and its area of specialty then we determine the type and characteristics of the interview and the analysis tools used. And we follow steps:

3.1 Data Collection

The interview, as its name suggests, involves an interviewee, a facilitator, and a discussion. The reasons that motivate the choice of device, the selection of partners, and the nature of the interactions will have an impact on the type of data to be collected and the way to analyze it. Let's take a closer look at those implications.

3.1.1 Preliminary Questions

The accumulation of methodological details observed in the works is induced by the addition, over the years, of motifs, nuances, objectives, diversified uses that the use of the interview has generated. One of the first responses is to determine why interviews must be conducted. This response will affect further data collection.

3.1.2 Characteristic of Sample

The number of people is 5 who have a medium and high knowledge in the field. These are 3 engineering students in industrial engineering, 1 industrial engineer with 5 years of experience in the automotive field, 1 student in the 2nd year of the master's degree in "Intelligent Industry and Digital Technology". As for the number of groups, no details are given. Economic proximity, available funds, availability of people, political function, role, impact or cultural environments, proximity to actors, traditions imperatives can mark out the constitution of the sample, which can be described as intentional.

3.1.3 Characteristic of Interview

This is an interview with 5 engineering students from a state school in Kenitra and Tangier in Morocco. We asked questions to extract information using the guide mentioned above. His interviews were individual and direct with a duration of 20 min.

3.1.4 Role of the Animator and Consideration of His Interventions

The role devolved to the animator is the subject of many details in the manuals and it is so heavy and contrasted that few would dare to venture into it without a long preliminary training.

Thus, the role of the facilitator is on two intertwined levels: maintaining communication and the socio-affective climate of the discussion and focusing on the cognitive tasks that the structuring of group thinking calls upon.

3.2 Data Processing

This second part deals with the preparation of the corpus of data and their analysis. Qualitative research involves a wide range of types of data analysis, regardless of the apparatus used to collect them.

3.2.1 Preparation Phase

In a qualitative approach to research, data preparation can involve several major steps, such as.

3.2.2 Re-Transcription

Audio and video recordings often need to be transcribed for easier analysis. To do this, you have to listen again and then write or use voice dictation tools.

3.2.3 Choose the Unit of Analysis

The unit of analysis is the element you choose for the analysis of your data in a qualitative approach. It can be words, sentences, paragraphs, sections of a text, or full transcripts of interviews. The choice of unit of analysis depends on study and the objectives of the research.

3.2.4 Prepare Tools for Coding

It is very useful to review the frames of reference, the research questions, the objectives, the hypotheses, which often emerged during the collection.

3.2.5 Analysis Phase

Coding, categorizing, describing, and then modeling or theorizing are actions that all researchers, regardless of the type of data they consider, must do to understand the phenomenon they are investigating. We will specify, in passing, the elements to be considered with regard specifically to group interviews while giving a brief overview of the operation itself.

3.2.6 Coding

Coding involves assigning codes or categories to transcribed data. It is a classification process that helps identify recurring themes and patterns in data.

3.2.7 Data Organization

The coded data should be organized in such a way that it can be easily consulted and analyzed.

3.2.8 Verification and Validation of Data

The results are checked and validated to ensure their reliability and validity. This may involve checking the transcript, clarifying ambiguous codes, and validating conclusions drawn from the data.

4 Results

During the analysis of the interviewees content, it was found that an employee has a different and more in-depth perspective on the subject in question compared to a student. This is due to their professional experience, which is more extensive and expertise in their field, making their answers more precise and focused on the practical applications of the subject. For the students with less professional experience, but a more theoretical knowledge of the subject, learned through their studies. This made their answers more focused on concepts and theories, without necessarily focusing on practical applications (Table 1).

Our results, show us that the most of the respondents answered that maintenance and manufacturing is the field most concerned with AI compared to other fields. To illustrate the results of this survey, a chart is presented that visualizes the fields mentioned by the interviewees.

To go more in depth, we try to identify the key for artificial intelligence techniques, we conducted a new analysis. As a result, we found that most of the interviewees mentioned machine learning, deep learning, and computer vision. These areas are considered essential for AI development and have continued to progress in recent years. Machine learning enables computers to learn and improve without explicit programming, while computer vision allows AI to interpret and interact with the visual world. This analysis highlights the significance of these technologies in AI development and shows us that their advancements will continue to have a major impact on both industry and society (Figs. 1 and 2).

During our interviews and when we asked each student about what they considered to be the key skills for successful AI application, each of the interviewees gave a different answer. Some people focused on the need for a solid understanding of statistics and machine learning, while others highlighted the importance of data understanding and the ability to manage it. Others still emphasized the need for a solid understanding of algorithms and the ability to implement them effectively. It is clear that to succeed in AI application, it is important to have a combination of technical skills and domain understanding (Fig. 3).

Our corpus was analyzed with NVIVO in order to measure the quality of results and choose the most frequently occurring words in each article. The results of our query revealed that the most important words were "Industry," "Manufacturing," "Data," and "learning," which means that the answers are within the context of the topic and the interviewees have more information about the subject. On the other hand, textual research is an approach

Table 1 List of interviews

Student	Gender	Diploma	Age
Student-1	M	Master's degree	23
Student-2	M	Industrial engineering	23
Student-3	F	Master's degree	22
Student-4	F	Master's degree	22
Student-5	M	Industrial Engineering	28

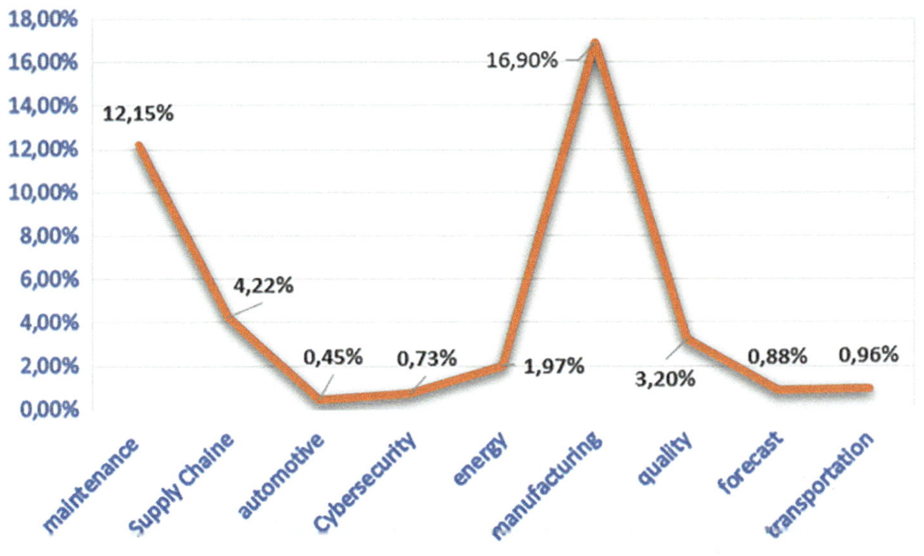

Fig. 1 Students' answers: AI application area

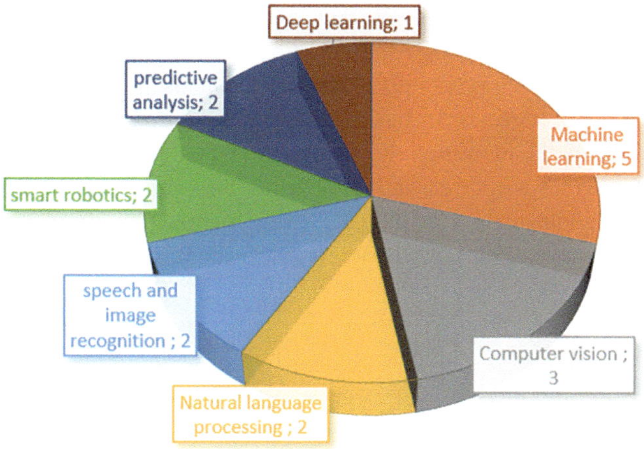

Fig. 2 Techniques of AI

Fig. 3 Word cloud

to research that focuses on the analysis of texts to understand the meanings and perspectives of authors. Our results in Nvivo based on this query can be used to generate conclusions and recommendations in order to make an important contribution to understanding the authors' perspectives and meanings (Table 2).

Because of their similar academic backgrounds, students in industrial engineering and M2 programs have strong similarities (above 0.8) according to the correlation analysis, while students at different educational stages have moderate correlations (0.7–0.8), which reflect both important differences and some commonalities. Finally, other students without experience and (student-5) with job experience have lower correlations (below 0.7), which emphasizes how viewpoints change once one enters the workforce. We can conclude that experience and educational attainment have a big impact on answers, with students expressing opinions that are more in line with those of seasoned professionals.

Table 2 Pearson correlation table

Interview A	Interview B	Pearson correlation coefficient	Observation
Student-2 (IE)	Student-1 (Master)	0.873895	Strong correlation
Student-3 (Master's)	Student-1 (Master)	0.869004	Strong correlation
Student-3 (Master's)	Student-4 (Master's)	0.848446	Strong correlation
Student-4 (Master's)	Student-1 (Master)	0.789884	Moderate correlations
Student-3 (Master's)	Student-2 (IE)	0.787734	Moderate correlations
Student-4 (Master's)	Student-2 (IE)	0.729912	Moderate correlations
Student-4 (Master's)	Student-5 (IE with Experience)	0.630309	Lower correlations
Student-3 (Master's)	Student-5 (IE with Experience)	0.626461	Lower correlations
Student-5 (IE with Experience)	Student-1 (Master)	0.620685	Lower correlations
Student-5 (IE with Experience)	Student-2 (IE)	0.604278	Lower correlations

5 Discussion

A discussion was conducted with the individuals who were interviewed to gain an understanding of the challenges and limitations of AI applications in Industry 4.0. The responses varied and provided a comprehensive outlook on the subject, highlighting the different perspectives and opinions on the potential and limitations of AI in the fourth industrial revolution. These insights will be useful in determining the future direction of AI in the industry and ensuring that its integration is both effective and efficient.

Here we can summarize the challenges and limitations of AI for Industry 4.0, based on the answers of the interviewees.

(Student1) highlighted the complex and costly integration of AI into existing industrial systems as a challenge, while limitations include the reliability and security of AI systems and the understanding of complex contexts and management of uncertainties. (Student2) listed improved efficiency and productivity, informed decision-making, and customization of products as challenges, while limitations include the impact on employment, security of systems, quality of data, regulation, and ethical issues. (Student3) emphasized the importance of high reliability of input data and clarity of processes and transparency. (Student4) mentioned predictive maintenance and optimization of production processes as challenges, while limitations include impact on employment, ethical issues, and the need to manage limitations to maximize benefits and minimize risks. (Student5) also discussed improved efficiency and productivity, informed decision-making, and customization of products as challenges, but highlighted limitations such as the quality of data used, regulation, impact on employment, and ethical issues.

During the interviews, students emphasized that the expected gain from the implementation of AI in Industry 4.0 is difficult to predict with precision. According to studies and research, AI offer many benefits for Industry 4.0, such as improving efficiency and productivity, reducing costs, increasing product quality, and better decision-making. The students also indicated that the gain would depend on many factors such as the industrial sector, the type of process targeted, the quality of available data, and the internal capabilities of the company. Specific use cases involved companies will also play an important role in determining the expected gain. Although the gain is difficult to quantify, it is possible to give a general idea by examining concrete examples, such as real-time data processing, which can bring significant time savings for data synthesis and sharing.

6 Conclusion

Based on our literature reviews and the interviews, we can conclude that it's evident that AI will continue to transform Industry 4.0 in significant and high ways. Our research highlights both the promising potential of AI through improvements in predictive maintenance, production efficiency, and decision-making and the critical challenges that must be addressed, including technological integration, workforce training, and data security. In parallel, during the interviews most of the answers from the interviewed participants offer valuable, real-world insights into how AI can bring tangible benefits to manufacturing and beyond, by driving innovation, optimizing processes, and enhancing overall operational effectiveness. Together, these findings illustrate a future where AI not only supports industrial advancements but also creates new opportunities for businesses and consumers. Ultimately, bridging the identified research gaps and overcoming current challenges will be essential for fully harnessing AI's capabilities, so ensuring that Industry 4.0 delivers on its promise of sustainable, efficient, and innovative growth.

Appendix

Interview guide
Duration: 20 Min
Student Diploma
Platform: Google Meet

Interview Structure

1st—Introduction and Presentation of the topic.
2nd—Explain how the interview will be recorded and analyzed and used in the study.
3rd—Question/Answer.

Main Interview Questions

- What do you know about AI?
- What is Industry 4.0?
- What is the impact of AI on Industry 4.0?
- Why should I implement AI solutions in Industry 4.0?
- What future in the face of the AI invasion?
- What are the AI techniques in Industry 4.0?
- What are the areas most affected by AI in Industry 4.0?
- What are the key skills for successfully applying AI in Industry 4.0?
- What are the challenges and application limits of AI for Industry 4.0?
- How much do you estimate the gain expected by the implementation of AI in Industry 4.0? (Table 3).

Table 3 Interview verbatim

	AI	Industry 4.0	The impact of AI on Industry 4.0	AI solutions in Industry 4.0	AI invasion
Student1	Field of computer science that aims to create computer programs that can perform tasks that normally require human intelligence	Term that describes the use of advanced technologies, such as automated manufacturing systems and automated control systems, to improve manufacturing processes	Enabling companies to automate, optimize manufacturing processes, demand forecasting, predictive maintenance, and quality management	Improved efficiency, informed decision making, customization of products, Predictive maintenance, process optimization	Numerous applications in different fields, but there are also concerns about its impact on employment and security
Student2	AI is a field of engineering focused on the development of intelligent machines that can perform tasks that typically require human intelligence, such as visual perception, speech recognition and translation between languages	Industry 4.0 refers to the fourth industrial revolution, which emphasizes the integration of advanced technologies, including artificial intelligence and the Internet of Things into manufacturing and other industries	AI has a significant impact on Industry 4.0 by enabling new levels of automation, efficiency, and decision-making	Implementing AI solutions in Industry 4.0 can lead to increased productivity, improved quality, and reduced costs	The future of AI is uncertain, but it is expected to continue to play a major role in various industries, including Industry 4.0
Student3	AI is a set of techniques that allow a machine to perform tasks that normally require human intelligence	It is the application of advanced technologies such as AI, Internet of Things, and 3D printers to improve the performance, quality, and flexibility	Data analysis, machine learning and the security of industrial systems	Make factories more efficient, more flexible, and smarter	It can offer many advantages for education, transport, and production
Student4	Is a branch of computer science that allows systems to learn and perform tasks normally associated with human intelligence	It is the use of advanced technologies to improve manufacturing processes and supply chains in industry	Automate repetitive tasks, improve quality, and increase the efficiency of their production processes	Machine monitoring, fault detection, analyzing production data and real-time decision-making	It can improve health care, road safety, communications, and businesses
Student5	It is a computer mathematical discipline that knows how to make the decision without human intervention	It is the 4th evolution of the industry which aims to use interconnected systems, and which allows us to communicate in real time	Revolutionizes and facilitates the way of data processing	To minimize time, mudas, maximize profit, reduce headcount	It's already been experienced, and the change has already started in the world ex: open Ai
	AI techniques	Application field	Key skills for successful application	Challenges and application limits	Expected gain
Student1	Machine learning computer vision Automatic natural language processing	Manufacturing processes, transportation, logistics and supply chains	Competencies in mathematical and statistical skills, programming skills and computer knowledge	Challenges: predictive maintenance and optimization of production processes. Limits: its impact on employment and ethical considerations, minimizing risks	Improved efficiency, productivity, lower costs, better product quality and better decision-making

(continued)

Table 3 (continued)

	AI techniques	Application field	Key skills for successful application	Challenges and application limits	Expected gain
Student2	Deep learning, computer vision, and natural language processing	Manufacturing, logistics and healthcare	Data science, machine learning, and programming	Challenges and limitations of implementing AI in Industry 4.0 include ethical concerns, data privacy, and the potential for job loss due to automation	The expected gain from implementing AI in Industry 4.0 is difficult to estimate, but it is likely to vary depending on the industry and the specific application
Student3	Robotics, machine learning, data analysis	Maintenance, manufacturing, energy, finance	Data competence Knowledge of the fields of application understand AI	Data quality limits AI complexity, requires employee training	It is difficult to give a precise estimate of the gain
Student4	Machine learning, speech and image recognition, intelligent robotics, predictive analytics	Predictive maintenance automated production planning optimization of supply chains cybersecurity	Understanding of AI and machine learning techniques ability to collect efficiently and reliably. Understanding of industrial processes and technologies	Challenges: risks related to the reliability and security of AI systems. limits: understand complex contexts, unforeseen situations, and manage uncertainties	It is difficult to give a precise figure to estimate the expected gain: because it will depend on many factors
Student5	Machine learning, deep learning, nature language processing NLP, treatment images	Market analysis, customer satisfaction, trend, forecast	Having an idea about AI and its prerequisites at the IT level	Challenges: reliability of data, clarity of process and transparency Limits: intervention of the human being for the control	The data processing will be in real time and the gain will be enormous in time for the synthesis

References

1. Zhang T, Li Q, Zhang CS, Liang HW, Li P, Wang TM, Li S, Zhu YL, Wu C (2017) Current trends in the development of intelligent unmanned autonomous systems
2. Banitaan S, Al-refai G, Almatarneh S, Alquran H (2023) A Review on artificial intelligence in the context of Industry 4.0. Int J Adv Comput Sci Appl 14(2):23–30
3. Kurkin AV, Giraev AV (2023) Prospects for the development of the digital economy based on advanced technologies of Industry 4.0: robots, big data and artificial intelligence. In: Advances in science, technology and innovation, vol Part F1, pp 385–389
4. Latif S, Driss M, Boulila W, Huma ZE, Jamal SS, Idrees Z, Ahmad J (2021) Deep learning for the industrial internet of things (IIoT): a comprehensive survey of techniques, implementation frameworks, potential applications, and future directions. Sensors
5. Marques M, Agostinho C, Zacharewicz G, Jardim-Gonçalves R (2017) Decentralized decision support for intelligent manufacturing in Industry 4.0. Sensors
6. Dinmohammadi F (2023) Adopting artificial intelligence in Industry 4.0: understanding the drivers, barriers and technology trends. https://doi.org/10.1109/ICAC57885.2023.10275230
7. Jan Z, Ahamed F, Mayer W, Patel N, Grossmann G, Stumptner M, Kuusk A (2023) Artificial intelligence for Industry 4.0: systematic review of applications, challenges, and opportunities. Expert Syst Appl 216
8. Gupta SS, Goyal R, Gupta D (2023) Artificial intelligence impacts on Industry 4.0: a literature-based study. In: Handbook of research on data science and cybersecurity innovations in Industry 4.0 technologies, pp 30–44
9. Chen W, He W, Shen J, Tian X, Wang X (2023) Systematic analysis of artificial intelligence in the era of Industry 4.0. J Manage Analytics 10(1):89–108
10. Alenizi FA, Abbasi S, Hussein Mohammed A, Masoud Rahmani A (2023) The artificial intelligence technologies in Industry 4.0: a taxonomy, approaches, and future directions. Comput Ind Eng 185
11. Ambadekar PK, Ambadekar S, Choudhari CM, Patil SA, Gawande SH (2023) Artificial intelligence and its relevance in mechanical engineering from Industry 4.0 perspective. Aust J Mech Eng
12. Tercan H, Meisen T (2022) Machine learning and deep learning based predictive quality in manufacturing: a systematic review
13. Rehse JR, Mehdiyev N, Fettke P (2019) Towards explainable process predictions for Industry 4.0 in the DFKI-Smart-Lego-Factor
14. Dalzochio J, Kunst R, Pignaton E, Binotto A, Sanyal S, Favilla J, Barbosa J (2020) Machine learning and reasoning for predictive maintenance in Industry 4.0: current status and challenges. Comput Ind
15. Peres RS, Jia X, Lee J, Sun K, Colombo AW, Barata J (2020) Industrial artificial intelligence in Industry 4.0-systematic review, challenges, and outlook
16. Marques M, Agostinho C, Zacharewicz G, Jardim-Gonçalves R (2017) Decentralized decision support for intelligent manufacturing in Industry 4.0
17. Wang Z, Li M, Lu J, Cheng X (2022) Business innovation based on artificial intelligence and Blockchain technology
18. Simeth A, Plapper P (2023) Artificial Intelligence based robotic automation of manual assembly tasks for intelligent manufacturing. In: Lecture notes in production engineering, vol Part F1162, pp 137–148
19. Calabrese A, Costa R, Tiburzi L, Brem A (2023) Merging two revolutions: a human-artificial intelligence method to study how sustainability and Industry 4.0 are intertwined. In: Technological forecasting and social change, p 188

Improving Battery Utilization and Lifespan of an Electric Vehicle: Advanced SOC Estimation Via Deep Neural Network

Elmahdi Fadlaoui, Hamza Hboub, and Noureddine Masaif

Abstract

Recent advances in artificial intelligence (AI) have revolutionized the capabilities of battery management systems (BMS), particularly in the critical task of state of charge (SOC) estimation for lithium-ion batteries (LIBs). While conventional methods struggle with dynamic operating scenarios, our research introduces a novel deep neural network (DNN) architecture designed to handle the complexities of rapid charge–discharge cycles, and thermal fluctuations. We focused on developing a real-time monitoring solution that overcomes the traditional challenges of SOC estimation, where battery behavior exhibits nonlinear characteristics across varying operational parameters. Our methodology leverages a comprehensive dataset derived from four driving profiles: US06, LA92, Highway Fuel Economy Test (HWFET), and Urban Dynamometer Driving Schedule (UDDS). The experimental design incorporates an 80–20 split for training and testing, with a unique validation approach using randomly combined drive cycles to assess real-world adaptability. The DNN processes three key input parameters, current, voltage, and temperature, to generate an accurate SOC prediction. Comparative analysis against the second-order RC Equivalent Circuit Model (ECM) with Extended Kalman Filter (EKF) demonstrates the superior performance of our approach. Our implementation achieves remarkable accuracy with a Root Mean Square Error (RMSE) below 3.3% and Mean Absolute Error (MAE) under 2.6%, representing a significant advancement in SOC estimation technology.

Keywords

Electric vehicles · Battery lithium-ion · State of charge estimation · Deep neural network · Equivalent circuit model · Extended Kalman filter

1 Introduction

The artificial intelligence (AI) domain has become a transformative force in various sectors in recent years due to its ability to provide solutions to a wide range of problems across various domains [1, 2]. Notably, AI has found extensive application in battery management systems, contributing significantly to the advancements in this field. The integration of AI technologies has facilitated more efficient and effective management of batteries, enhancing their performance and safety in various applications. Energy storage systems (ESS) have gained critical importance in recent years as the world shifts toward renewable energy sources like solar and wind, which are inherently intermittent. ESS helps stabilize the power grid by storing surplus energy during peak generation and releasing it when demand surges, which reduces dependency on fossil fuels. These systems are also key to decentralized energy management, supporting electric vehicles, smart grids, and household backup power, thereby promoting energy security and sustainability.

The lithium-ion batteries are becoming increasingly popular as a new source of clean energy due to environmental pollution and the energy crisis. Their adoption in electric vehicles (EVs), plug-in hybrid electric vehicles (PHEVs), and hybrid-electric vehicles (HEVs) is driven by their exceptional characteristics: superior energy retention, high energy, and power density metrics, and extended operational lifespan [3]. While these energy storage systems become more prevalent, the researchers focus specifically on the critical challenge of state of charge and state of health estimation, capacity

E. Fadlaoui (✉) · H. Hboub · N. Masaif
Laboratory of Electronic Systems, Information Processing, Mechanics and Energy, Ibn Tofail University, Kenitra, Morocco
e-mail: elmahdi.fadlaoui@uit.ac.ma

A. Moumen et al. (eds.), *Artificial Intelligence and Data Analytics for Innovative Applications in Engineering, Sustainability and Technology*, Sustainable Artificial Intelligence-Powered Applications, https://doi.org/10.1007/978-3-031-90318-2_3

prediction, and safety [4]. Among these factors, accurately estimating the SOC is crucial [5].

The SOC which represent the remaining capacity available on the battery is a key indicator that directly impacts operational safety and user experience. The accurate estimation of the SOC of the lithium-ion battery can prevent over-discharge and overcharge, which improves the battery's service life, performance, and reliability. However, the complexity of SOC estimation lies in its indirect nature, requiring sophisticated interpretation of measurable parameters including voltage, current, and thermal readings [6, 7].

Several researches have been proposed to obtain an accurate estimation of the battery state of charge of electric vehicles. These methods can be classified into four groups: the conventional Coulomb counting method (CCM), which tracks current integration over time but it cannot handle measurement noise and inaccurate initial SOC; the open circuit voltage method (OCVM), is easy to implement and has high precision for SOC estimation, but is not appropriate for online applications; model-based methods (MBM), which leverage mathematical representations of battery behavior, it takes into account the physical properties and behavior of the battery [8]; and data-driven approaches that employ machine learning (ML) to learn the nonlinear relationships between measurement parameters (current, voltage and temperature) and SOC, which consider the battery as black box and learn its internal characteristics by training the ML model with measurement factors [9, 10]. Notably, these methods do not need any information about the physical characteristics of the battery. Besides that, model-based methods (MBM) are based on developing a model that can simulate the battery's behavior under operation [11]. Recently, artificial NN- based methods received more attention from researchers all around the world. Thanks to the increasing computing power of GPUs, the network training time has been significantly reduced from months to hours. Additionally, researchers can gather vast amounts of training data by collecting field data from BMSs.

Installed on-site and transferring it to the laboratory to perform tests with different dynamic loading profiles. With a pre-trained model, the estimation can be completed in a few milliseconds, which is sufficient for most onboard EV applications. This paper examines the implementation of Deep Neural Networks (DNN) for SOC estimation, benchmarking its performance against the established second-order RC Equivalent Circuit Model with Extended Kalman filter approach. The validation process employs mixed drive cycle testing under varying thermal conditions, with performance evaluated through RMSE and MAE metrics. The rest organizational structure of this paper is organized as: Sect. 2 presents the establishment of the 2nd order RC Equivalent circuit model with an Extended Kalman filter (ECM-EKF) and the framework utilized for SOC estimation. Section 3 illustrates the DNN architecture proposed for the estimation. Section 4 shows the implementation and the comparison of the two methods, and finally, Sect. 5 presents the conclusion.

2 Second-Order RC Equivalent Circuit Model and Extended Kalman Filter ECM-EKF

In this section, we develop the ECM-EKF model-based method [12]. First the 2nd order RC equivalent circuit model as shown in Fig. 1 contains two parallel RC branches connected in series with resistance and input voltage source, after that the EKF algorithm is implemented to estimate the battery SOC.

The model's output $V(k)$ equation establishes a connection between the open circuit voltage V_{OCV} in function of SOC and the voltage drops across the elements, according to Kirchhoff's laws, the output voltage is calculated as follows:

$$V(k) = V_{OCV}(SOC(k)) - V_1(k) - V_2(k) - R_0 I(k) \quad (1)$$

where

- R_0: cell internal resistance
- R_1: resistance of the electrochemical
- R_2: concentration polarization
- C_1, C_2: the polarization process of the battery dynamics.

The state space equation is written as:

$$\begin{bmatrix} SOC[k+1] \\ V_1[k+1] \\ V_2[k+1] \end{bmatrix} = A_k + \begin{bmatrix} SOC[k] \\ V_1[k] \\ V_2[k] \end{bmatrix} + B_k I[K] \quad (2)$$

where

$$A_k \begin{bmatrix} 1 & 0 & 0 \\ 0 & \exp(-\Delta t/\tau_1) & 0 \\ 0 & 0 & \exp(-\Delta t/\tau_2) \end{bmatrix}; \text{ and}$$
$$B_k \begin{bmatrix} -\Delta t/(Q_N) \\ R_1(1-\exp(-T/\tau_1)) \\ R_2(1-\exp(-t/\tau_2)) \end{bmatrix} \quad (3)$$

In order to obtain the parameters of the ECM, a parameter identification procedure was carried out using the hybrid pulse power characterization (HPPC). Table 1 shows the mean values of the parameters, for more information see the ref [13].

The SOC-OCV relationship was also taken into account in the ECM. The OCV curve was obtained by fully charging the

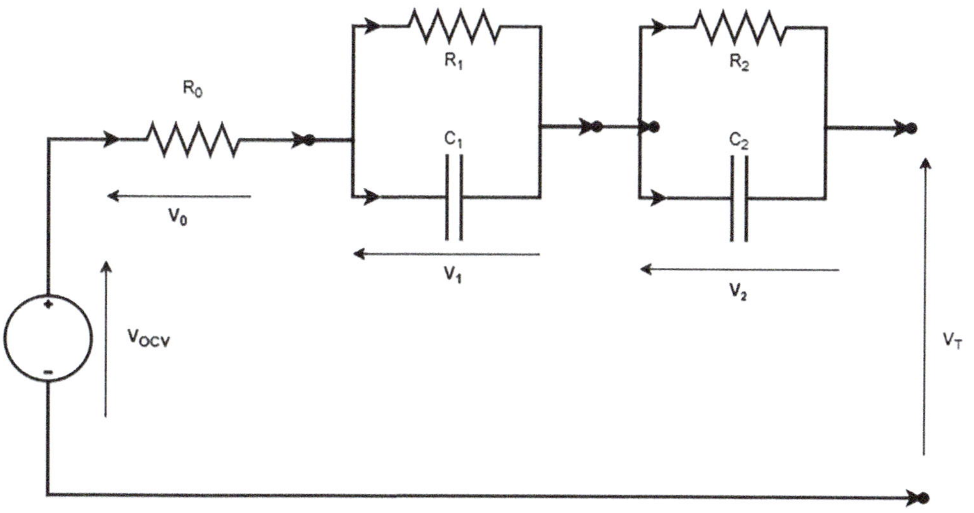

Fig. 1 2nd-order RC equivalent circuit representation of a cell model (ECM)

Table 1 Battery internal parameters mean values

Parameters	R0	R1	R2	C1	C2
Values	0.01	0.0005	0.0005	20,108	200,448

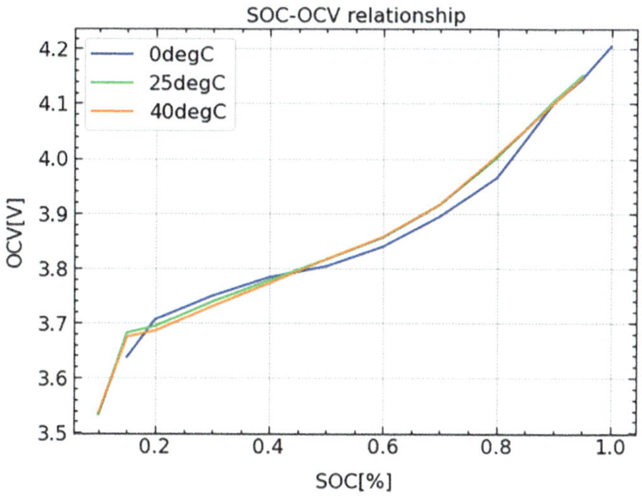

Fig. 2 SOC-OCV relationship at different temperatures

battery and then allowing it to rest for several hours before measuring the OCV [14]. The SOC was then estimated by interpolating the OCV curve beside the Extended Kalman Filter (EKF). Figure 2 shows the SOC-OCV relationship at various temperatures (0,25,40).

The widely used EKF algorithm [15], known for its excellent tracking performance of a system state, such as the state of charge of Li-ion batteries, by combining a mathematical model with measurements of inputs and outputs. The EKF utilizes the first-order Taylor expansion to the measured data and state function To achieve local linearization. This allows for the description of a nonlinear system using the following state-space equations,

$$\begin{cases} X_k = f(x_{k-1}, u_{k-1}) + w_{k-1} \\ y_k = h(x_k, u_k) + v_k \end{cases} \tag{4}$$

where X_k denotes the state variable of the system:

$$X_k = \begin{bmatrix} \mathrm{SOC}_k \\ V_{1,k} \\ V_{2,k} \end{bmatrix} \tag{5}$$

y_k is the observation of the system Eq. 1. f and h are the nonlinear functions of the state and measurement respectively; w_{k-1} and v_k are Gaussian process noise and measurement process noise; also R and Q are measurement and process noise covariance respectively. The standard calculation steps of EKF algorithm are given in Table 2.

Table 2 Extended Kalman filter algorithm

Predict
Step 1 State prediction: $X_k = f(X_k, u_k)$
Step 2Error covariance prediction: $P_k = F_k P_k F^T_k{}_k + Q_k$
Update
Step 3 Innovation or measurement residual: $\tilde{y} = z_k - h(x_k)$
Step 4 Innovation (or residual) covariance: $S_k = H_k P_k H^T_k{}_k + R_k$
Step 5 Near-optimal Kalman gain: $K_k = P_k H^T_k S^{-1}_k$
Step 6 Updated state estimate: $\hat{x}_k = \hat{x}_{k-1} + k_k \tilde{y}$
Step 7 Updated covariance estimate: $P_k = (I - k_k H_k) P_k$

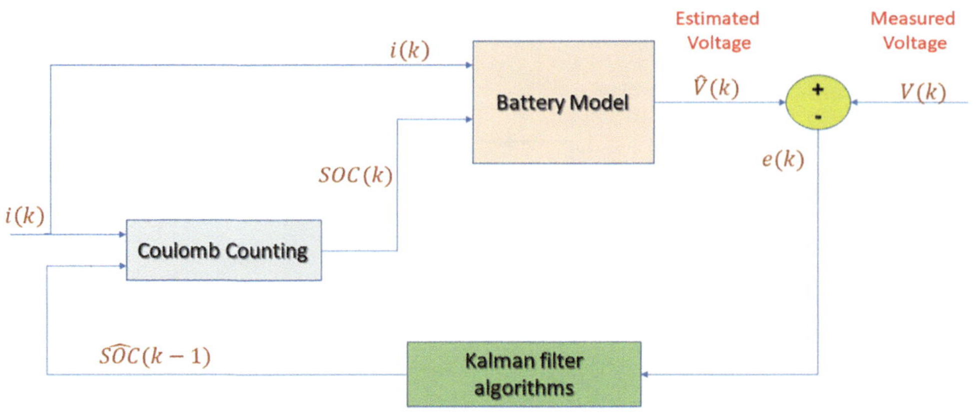

Fig. 3 ECM-EKF framework of the battery SOC estimation

where the following Jacobians are used to define the state transition and observation matrices:

$$\begin{cases} F_k = \frac{\partial f}{\partial x}\Big|_{(x_k,u_k)} \\ H_k = \frac{\partial h}{\partial x}\big|_{(x_k)} \end{cases} \quad (6)$$

The implementation architecture shown in Fig. 3 consists of a dual-stage process for SOC estimation. The primary stage employs a battery model that processes current and temperature measurements to generate terminal voltage predictions. Subsequently, the EKF stage analyzes the difference between predicted and actual terminal voltage measurements to compute the SOC estimation, creating a continuous feedback loop for enhanced accuracy.

3 Deep Neural Network

Our innovative approach to SOC estimation leverages deep learning capabilities, eliminating the need for traditional battery modeling [16]. Instead of relying on conventional circuit representations, we developed a data-driven framework that captures battery dynamics directly from operational measurements voltage, current, and temperature (V, I, T) readings. This methodology treats the battery system as an encapsulated entity, deriving its predictive accuracy from comprehensive charge–discharge cycling data. The core of our implementation utilizes Deep Neural Networks (DNNs), which represent an advanced evolution of conventional Artificial Neural Networks (ANNs). While traditional neural architectures typically employ minimal hidden layer structures, our DNN implementation incorporates multiple processing layers, enabling sophisticated feature extraction and complex pattern recognition. This enhanced architectural depth allows for nuanced mapping of the intricate relationships between input parameters and SOC estimates. The proposed DNN architecture, illustrated in Fig. 4, processes a three-dimensional input vector $X_k = [I_k, V_k, T_k]$ comprising instantaneous current, voltage, and temperature measurements. Our design employs a fully connected topology with three strategically sized hidden layers, culminating in the output layer that generates the SOC prediction $y_k = [\text{SOC}_k]$.

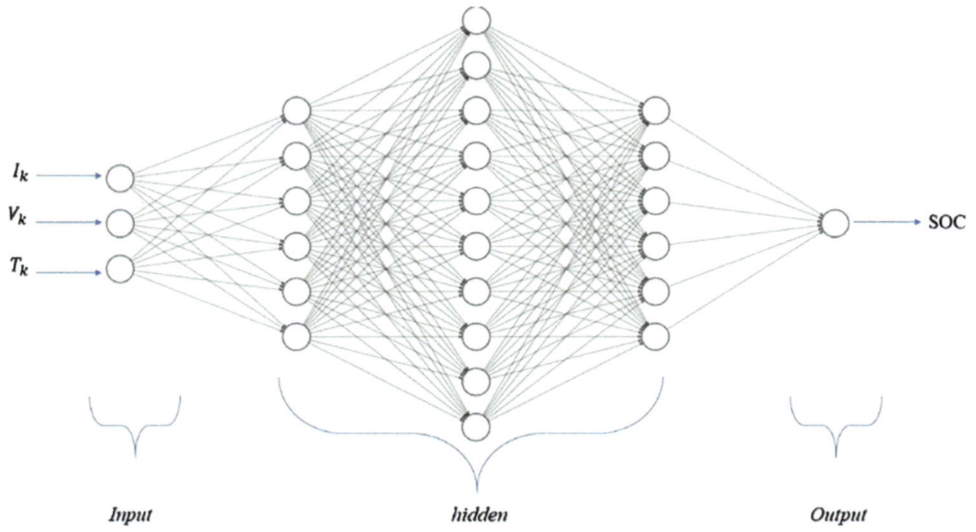

Fig. 4 Architecture of the proposed deep neural network

To rigorously validate our approach, we implemented a dual-metric evaluation framework: Root Mean Square Error (RMSE), Eq. 7: Quantifies prediction accuracy by measuring the standard deviation of estimation errors. Mean Absolute Error (MAE), Eq. 8 provides a direct measure of estimation accuracy through absolute deviation analysis.

$$\text{RMSE} = \sqrt{\sum_{i=1}^{n} \frac{(y_i - \hat{y}_i)}{n}} \tag{7}$$

$$\text{MAE} = \frac{1}{n} \sum_{i=1}^{n} |y - \hat{y}_i| \tag{8}$$

This comprehensive evaluation methodology enables detailed performance assessment, validating both the accuracy and reliability of our SOC estimation approach across diverse operational conditions.

4 Results and Discussion

4.1 SOC Pre-Estimation with DNN

The study utilized Turnigy Graphene 5000mAh 65C Li-ion Battery [17]. Table 3 presents the essential parameters of this battery.

Three dynamic loading profiles were employed to simulate real-world driving conditions: US06, UDDS HWFET, and LA92. These drive cycles represent diverse discharge/charge patterns commonly encountered in electric vehicles. Approximately 320,000 data points were collected from these tests. To prevent overfitting and enhance model generalization, a sampling rate of 10 was applied, resulting in one data point for every 10 original samples. Before training, the data were scaled to [1] to accelerate convergence and improve parameter optimization. Figure 5 illustrates the current profiles of the US06, LA92, HWFET, and UDDS drive cycles used for training. A mixed drive cycle, composed of a combination of these four profiles, was created to provide a more representative and challenging testing environment for the battery model.

Table 3 Battery main specification

Chemistry	LiPO
Nominal capacity	5 Ah
Nominal voltage	3.7 V
Energy density	134 (Wh/Kg)
Discharge	2.8 V end-voltage, 20A MAX continuous current
Charge	4.2 V, 50 mA end-current (CC-CV) fast

First, a DNN with three hidden layers is trained with four drive cycles at (0, 25, and 40 °C). (V, I, T) are used as input

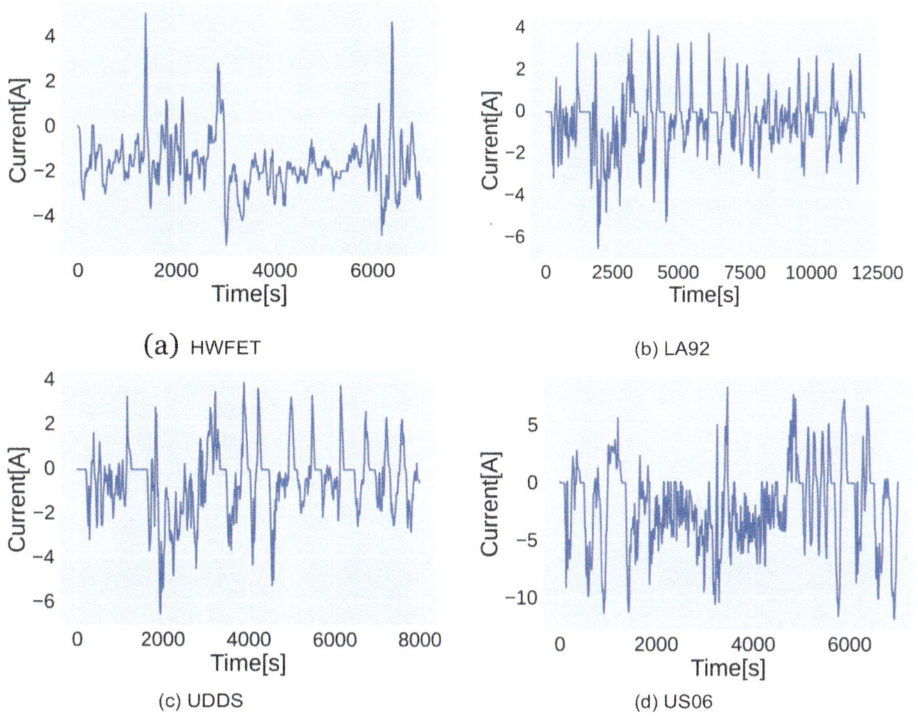

Fig. 5 Fourth current profiles: HWFET, LA92, UDDS, and US06

of the neural network with (tanh) as the activation function, and SOC as output with RELU as the activation function. The neural network is trained with 80% of data, while the 20% rest is chosen for validation. The mixed data at different temperatures of the four drive cycle are used for testing.

In this study, the DNN architecture employed consists of 5 layers, the input and the output layer with three hidden layers. The first and third hidden layers each contain 64 neurons, while the second layer has 128 neurons with (PRelu) activation, and the output layer with (RELU) activation. The model was trained for 200 epochs using the Adam optimizer, set at a learning rate of 0.01, and optimized with a mean squared error loss function, with a batch size of 1000.

For the comparison, the EKF method is used, and the most suitable covariance matrices values of this method are concluded after multiple experiments illustrated in Table 4.

Table 4 Extended Kalman filter covariance matrices values

Parameters	Values
R_x	$2.5e^{-5}$
P_x	$\begin{bmatrix} 0.025 & 0 & 0 \\ 0 & 0.01 & 0 \\ 0 & 0 & 0.01 \end{bmatrix}$
Q_x	$\begin{bmatrix} 1e^{-6} & 0 & 0 \\ 0 & 1e^{-5} & 0 \\ 0 & 0 & 1e^{-5} \end{bmatrix}$

4.2 Comparison of DNN and EKF Method Under Different Temperatures

Figure 6 shows the state of charge estimation test results of the DNN and EKF for a dataset of mixed drive cycles at (0, 25,

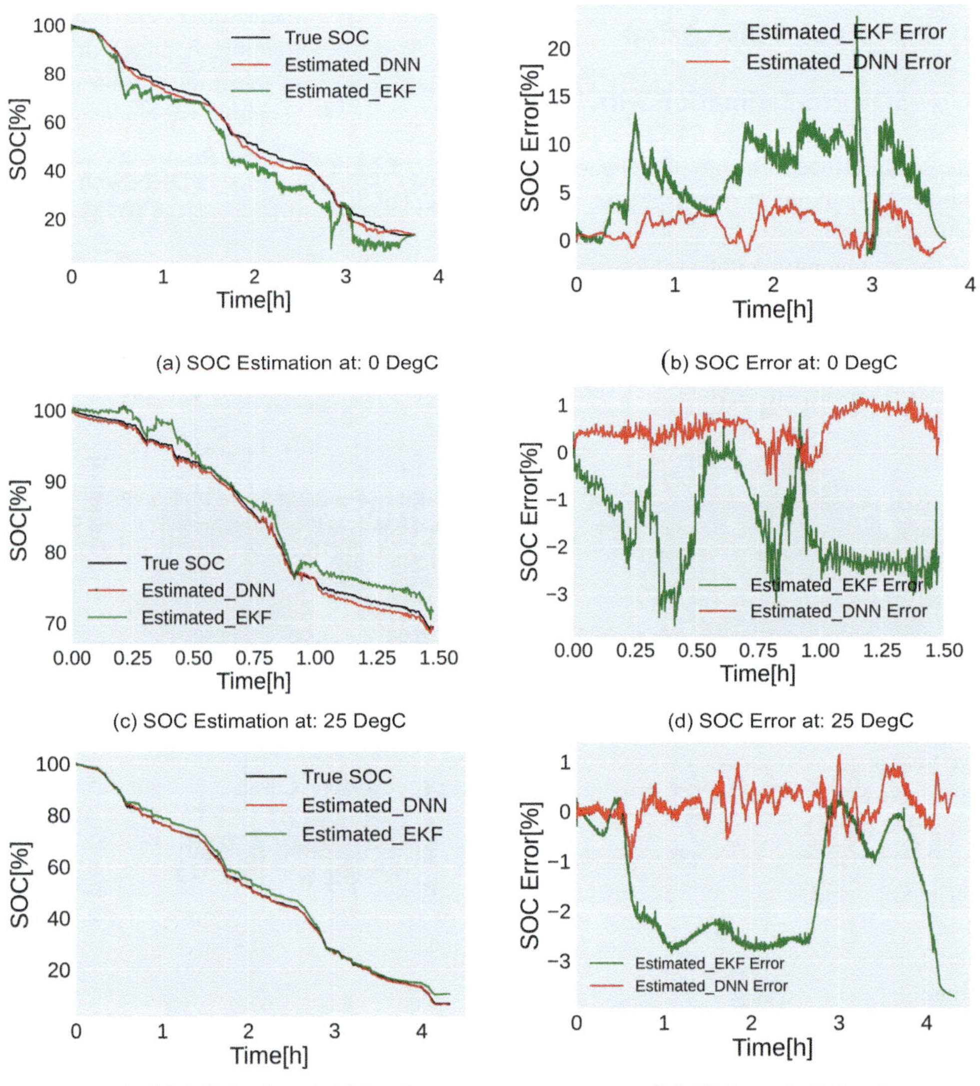

Fig. 6 SOC estimation with mixed drive cycle at different temperatures

(a) SOC Estimation at: 0 DegC
(b) SOC Error at: 0 DegC
(c) SOC Estimation at: 25 DegC
(d) SOC Error at: 25 DegC
(e) SOC Estimation at: 40 DegC
(f) SOC Error at: 40 DegC

Fig. 7 RMSE and MAE of the SOC estimation of the mixed drive cycle

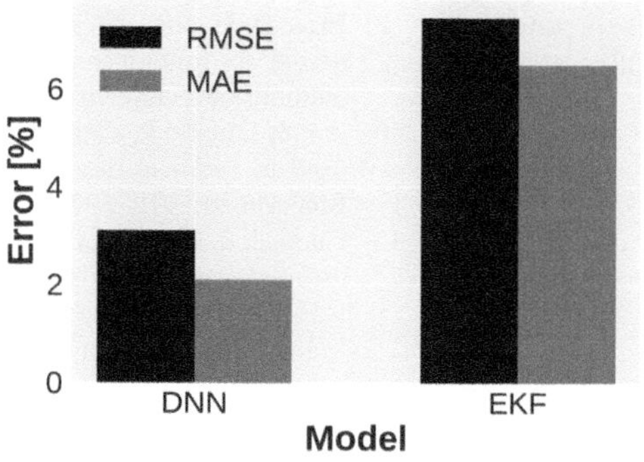

(a) RMSE and MAE at: 0 DegC

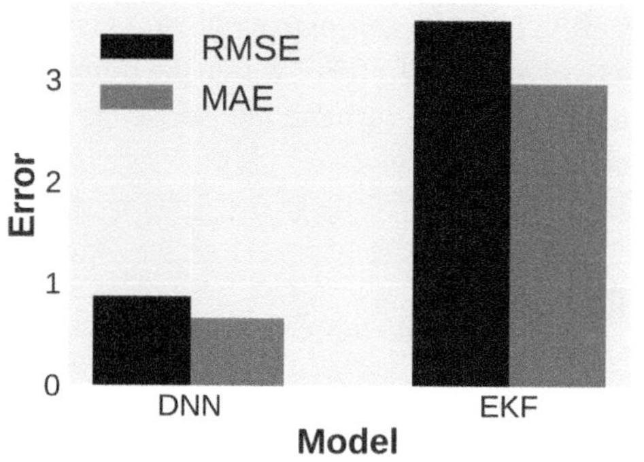

(b) RMSE and MAE at: 25 DegC

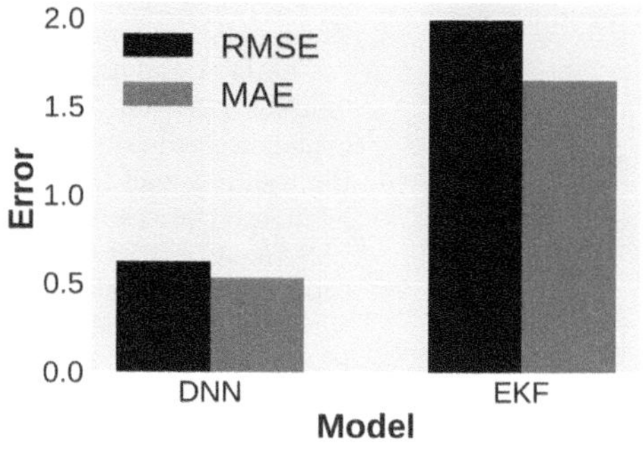

(c) RMSE and MAE at: 40 DegC

and 40 °C) with their corresponding estimation errors; the true SOC indicates the SOC obtained using the coulomb counting method, which was obtained by discharging the battery from 100% SOC with well-calibrating current sensor to ignore the integral error. This figure shows that the DNN method is more accurate and stable in SOC estimation compared to EKF method for different temperatures.

These bar charts in Fig. 7 shows the RMSE and MAE of SOC estimation results of the two models for different temperature, for each figure that we see, the first two bars illustrate the RMSE and MAE respectively of the DNN model and the last two bars for the EKF model, we can observe that DNN give good results in different temperatures.

In comparison with the EKF method for SOC estimation, the proposed deep neural network approach presents notable superiority.

While deep neural networks have demonstrated impressive performance in battery state of charge estimation, they inherently suffer from limitations in terms of interpretability compared to physics-based models. DNNs are complex black-box models that learn patterns from data without explicit consideration of underlying physical principles. This lack of interpretability can hinder our understanding of how the model arrives at its predictions, making it difficult to assess the reliability and robustness of the results. In contrast, physics-based models are grounded in well-established physical laws, providing a transparent and understandable framework for SOC estimation. This interpretability allows for a deeper understanding of the battery's behavior, enabling more reliable predictions and better diagnostics in various operating conditions.

5 Conclusion

Lithium-ion batteries continue to play a pivotal role in powering electric vehicles, and the necessity for an accurate and dependable battery management system becomes paramount. This study introduces a deep neural network for SOC estimation, utilizing measurements of current, voltage, and temperature. The mixed drive cycle at different temperatures is used for testing the model's reliability across various driving scenarios. Comparative analyses demonstrate the superiority of the proposed deep neural network method over traditional approaches like the 2nd order RC equivalent circuit model and the Extended Kalman Filter estimation method. Impressively, the experimental results showcase that the RMSE is less than 3.3% and the MAE is under 2.6%.

The integration of artificial intelligence techniques into battery management systems marks a significant advancement. This innovation facilitates the accurate determination of SOC, even in demanding scenarios characterized by rapid charging or discharging, temperature fluctuations, and battery degradation. Consequently, the heightened precision in SOC estimation contributes to optimizing battery capacity utilization, prolonging battery lifespan, and enhancing overall safety. This transformative role of AI underscores its potential as a revolutionary force within the domain of battery management, poised to reshape the landscape of energy storage and electric mobility.

References

1. El Mountassir Y (2020) Economic Intelligence system and decision making: proposal of a theoretical model. Moroc J Quant Qual Res 2:137–152
2. Imane S, Mohamed O, Said H, Mohammed Z, Jamal C, Larbi S (2023) New GIS approach using machine learning algorithm for early floods detection. Moroc J Quant Qual Res 5
3. Shrivastava P, Soon T, Idris M, Mekhilef S (2019) Overview of model-based on-line state-of-charge estimation using Kalman filter family for lithium-ion batteries. Renew Sustain Energy Rev 113:109233
4. Jia J, Liang J, Shi Y, Wen J, Pang X, Zeng J (2020) SOH and RUL prediction of lithium-ion batteries based on Gaussian process regression with indirect health indicators. Energies 13:375
5. Wang Y, Tian J, Sun Z, Wang L, Xu R, Li M, Chen Z (2020) A comprehensive review of battery modeling and state estimation approaches for advanced battery management systems. Renew Sustain Energy Rev 131:110015
6. Lin X, Tang Y, Ren J, Wei Y (2021) State of charge estimation with the adaptive unscented Kalman filter based on an accurate equivalent circuit model. J Energy Storage 41:102840
7. Tian Y, Lai R, Li X, Xiang L, Tian J (2020) A combined method for state-of-charge estimation for lithium-ion batteries using a long short-term memory network and an adaptive cubature Kalman filter. Appl Energy 265:114789
8. How D, Hannan M, Lipu M, Ker P (2019) State of charge estimation for lithium-ion batteries using model-based and data-driven methods: a review. IEEE Access 7:136116–136136
9. Anton J, Nieto P, Viejo C, Vilán J (2013) Support vector machines used to estimate the battery state of charge. IEEE Trans Power Electron 28:5919–5926
10. Chemali E, Kollmeyer P, Preindl M, Emadi A (2018) State-of-charge estimation of Li-ion batteries using deep neural networks: a machine learning approach. J Power Sources 400:242–255
11. Wang Z, Feng G, Liu X, Gu F, Ball A (2022) A novel method of parameter identification and state of charge estimation for lithium-ion battery energy storage system. J Energy Storage. 49:104124
12. Maheshwari A, Nageswari S (2021) Sunflower optimization algorithm based filtering method for state of charge estimation of batteries in electric vehicle
13. Khanum F, Louback E, Duperly F, Jenkins C, Kollmeyer P, Emadi A (2021) A Kalman filter based battery state of charge estimation MATLAB function. In: 2021 IEEE transportation electrification conference and expo (ITEC), pp 484–489
14. Elmahdi F, Ismail L, Noureddine M (2021) Fitting the OCV-SOC relationship of a battery lithium-ion using genetic algorithm method. E3S Web Conf 234:00097
15. Cui Z, Hu W, Zhang G, Zhang Z, Chen Z (2022) An extended Kalman filter based SOC estimation method for Li-ion battery. Energy Rep 8:81–87
16. Zhang D, Zhong C, Xu P, Tian Y (2022) Deep learning in the state of charge estimation for Li-ion batteries of electric vehicles: a review. Machines 10:912
17. Kollmeyer P, Skells M (2020) Turnigy graphene 5000mAh 65C li-ion battery data. Mendeley Data 1:10–17632

Impact Evaluation of School Transport Program on Schooling in Rural Areas in Morocco

Fatima Azdagaz, Mariem Liouaeddine, and Omar Zirari

Abstract

Improving education in rural areas and reducing school dropout rates continue to be significant challenges for Morocco's education system. In this regard, this study seeks to evaluate the effectiveness of the social support program for schooling, specifically the school transportation program, on school enrollment and reduction of school wastage among students in rural schools. The research analyzes evidence based on data from the national social support survey implemented by the National Human Development Observatory in 2018 through a representative sample of 3,039 rural and urban households. Through the micro-econometric strategy in the propensity score matching (PSM) method, the evidence indicates that school transportation provision significantly enhances the level of school enrollment but, simultaneously, reduces dropout and repetition rates of students in rural schools. Additionally, the analysis takes into consideration the implementation of this public policy, its targeting strategy, and the identification of key success factors. This public policy analyzes its targeting strategy, and identifies the key factors for achieving a high-performance education system with universal access and quality.

F. Azdagaz (✉) · M. Liouaeddine
Laboratory of Economics and Public Policy, Faculty of Economics and Management, Ibn Tofaïl University, Kenitra, Morocco
e-mail: fatima.azdagaz@uit.ac.ma

M. Liouaeddine
e-mail: mariem.liouaeddine@uit.ac.ma

O. Zirari
Laboratory of Research in Theoretical and Applied Economics, Faculty of Economics and Management, Hassan 1th University, Settat, Morocco
e-mail: omar.zirari@uhp.ac.ma

Keywords

Social support program · Public policy evaluation · School transportation · Education · Propensity score matching (PSM)

1 Introduction

Public education policies are significant in increasing the enrollment of students while reducing wastage of human educational potential at the same time. The policies aim to promote equal access to educational opportunity, improve instructional quality, and aid disadvantaged students. We have explained in this discussion the implication of these policies and the practices that have been put in place to improve academic performance.

Most significantly, education public policy has a key role in the promotion of universal access to education opportunities. Through the encouragement of the passage and enforcement of laws making free and compulsory education, government institutions strive to make children from all socio-economic backgrounds enroll in education programs. Inequalities in education access are reduced and social inclusion enhanced through such policies.

Additionally, public education policy priority is largely focused on improving the quality of teaching. This can be achieved through a variety of ways, such as initiating teacher training programs, promoting pedagogical skill development programs, and endorsing innovative policies in teaching approaches. Public policies enable the provision of improved educational opportunities to all learners by availing resources for teacher professional development and by equipping schools with necessary tools.

Moreover, public educational policies play a significant role in preventing school dropouts or early school leavers.

A. Moumen et al. (eds.), *Artificial Intelligence and Data Analytics for Innovative Applications in Engineering, Sustainability and Technology*, Sustainable Artificial Intelligence-Powered Applications, https://doi.org/10.1007/978-3-031-90318-2_4

The phenomenon typically happens as a result of a set of intricate factors, including socioeconomic disadvantages, social inequalities, academic problems, or lack of family support. Public policies can properly respond to the issues by offering support programs to dropout students, implementing prevention initiatives against early school leaving, and initiating programs with the aim to increase parents' involvement in educating their children.

In addition, there is a necessity that social justice and equity within the education system be enhanced through public education policy. Through adopting policies meant to narrow the gap of achievements for various student populations, including ethnic minorities, financially challenged students, and disabled pupils, governmental organizations can play their role in generating a more equitable and diverse schooling system. A few of the policies would be the distribution of additional resources to schools serving low-income areas, the implementation of mentorship programs for students at risk, and the promotion of diversity and inclusiveness in the curriculum.

Public education policy is effectively a tool for promoting student enrollment and reducing school wastage, particularly in rural and remote areas. Through equal access to learning facilities, improvement in the quality of teaching, school wastage reduction, and promotion of equity and social justice, public education policy provides a learning environment conducive to educational achievement for all students. Thus, it is crucial that government departments continue to invest in these policies and develop good systems to deliver education and achievement to every child.

In light of the above background, the aim of the current study was to quantify the effect of the public school transport policy in rural communities for increasing enrollment rates among the affected population and avoiding school wastage.

2 Literature Review

2.1 Theoretical Framework

State intervention and education policies have been the source of great debate and controversy. The state's role in making schooling possible socially lies at the center of this controversy. Different theories express various viewpoints, ranging from proposing a large extent of state intervention in the management of the economy to no intervention at all. This essay discusses five such theoretical frameworks, their underlying principles, their major proponents, and their implications for educational policy. In particular, we will discuss the perspectives of socialism, distributive justice, and socio-liberalism.

Socialists, as exemplified by Engels [1], are prone to uphold an active state intervention in most spheres of human existence, including education. There is a fundamental assumption in this model that the state has to play an active role in ensuring equitable access to quality education for all, especially with the aim of narrowing socioeconomic disparities. To the supporters of socialism, education is a right and need not be determined by economic affordability but by need and merit.

Within a socialist context, it's up to the state to secure the provision of free and equal public education financed from public sources and managed by the government. Its primary goal is to equip each individual with an equal prospect for accessing schooling services and facilities in comparison with his or her fellow citizens regardless of their place and socioeconomic background. Through investing in education, the state can be in a position to facilitate the development of a more democratic and equitable society where every individual is afforded the opportunity to achieve their fullest potential.

The distributive justice theory, as presented by Rawls [2], examines the fair distribution of resources and opportunities in a social setting. Rawls, a key exponent of this theory, argues that the state has to have a role in making quality education accessible to all individuals, especially the disadvantaged. Rawls also devised the difference principle, which holds that economic inequalities can be justified only if they enhance the situation of the worst off in society.

In education, this implies that the state enacts policies and programs aimed at reducing educational disparities along various socioeconomic strata. These may encompass tutorial programs, students' financial aid to needy students, and initiatives aimed at enhancing the quality of schools in poor neighborhoods. Through equalizing access to quality education, the state assists in fostering a more equal and just society for all individuals.

The social liberalism perspective, adopted by authors such as [3, 4], emphasizes the role of individual competence and freedom. State intervention in designing an educational system that assures equal opportunities for everyone to utilize their potential and capacities in harmony with justice and equity principles, the social liberals believe.

Within a social liberal framework, the state is responsible for guaranteeing equal access to quality education within the framework of heterogeneity of individuals' abilities and needs. This can include policies addressing individual students' needs, such as differentiated instruction, inclusive education initiatives, and special provision for students with particular needs. By investing in each person's potential development, the state can help to create a more diverse and inclusive society.

Generally, the theoretical approaches to government intervention in education offer a range of views on the extent to which the state should facilitate education at a social level. Some views, such as those of socialism, promote active intervention by the state with the aim of promoting equality of access to quality education. Other views, such as those of neoliberalism, support a less interventionist, market-oriented approach. Each has distinct implications for education policy and resource distribution in society.

2.2 Empirical Framework

In the past ten years, efforts have been increased to curb student absenteeism as a means to enhance overall school performance. Numerous studies have reinforced the fact that school attendance has a strong association with student achievement and postsecondary education outcomes, whereas absenteeism is associated with decreased student performance [5–8].

Within this framework, public policies on education are instrumental in facilitating student enrollment and controlling school waste, through absenteeism and dropout rates. Public social programs aimed at school enrollment are governmental initiatives that are developed to increase education access, encourage academic achievement, and eliminate barriers that may limit students' educational attainment. The goal of such initiatives is to ensure that every child, regardless of their background or socioeconomic status, has the opportunity to receive a quality education. Such programs take various forms and focus on different aspects of schooling, including affordability, quality of instruction, student well-being, and equity in the school system.

From this perspective, four aspects of social support for school were addressed. First, financial access is often facilitated by scholarship programs, grants, or direct financial aid to families to cover tuition fees, school supplies, and transportation. These initiatives aim to remove financial barriers that might prevent some children from attending school. Moreover, the quality of education can be enhanced through in-service training for teachers, curriculum reforms, investments in educational infrastructure, and efforts to strengthen teaching skills. These actions are intended to create a stimulating and inclusive learning environment for all students. Additionally, student well-being is supported through school health programs, nutritious meals, psychological counseling, and extracurricular activities. These programs are designed to ensure that students maintain good physical and mental health, which is crucial for their academic success.

Lastly, equity in the education system is fostered through initiatives aimed at reducing social, economic, and gender disparities. This could include policies supporting disadvantaged groups, programs promoting gender equality, measures to combat discrimination, and policies ensuring equal access to education for all children.

In conclusion, public social programs that support schooling are a vital part of educational policies designed to guarantee equitable access to education and foster academic success for all children. By investing in these programs and implementing effective policies, governments can create an inclusive and equitable educational environment that enables all children to reach their full potential.

In line with the focus of this study on the effect of providing school transport on school attendance for Moroccan students in rural areas, existing literature underscores the critical role of such public social support schemes in reducing absenteeism and combating school wastage. Specifically, [9] conducted a study based on 26 research projects from the United States, Canada, Croatia, England, Nepal, Pakistan, Portugal, South Africa, and Spain, concluding that transport plays a key role in academic success, particularly for students from low-income families. In terms of public intervention, an efficient and integrated transport system plays an important role in enhancing accessibility and improving the quality of education [10]. Education is one of the main pillars of human resource development, and easy and affordable access to educational institutions is one of the main factors that influence student participation and success [11].

It is important to note the importance of transport demand for educational travel to consider educational access factors. Primarily, in rural areas, school attendance and enrolment are indicators of access to education and the success of government educational programs [12].

Physical distance from schools has a negative impact on enrolment and educational success [13, 14]. Similarly, several previous studies based on survey data found that school attendance is positively associated with school accessibility and negatively related to poverty and household size. According to [15], by aggregating distance as an indicator of accessibility at the district level and from a cross-sectional analysis, he found that the gender difference in school attendance decreases with improving school accessibility.

For his part, [16] presented the multidimensional impact of rural road accessibility and enrolment gains at the primary level of education evident in rural India. Previous studies have used distance to schools and public transport to measure accessibility indicators. However, there is a need for thorough investigation of the supply of opportunities in the region, which is crucial for planning and policy decisions. A supply based measure of accessibility is needed in rural areas, as they are sparsely populated, supply is even sub-standard, and private service providers have little or no interest in them.

In his work measuring the impact of school busing on student outcomes in the Michigan region of the USA, [17] found that eligibility for transportation increased attendance rates and reduced the likelihood of chronic absence, with these effects being strongest for economically disadvantaged students.

There is evidence that school bus eligibility increases attendance, although eligibility may not be sufficient in contexts where school choice is extreme and few pupils are eligible or take the bus [18].

Investigating the relationship between school transport services and attendance, [19] used a nationally representative dataset of kindergarten children in the United States, the longitudinal study mobilized showed that children who took the bus to school were 3% points less likely to be chronically absent. Furthermore, another study [20] found that rural communities have unique transportation needs; for rural students, taking the school bus correlated with reduced school absences on average, and reduced absences even further for rural students who have siblings and whose fathers work full-time. These findings are in line with previous studies showing that school buses can be an important routine-setting mechanism for kindergarten students and their families. Although not generally considered an attendance intervention, the school bus could be an important mechanism for promoting positive attendance behaviors among kindergarten children.

For their part, [21] asserted that school buses are essential, if underappreciated, components of American schools' academic infrastructure. Nationally, more than half of the 49.5 million K-12 students use school buses to get to school each day, at an average cost of around $1,000 per student. If the environment views school buses simply as a means of transport between school and home, this leads to an underestimation of their role in American schools and society, while school buses facilitate reforms such as desegregation and district consolidation and enable students to access schools that are better suited than their neighborhood school.

At the national level, there is limited research on the impact of school transportation on student enrollment and the reduction of school wastage. In this context, [22], in a qualitative study, confirmed that the availability of school transport and extracurricular activities motivates student retention and attention in class, and therefore, the fight against school wastage. However, the inadequacy of these factors is a key determinant of a negative perception of school climate and, therefore, educational performance. In the same context, the study by Azdagaz et al. [23] shows that benefiting from social support programs for schooling in Morocco, such as school canteens and transport, improves school attendance and combats school drop-out in rural areas.

Within this framework, the present work aims to evaluate, using a micro-econometric approach, the impact of the school transport program in rural areas of Morocco on children's school attendance and the fight against school wastage.

3 Methodological Framework and Data

3.1 Propensity Score Matching Approach

This empirical study aimed to assess the impact of the school transport program as a social support program on school enrolment in rural areas. With this in mind, the method used to assess the causal effect of this program is propensity score matching (PSM), which makes it possible to evaluate the impact of the program on pupils benefiting from school transport in comparison with non-beneficiary pupils. Measuring the effect of the social support for schooling program comes down to asking the following question: What is the impact of the school transport program on school enrolment and the fight against school wastage among pupils in rural areas in Morocco?

The aim of this study is to quantify the impact of the school transport program on the enrolment of pupils in rural areas (ΔATT) by measuring the difference between beneficiaries (treatment group) and non-beneficiaries (comparison group). We applied the standard matching approach formalized by Rubin [24], which involves treating the causal link by comparing the two groups. This approach reduces selection bias because it is impossible to attribute the differences in school enrolment and wastage observed between the two groups of pupils solely to the fact that social support for schooling is provided (school transport).

According to the mathematical expression, the treatment effect Δ_i for pupil I at time t is defined as the difference between the potential outcome Y_i^T if the pupil receives school transport and the potential outcome Y_i^C if they do not. In this context, T represents the treatment group, while C denotes the control group. This relationship can be formally expressed as follows:

$$\Delta_i = Y_i^T - Y_i^C \tag{1}$$

However, starting with a direct comparison of potential outcomes could lead to a bias. In this respect, it makes sense to compare the average effect of a school transport program on a student randomly selected from the study population. This effect is called the average treatment effect on the entire population (ATE) and can be expressed as follows:

$$\text{ATE} = E\left(Y_i^T\right) - E\left(Y_i^C\right) = E(Y_{i1}|T_i = 1) - E(Y_{i0}|T_i = 0) \tag{2}$$

The ATE estimator (naive estimator) assumes a simple difference in mean outcomes between the treatment and control groups. However, does this estimator really correspond to what we are trying to measure—that is, the difference between the average results of pupils in the presence of school transport (treated pupils) and the average results of the same pupils in the absence of school transport? Thus, the most suitable estimator is the Average Treatment Effect on the Treated (ATT), which measures the mean impact of the treatment on pupils who actually receive it and is expressed as follows:

$$\text{ATT} = E\left(Y_i^T - Y_i^C | T_i = 1\right) = E(Y_{i1}|T_i = 1) - E(Y_{i0}|T_i = 1) \quad (3)$$

The challenge in this context is that we cannot directly observe how students in the control group would have behaved had they received school transport. Thus, the second component of the expression $E(Y_{i0}|T_i = 1)$ is hypothetical and cannot be directly observed. To address this, we need to identify a suitable proxy for this unobservable scenario, leading to the following valid equation:

$$E(Y_{i0}|T_i = 1) - E(Y_{i0}|T_i = 0) \quad (4)$$

This equation illustrates that the expected mean outcome must be the same for both groups, provided they have an identical distribution of observable characteristics. Consequently, identifying a counterfactual to estimate the ATT becomes a key challenge in causal inference. To address this issue, [25] proposed a solution utilizing a propensity score matching (PSM) approach.

The intuition behind this method is simple: for each company in the treatment group, a statistical twin with observable characteristics similar to those of the treated company must be found in the control group, so that the sample can be considered randomly selected.

PSM creates a statistical comparison group (or counterfactual) by estimating the probability of receiving treatment T, given the observed characteristics X. This probability is known as the propensity score, and is represented as:

$$P(X) = P(T = 1|X) \quad (5)$$

The application of this matching method is based on several key assumptions [26].

The first is called the conditional independence assumption, which assumes that all control variables X contain all the information needed to characterize the potential outcomes.

$$\left(Y_i^T, Y_i^C\right) \perp T|X \quad (6)$$

The second assumption addresses the issue of dimensionality: as the number of characteristics to match increases, the likelihood of failing to find enough corresponding units for each program participant also grows.

Matching is performed through several methods, including Nearest Neighbor Matching, where students with the closest scores are paired; Kernel Matching, which estimates the counterfactual outcome by using weighted averages of all students in the control group; and Stratification Matching, which organizes students with similar covariates into strata (blocks); and also radius matching, which sets a tolerance level for the maximum distance between the propensity scores of treated and untreated students. These estimation methods can be used together, as recommended by Smith and Todd [27], and typically yield similar or closely aligned results. Given the lack of consensus on the most effective estimation method, we opted to use various matching techniques.

3.2 Data Description

To address our research question, we will utilize data from the 2018 National Social Support Survey conducted by the National Human Development Observatory, which surveyed a sample of 3039 households. The survey geographically covered all regions of Morocco.

It is important to note that two sub-samples from the survey will be used to evaluate the impact of the post-matching school transport program, consisting of 5704 individuals for whom data on social support programs for schooling are available. The first sub-sample includes beneficiaries ($T = 1$), while the second consists of untreated individuals ($T = 0$), as shown in Table 1.

Based on the empirical literature and depending on the availability of the National Social Support Survey data, the variables selected are presented in Table 2.

4 Empirical Estimates

Before proceeding with the first stage of estimating the program's impact, we conducted a student's t-test analysis to assess the similarity in observable characteristics between the two groups (Table 3) by comparing the responses of the control group and the treatment group. The objective was to evaluate the quality of the matching. Our analysis reveals that,

Table 1 Description of the treatment variable

	Treatment	Freq.	Percent	Cum.
School transport	$T = 1$	5353	93.85	93.85
	$T = 0$	351	6.15	100.00
	Total	5704	100.00	–

Table 2 Description of variables of interest and control

	Variables	Description
Result variables	Schooling (dropping out)	1: Studying; 0: Gave up studies
	Repetition	0: Never repeated 1: Repeated once 2: Repeated twice 3: Repeated three times 4: Repeated four times
Control variables	Gender	1: Male; 0: Female
	Age	Individual's age
	(Area) area of residence	1: Urban; 0: Rural
	Preschool	1: Pre-schooled; 0: Not pre-schooled
	(Level) School level of student	1: Primary 2: College 3: High school 4: Higher education
	(FatherLevel) father's level of education	0: No level 1: Preschool or Koranic school 2: Primary school 3: College 4: High school 5: Higher education
Control variables	(MotherLevel) mother's level of education	0: No level 1: Preschool or Koranic school 2: Primary school 3: College 4: High school 5: Higher education
	(ClassSize) class size	Number of students in class
	(Tayssir) Tayssir beneficiary	1: Tayssir beneficiary; 0: Non-beneficiary

Table 3 Mean test of control variables by treatment

Variable	N. treated	N. control	T-stat.	P-value
Preschool	5367	351	0.0044	0.8543
Level	5367	351	– 0.0483***	0.0000
Area	5367	351	0.681***	0.0000
Gender	5367	351	– 0.0280	0.3074
Age	5367	351	0.3627	0.2061
FatherLevel	5367	351	– 0.1438*	0.0589
MotherLevel	5367	351	– 0.2934	0.6074
ClassSize	2985	279	–3.0926***	0.0000
Tayssir	5367	351	– 0.1128	0.5682

Note Test of equality of means at (treated) and (control)
With: diff = mean (control)—mean (treated) H0: diff = 0

Table 4 Estimation of propensity scores (Probit model)

Variable	Effet marginal dy/d*x*	Std. err.	*P* > *z*
Preschool	0.0010	0.00838	0.899
Level	0.0529	0.00780	0.723
Area	-0.0993^{***}	0.00716	0.000
Gender	− 0.0045	0.00734	0.537
Age	0.0109^{***}	0.00113	0.000
FatherLevel	0.0025	0.00278	0.352
MotherLevel	0.0028	0.00362	0.430
ClassSize	0.0022^{***}	0.00048	0.000
Tayssir	-0.0253^{***}	0.00721	0.002

Robust standard errors in parentheses
$^{***}p < 0.01$, $^{**}p < 0.05$, $^{*}p < 0.1$

on average, the two groups are comparable across all variables except for education level, environment, father's education level, and class size.

The first step is to estimate the probit model (Table 5), which calculates the propensity scores or probabilities of participation associated with each individual in the sample, based on *X* characteristics.

Table 4 shows the results of the marginal effects after estimating the Probit model.

What is most important in the above table is the sign and significance, which are interpretable and therefore determine the probability of benefiting from the school transport program. From the table, it seems that the probability of participating in the program decreases if the pupil is from a non-rural environment and when the pupil benefits from another social support program for schooling, such as the Tayssir program. On the other hand, it increases as the pupil's age increases and as the class size increases.

Table 5 Estimation of ATT by the 4 approaches on the outcome variable: schooling

Matching approaches	Independent variable: schooling				
	Treated	Control	ATT	Sd.E	T.St
Nearest neighbor	342	2542	0.02^{***}	0.01	2.068
Kernel	342	4745	0.23^{***}	0.02	10.96
Radius	342	4745	0.24^{***}	0.02	11.33
Stratification	342	4745	0.19^{***}	0.02	9.075

Robust standard errors in parentheses
$^{***}p < 0.01$, $^{**}p < 0.05$, $^{*}p < 0.1$

The second step was to test the common support hypothesis, which allowed us to ensure that students were sufficiently similar in terms of observed characteristics unaffected by participation.

Figure 1 demonstrates that the two curves overlap. This indicates that there is common support for matching beneficiaries and non-beneficiaries. Common support ensures that we can identify non-beneficiary students with propensity scores nearly identical to those of the beneficiary students.

Common support refers to the range where the propensity score curves of beneficiaries and non-beneficiaries perfectly overlap. Figure 2 displays the common support range obtained in this study, which spans from [0.00538486 to 0.18360874].

Figure 2 illustrates the distribution of scores within this common support region, categorized by whether the student participated in the program.

After identifying the propensity score and common support region, the treatment group units can be matched with the closest counterparts in the comparison group based on their scores.

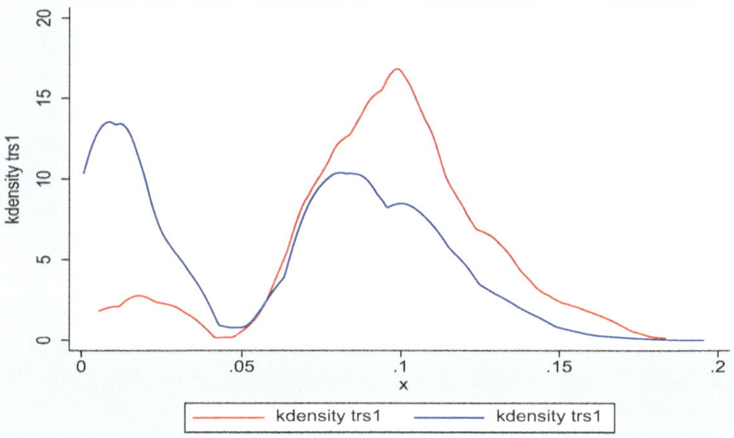

Fig. 1 Distribution of propensity scores before matching. *Source* Author's

Fig. 2 Distribution of propensity scores within the common support region. *Source* Author's

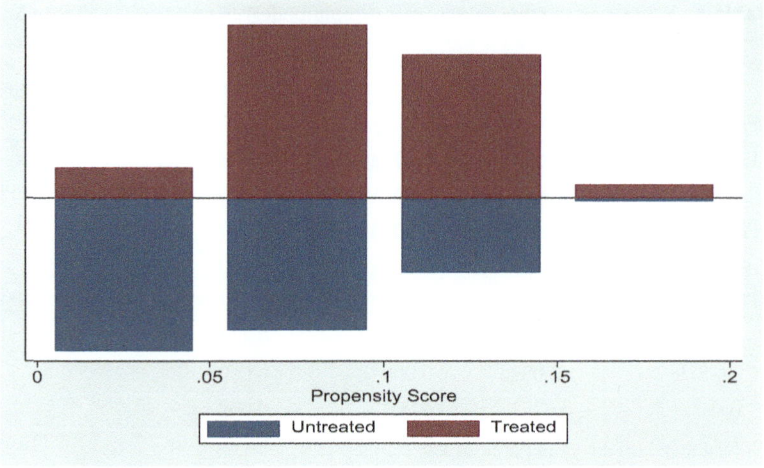

The following table presents the estimated effect of program participation on the two outcome variables: enrollment and repetition.

Table 5 shows the estimated effect of program participation on student enrolment in rural areas. An initial reading of the results shows that the school transport program has a knock-on effect on enrolment. Limiting ourselves to the significant results of public action in the field of school transport, the results show that the effect of both programs on school enrolment is positive, and inversely, when the discussion is about dropping out of school, this impact is strictly significant. In other words, all four PSM approaches show positive and significant coefficients.

In fact, we can see that benefiting from the said program helps to combat school dropout and, thus, to enroll beneficiaries in school.

For example, an estimation using the kernel approach shows that the school transport program leads to a 23% point improvement in school enrolment. In other words, on average, the proportion of individuals in the treatment group was 23% higher than that in the control group. This result corroborates the reality of the school system in disadvantaged environments, especially in rural areas, where schools are far from villages and douars, where dropout rates are higher. Therefore, the introduction of the school transport program helps reduce school dropout and improve children's enrolment.

Looking at another variable that determines school waste, Table 6 presents an estimate of the effect of participation in the school transport program on the repetition rate of students benefiting from the program. The results show that this scheme has a significant impact on the variables of interest. It should be pointed out that the results show that the effect of all four programs on repetition is negative and strictly significant. In other words, all four PSM approaches show negative and significant coefficients. Thus, benefiting from this public measure reduces the number of times students repeat a year.

Table 6 Estimation of ATT by the 4 methods on the outcome variable: repetition

Matching approaches	Independent variable: repetition				
	Treated	Control	ATT	Sd.E	T.St
Nearest neighbor	342	2473	– 0.06	0.05	– 1.03
Kernel	342	4746	– 0.09***	0.03	– 2.56
Radius	342	4746	– 0.10***	0.06	– 4.73
Stratification	342	4746	– 0.08***	0.04	– 1.99

Robust standard errors in parentheses
*** $p < 0.01$, ** $p < 0.05$, * $p < 0.1$

In general, estimations using one or more of the four approaches show that the transport program leads to a reduction in pupil repetition by 11, 10%, and 7%, respectively. On average, the proportion of individuals in the treatment group was 11, 10, and 7% lower than that in the control group. The context of the sector confirms the results obtained insofar as school waste, particularly repetition, increases in rural areas due to factors such as lateness and absence, lack of support and monitoring by households, and several other factors. The widespread use of social support programs for school enrolment, such as school transport, can reduce the scale of these factors, and thus reduce school wastage.

Looking at the same question but measuring its impact in relation to gender, place of residence, and level of education (Table 7).

In terms of gender, the results show that the effect of the program is insignificant for female beneficiaries. On the other hand, the program had a positive and significant impact at the 10% threshold on the enrolment of male beneficiaries compared to non-beneficiaries. The proportion of boys benefiting from the school transport program was 2.7% higher, on average, than that of non-beneficiaries.

Regarding place of residence, the estimation of the average treatment effect using the "NN nearest neighbor" approach

Table 7 Estimated ATT by gender, area and level on schooling

		Independent variable: schooling			
		Coeff.	Std. err.	Z	P > Z
Gender	Male	0.0171679	0.0134565	1.28	0.202
	Female	0.0272816*	0.0143719	1.90	0.058
Area	Rural	0.02381**	0.0106347	2.24	0.025
	Urban	0.00000017***	1.15e−17	3.61	0.000
Level	Primary	0.00000024***	2.84e−17	7.62	0.000
	College	− 0.004902	0.0048899	− 1.00	0.316
	High school	0.0614219	0.0395081	1.55	0.120
	Higher education	0.2269841**	0.0979681	2.32	0.021

Robust standard errors in parentheses
*** $p < 0.01$, ** $p < 0.05$, * $p < 0.1$

Table 8 Estimated ATT by gender, area, and level on repetition

		Independent variable: repetition			
		Coeff.	Std. err.	Z	P > Z
Gender	Male	− 0.0720089	0.0582737	− 1.24	0.217
	Female	− 0.0901863	0.0643154	− 1.40	0.161
Area	Rural	− 0.0773707*	0.0457681	− 1.69	0.091
	Urban	− 0.1304945	0.1439164	− 0.91	0.365
Level	Primary	− 0.0660008	0.0539943	− 1.22	0.222
	College	0.0515139	0.0588843	0.87	0.382
	High school	− 0.0587461	0.0996318	− 0.59	0.555
	Higher education	0.0896825	0.1376192	0.65	0.515

Robust standard errors in parentheses
*** $p < 0.01$, ** $p < 0.05$, * $p < 0.1$

indicates that the program yields a positive and significant result at the 10% level for individuals in the treatment group living in both environments. The impact of the transport program is particularly significant for those residing in rural areas. Indeed, the proportion of rural pupils benefiting from school transport was 2.3% points higher than that of rural pupils who did not participate in the transport program.

Based on education level, the results show that the effect of the school transport program is significant for beneficiaries in primary and higher education, with a positive and significant impact at the 5% threshold on the enrolment of beneficiaries in primary education and those with higher education compared with non-beneficiaries. The proportion of primary school beneficiaries is slightly higher than that of non-beneficiaries in the school transport program, while the proportion of higher education beneficiaries is very high, at 22.69% higher on average than that of non-beneficiaries in the program.

Analysis of the impact of the school transport scheme on grade repetition reveals that, in terms of gender, the program had no significant impact on grade repetition for boys and girls who benefited from it compared with non-beneficiaries (Table 8).

Regarding place of residence, the results show that the transport program has a negative and significant result at the 10% threshold for individuals in the treatment group living in rural areas. Indeed, the proportion of rural pupils benefiting from school transport was 7.7% points lower than that of rural pupils who did not participate in the program.

The current context of the Moroccan economy is based on several constraints that hamper the success of several social policies, particularly in the education sector.

Our study shows that access to school transport in rural areas contributes significantly to improving school enrolment and reducing dropout and repetition rates. These results are in line with findings from similar studies worldwide that explore the relationship between school transport and academic success. In this context, [28] found that, in the Brazilian context, the majority of pupils benefiting from a school transport program were very satisfied, with regularity being the most highly valued criterion, while teachers felt that punctuality had improved, dropout had decreased, and academic performance had increased. The state must guarantee the right to education for all children and adolescents, especially those living in rural areas. To guarantee this right, governments are also required to provide the necessary resources to ensure access to schools and regular attendance. The provision of free and efficient school transport for all pupils in the public school network stands out among these resource needs [29].

Admittedly, the results of the study revealed a significant impact of the social support program for school transport on improving school enrolment and combating school wastage. However, if we focus on the number of beneficiaries of this public action to support schooling, given the various problems in the education sector other than the generalization of schooling (quality of the system). The success of social programs in Morocco depends on the development of demand-driven elements of economic inclusion, in particular, by complementing well-targeted cash transfer and support programs with a range of social services aimed at improving school enrolment and combating school wastage, but primarily to improve the quality and efficiency of education. To ensure that public policies to support school enrolment have a maximum impact on the target population, public authorities should pay greater attention to all the factors linked to the national context, which has a direct and indirect impact on inclusive development and the improvement of the education system as a whole.

The vast sector is weakly protected by a poorly structured system, based on generalization and deteriorating quality. The vast majority of education funding was provided by the state. It allocates almost 22% of its budget or 7% of GDP. This public effort is weighed down by a very high wage bill to the detriment of investment in training programs, research and development, human resources training, and the renewal of the training provided, while taking into account the various forms of inequality in the sector.

5 Conclusion

The goal of this research is to assess the impact of the public school transport program in rural areas on enhancing school enrollment among the target population and reducing school wastage. Using data from the National Social Support Survey conducted by the National Human Development Observatory, we employed a propensity score matching (PSM) approach.

The results of the estimations show that there is an overall improvement in school enrolment of school transport beneficiaries compared to non-beneficiaries with the same characteristics. Using different matching techniques, the program's effect is positive and significant on enrolment and negative and significant for grade repetition. For example, the Tayssir conditional cash transfer program in Morocco helped reduce school dropout rates overall, although it did not encourage the poorest children to attend school [30]. The problems of absenteeism and dropout have an impact on the effectiveness and ability to capitalize on the skills and competencies targeted by public education programs. In short, all the limitations that hamper the effectiveness of educational programs are system-related constraints and socio-economic constraints due to the precariousness of the rural population [31]. Social support for school enrolment in rural areas significantly improves quantitative indicators of school enrolment and significantly reduces school dropout, despite a massive increase in repetition and backwardness [32].

To reinforce the impact of our various programs, we need to focus on spending and efforts, primarily on those in precarious situations. The challenge lies in determining the type of action needed to reduce the determinants of social inclusion and exclusion; hence, the importance of the new Single Social Register (RSU) system, which aims to strengthen the ability to target the poor and vulnerable populations most affected by social action.

In general, this work has shown that improving educational indicators, and thus the success of the pillars of an education system, is possible but complex; in particular, financial aid alone is not enough. They must be accompanied by measures to strengthen the supply of education on the one hand, and actions on the demand side on the other. Drawing on international experience, we recommend, for example, the implementation of a number of complementary measures aimed at improving the quality of educational provision and not just the question of generalization. In this respect, we refer to the Indian experience of "Multilingual Education Intervention," aimed at girls with no command of languages other than those spoken at home. Once at school, these girls are faced with a handicap due to their lack of command over the language of instruction. This example illustrates a better match between the reason for dropping out of school (the language of instruction) and the measures introduced to remedy it. This logic could be transposed to the Moroccan context, where some children from Berber families in rural areas face a real obstacle linked to classical Arabic used as the language of instruction [33]. The study also revealed that despite the state's efforts in the area of school enrolment, repetition and dropout rates are still high, legitimizing the need to back up these financial strategies with non-monetary measures such as tutoring for pupils with learning difficulties. The strategic choice of boarding schools to accommodate pupils, combat absenteeism and lateness due to school transport difficulties, and raise awareness and inform households about the importance of girls' schooling, especially in rural areas, and spread the culture of education to the public. In terms of research perspectives, it is important to strive to understand parents' underlying motivations and interests in their children's education while ideally addressing gender issues.

The success of social and educational reform, especially in Morocco, depends on taking into account the real needs of the poor and vulnerable, making optimal use of the means and resources available to the state, and involving all stakeholders in the sector.

References

1. Engels F (2010) Socialism: utopian and scientific. In: Capaldi N, Lloyd G (eds) The two narratives of political economy. https://doi.org/10.1002/9781118011690.ch25
2. Rawls J (1999) A theory of justice: revised edition. Harvard University Press. https://doi.org/10.2307/j.ctvkjb25m
3. Sen A (2000) Social exclusion: concept, application, and scrutiny. Asian Development Bank. https://www.adb.org/sites/default/files/publication/29778/social-exclusion.pdf
4. Nussbaum M (2009) The capabilities of people with cognitive disabilities. Metaphilosophy 331–351. https://doi.org/10.1111/j.1467-9973.2009.01606.x
5. Aucejo EM, Romano TF (2016) Assessing the effect of school days and absences on test score performance. Eco Educ Rev 70–87. https://doi.org/10.1016/j.econedurev.2016.08.007
6. Gershenson S, Jacknowitz A, Brannegan A (2017) Are Student absences worth the worry in U.S. Primary Schools? Educ Financ Policy 12(2):137–165. https://doi.org/10.1162/EDFP_a_00207
7. Kirksey JJ (2019) Academic harms of missing high school and the accuracy of current policy thresholds: analysis of preregistered

administrative data from a California School District. AERA Open 5(3). https://doi.org/10.1177/2332858419867692
8. Gottfried MA (2019) Chronic absenteeism in the classroom context: effects on achievement. Urban Educ 54(1):3–34. https://doi.org/10.1177/0042085915618709
9. Hopson LM, Lidbe AD, Jackson MS, Adanu E, Li X, Penmetsa P, Abura-Meerdink G (2022) Transportation to school and academic outcomes: a systematic review. Educ Rev 76(3):648–668. https://doi.org/10.1080/00131911.2022.2034748
10. Deakin M, Reid A (2018) Smart cities: under-gridding the sustainability of city-districts as energy efficient-low carbon zones. J Clean Prod 39–48. https://doi.org/10.1016/j.jclepro.2016.12.054
11. Chen P, Jiao J, Xu M, Gao X, Bischak C (2018) Promoting active student travel: a longitudinal study. J Transp Geog 265–274. https://doi.org/10.1016/j.jtrangeo.2018.06.015
12. Kingdon GG (2007) The progress of school education in India. Oxf Rev Eco Pol 168–195. https://doi.org/10.1093/icb/grm015
13. Haepp T, Lyu L (2018) The impact of primary school investment reallocation on educational attainment in rural China. J Asia Pac Econ 23(4):606–627. https://doi.org/10.1080/13547860.2018.1515004
14. Mbiti IM (2016) The need for accountability in education in developing countries. J Eco Persp 109–132. https://doi.org/10.1257/jep.30.3.109
15. Jayachandran U, Socio-economic determinants of school attendance in India. Working papers 103, Centre for Development Economics, Delhi School of Economics. http://www.cdedse.org/pdf/work103.pdf
16. Aggarwal S (2018) Do rural roads create pathways out of poverty? Evidence from India. J Dev Econ 133:375–395. https://doi.org/10.1016/j.jdeveco.2018.01.004
17. Edwards DS (2024) Another one rides the bus: the impact of school transportation on student outcomes in Michigan. Educ Financ Policy 19(1):1–31. https://doi.org/10.1162/edfp_a_00382
18. Blagg K, Chingos M, Corcoran SP, Cowen J, Denice P, Gross B, Valant J (2017) Student transportation and educational access: how students get to school in Denver, Detroit, New Orleans, New York City, and Washington-DC. Urban Institute Student Transportation Working Group. https://www.urban.org/sites/default/files/publication/88481/student_transportation_educational_access_0.pdf
19. Gottfried MA, Kirksey JJ (2017) "When" students miss school: the role of timing of absenteeism on students' test performance. Educ Res 46(3):119–130. https://doi.org/10.3102/0013189X17703945
20. Gottfried MA, Ozuna CS, Kirksey JJ (2021) Exploring school bus ridership and absenteeism in rural communities. Earl Childh Res Quart 236–247. https://doi.org/10.1016/j.ecresq.2021.03.009
21. Cordes SA, Rick C, Schwartz AE (2022) Do long bus rides drive down academic outcomes? Educ Eval Policy Anal 44(4):689–716. https://doi.org/10.3102/01623737221092450
22. Alladatin J, Barjit R (2022) Analysis of students' perceptions of the school climate: comparative case study of 2 schools in Morocco. In: Langran L, Henriksen D (eds) Proceedings of SITE interactive conference, p 168. Online: Assoc Adv Comput Educ. https://www.learntechlib.org/primary/p/221649/
23. Azdagaz F, Liouaeddine M, Zirari O (2025) Evaluation of the impact of school canteen programs on schooling and combating school wastage of students in rural schools in Morocco. Dev Stud Res 12(1). https://doi.org/10.1080/21665095.2025.2449935
24. Rubin DB (1974) Estimating causal effects of treatments in randomized and nonrandomized studies. J Educ Psych 66(5):688–701. https://doi.org/10.1037/h0037350
25. Rosenbaum PR, Rubin DB (1985) The bias due to incomplete matching. Biometrics 41(1):103–16. https://doi.org/10.2307/2530647
26. Gertler PJ, Martinez S, Premand P, Rawlings P, Vermeersch CM (2016) Impact evaluation in practice. W Bank Pub. https://doi.org/10.1596/978-1-4648-0779-4
27. Smith JA, Todd PE (2005) Does matching overcome LaLonde's critique of nonexperimental estimators? J Economet 305–353. https://doi.org/10.1016/j.jeconom.2004.04.011
28. Nascimento MVLDA, Andrade MOD (2021) School transportation program as means to improve public education in a minor rural town in Northeastern Brazil. Ensaio: Avaliação e Políticas Públicas em Educação 30:182–206. https://doi.org/10.1590/S0104-40362021002903093
29. Carvalho WL, Yamashita Y, Aragao JJQD (2016) Rural school transportation in brazil as an essential factor for the education policy 2(1):263627. https://www.ijntr.org/download_data/IJNTR02010004.pdf
30. Mourji F, Ikira M, Ricard C (2021) Le soutien à la scolarisation au Maroc: Portée et limites du programme Tayssir. RefEco, Billet de blog n°16. https://refeco.org/wp-content/uploads/2021/07/Billet_16.pdf
31. Bouzakria B (2020) Le programme de post alphabétisation au Maroc: entre cohérence et divergence. Moroccan J Quant Qual Res 2(3):92–112. https://doi.org/10.48379/IMIST.PRSM/mjqr-v2i3.23893
32. ONDH (2019) Évaluation d'impact des programmes d'appui social à la scolarisation. https://www.men.gov.ma/Fr/Documents/evaluation%20appui_social_ondh.pdf
33. Sylva K (2014) The role of families and pre-school in educational disadvantage. Oxf Rev Educ 40(6):680–695. https://doi.org/10.1080/03054985.2014.979581

Detection of Facial Emotion with Deep Learning Models and Contribution of Inception-V3

Fouad Lehlou, Adil El Makrani, and Jalal Laassiri

Abstract

Facial expression recognition (FER), also known as facial emotion detection, is an emerging domain within computer vision and machine learning. This field has broad implications to education, psychology human–computer interaction and market analysis, etc. Facial expressions recognition need to correctly identify facial expressions in order to solve requests of these areas. In this work, we investigate emotion from facial expression detection in FER-2013 dataset by just extracting label. It investigates the accuracy of two different Convolutional Neural Network (CNN) architectures are: Inception-V3 and Sequential model. So, we have to classify facial expressions (Accounting for 7 classes: anger, fear, disgust, happiness, surprise, sadness, and a neutral category). Empirical results also indicates that the Inception-V3 model fine-tuned performs better than the one most sequential model for predicting the labels of FER-2013.

Keywords

Machine learning · Deep learning · Transfer learning · Inception-V3

1 Introduction

Non-verbal communication works through the face (of man as a rule) in a very close medium for feelings expression. Facial expressions can send an extensive array of not-so-subtle emotions. A smile as an example implies happy and a frown of the brow sadness negative. Similarly, widened eyes probably suggest surprise and a mouth that is curled expresses disgust [1]. Making these facial cues work at one with machines will go a long way to make human–machines interactions more fluent.

Facial Expression Recognition (FER) constitutes a paramount pillar of non-verbal communication in this paper, as it helps Human–Machine Interaction (HMI) system interrelate an emotion from the facial appearance of an observer and understand his/her intention.

FER (Facial Expression Recognition) is a key pillar of Haptic communication that permits the Human–Machine Interaction (HMI) systems to comprehend the emotions and intentions of you, human. This technology is used in a diverse range of industries like education, security surveillance, gaming, and even social robotics.

FER contributes to the analysis of social features (age, sex, cultural background, etc.) in behavioral science and, in the medical area it is an extremely important a tool in monitoring psychological and physiological conditions that warrant pain detection depression recognition, and anxiety determination; support the treatment of cognitive impairments.

Humans have an almost instinctive ability to interpret most facial expressions but until recently machines have struggled to recognize faces reliably even when they are expressing themselves. From the FER obstacles, since many of these factors including pre-processing, feature extraction, and classification need to work seamlessly in all configurations the complexity and range of facial variation becomes apparent i.e. varying input conditions, head poses and environments {lighting condition).

F. Lehlou (✉) · A. E. Makrani · J. Laassiri
Laboratory of Research in Informatics, FS, UIT, Kenitra, Morocco
e-mail: Fouad.lehlou@uit.ac.ma

A. E. Makrani
e-mail: adil.elmakrani@uit.ac.ma

J. Laassiri
e-mail: Laassiri@uit.ac.ma

A. Moumen et al. (eds.), *Artificial Intelligence and Data Analytics for Innovative Applications in Engineering, Sustainability and Technology*, Sustainable Artificial Intelligence-Powered Applications, https://doi.org/10.1007/978-3-031-90318-2_5

Even with deep learning being applied to FER, the problem is not totally solved, regarding its feature learning ability. Deep neural networks must be trained with large amount of data to prevent overfitting. Nevertheless, the available facial expression databases are not enough to adequately train the neural networks that are implemented in the known deep frameworks in various tasks at this stage where access to object recognition.

However, we have the personal factors of background, gender, and age to say nothing with ethnic or an expression intensity that contribute more for inter-subject differences. Furthermore, expression recognition is also difficult due to pose variability, occlusions and lighting variability in unconstrained facial expression scenarios. These, although not exclusive to facial expressions create difficulty for neural networks to handle and learn the same, ultimately aiding in the reduction of intra-class variation and accurate training of an expression recognition model.

This study describes and purports to advance the approach in addressing these challenges. Our transfer learning Inception-V3 architecture (compared to a Sequential model) enhances the performance of FER. Deep learning methods such as Convolutional Neural Networks (CNNs), for example yield great results on feature extraction and classification. CNNs which was inspired by biological model can learn hierarchical features automatically. This work requires transfer learning techniques because of the data sparsity and of time-saving to train the model (you cannot do a new one), rather than doing it from scratch. In our end-to-end framework, we use Convolutional Neural Networks for facial expression labelling, where the input is in image form which contains neutral, happiness, anger, surprise, sadness, and disgust. We evaluate the models in detail.

2 Facial Expression Background and Overview

Deep learning is a subfield of machine learning that learns representations from data through stacked feature abstraction layers (e.g., text, images, or sounds) and makes predictions directly on raw data. How it is done: via neural network architecture. This deep is used in the fact that the more layers a neural network has, the deeper it is (the number of layers within a network is deep).

The aforementioned algorithms are based on a learning in loop fashion, where with each execution of a cycle model adjusts a little bit its predictions by making very small modifications. This technique enables a computer to be molded so that it learns (in the manner of humans) and adapts by observing examples [2].

Deep learning is widely used in few aspects: mainly self-driving automobile sector that enables pedestrian detection and sign-board recognition [3]. The accuracy of these models has been noticed to be so remarkable, sometimes surpassing humans in certain tasks.

In order to get this level of performance, deep learning models take tremendous amounts of labeled data for training. They use neural network architectures that are typically multiple layered in nature to find relevant patterns without the need for an expert feature selection [4].

2.1 Convolution Neural Network

Perhaps most widely known is a Convolutional Neural Network (CNN) which is a deep learning architecture which has achieved outstanding results on visual data sample analysis and interpretation. Their technique of automatically lifting and learning hierarchical image features has become the heart of contemporary computer vision implementations. They have worked extremely well across several computer vision tasks, demonstrating strength in object detection/localization within natural images [5, 6]. Current State of the Art: The literature on automatic facial expression analysis [7–9] has already formalized standard algorithmic approaches involving Facial Expression Recognition (FER). They are mainly in the domain of conventional approaches, yet without having adequately discussed deep learning techniques as well.

In recent days there is rising demand of deep learning based Facial Expression Recognition (FER) [10]. Nowadays, most of the works in this regard are not fully formed to the extent they do not specifically dive into the characteristics of FER datasets and the details behind technical heavy deep FER. This work thus investigates deep learning for FER in a more systematic way, including videos (i.e. image sequences) and static images. For those new to this space, we aim to provide an in-depth presentation of the modular architecture and cutting edge techniques of deep FER.

Common architecture it contains layers like input, output, and intermediate such as convolutional, fully connected, max-pooling, etc. Depending on the network these will can be arranged differently across different CNN architectures.

2.2 Transfer Learning

Transfer learning is the process of applying knowledge to new but related tasks learned from previous one [11]. This lets deep learning models re-use the feature learnt in previous tasks through parameter sharing to change their objectives but train these at same layer/features (as performed by various deep object detectors). In neural networks, we update some layers selectively even while keeping others as they are. Those frozen layers keep their learnt feature maps so that only well aligned parts of model will be trained. This approach is especially useful with pre-trained models on large datasets since it minimizes computational effort by directing learning action toward selected portions of the architecture [12].

2.3 Related Work

Some of recent developments in facial expression classification are really good, where researchers demonstrated a lot of efforts in this area [13–15]. Numerous researchers have published deep learning techniques, such as [1], Convolutional neural networks (CNNs) to solve for the representation of facial expression.

We introduced Original CNN architecture [16] and the CNN architecture used for developing Facial Action Coding System (FACS) model as one of the similarities. Both break the input data down to individual features and data goes through each layer of the these models in constructing increasingly complex representations based on the features that are selected (extracted) in proceeding layers in helping the classification. That being said CNNs are different from FACS model [17]. And you can even train CNNs with enough training data to recognize some parts overall (e.g., smiling) without having to define structural representations for every visual component.

In the EmotiW 2015 challenge for Facial Expression Recognition (FER) using CNNs, Zhang and Yu. With the aid of an extensive [18] CNN model and random image partitions they achieve some 2–3% better performance. Kahou et al. Both experiments used CNNs [19] for two separate investigations. In meal one, they trained a default CNN by using the Acted Facial Expression in the Wild(fcn18) dataset [15] and in the following one used Toronto Face dataset as second-experiment. In both experiments, same called primary and neutral expression with 6 universal facial expression to be more accuracy with networks.

Hamester et al. [20] used CNNs for facial images in their task and showed a pixel-level accuracy of 94.4%, surpassing other methods by combining information from the network.

This work developed a common CNN algorithm for average facial expressions prediction. The technique combines classical solutions, Transfer Learning with convolutional neural networks (Inception-V3) and Sequential model for sequence of datasets. We improve the state-of-the-art by adding layers for classification, albeit cross comparing two models.

3 Methodology

This part of the paper introduces the approach we have used in this research, divided into three main phases: Data preprocessing, Feature extraction, and Classification. Figure 1 is the visual of our methodology.

3.1 Face Expression Dataset

The Fer2013 dataset contains approximately 30,000 RGB facial images showing a range of expressions, all standardized to a size of 48×48 pixels. The main labels in this dataset fall into seven distinct categories: 0 for Angry, (1) for Disgust, (2) for Fear, (3) for Happy, (4) for Sad, (5) for Surprise, and (6) for Neutral. It is noteworthy that the Disgust category has the fewest images, with a total of 600, while the other labels are more balanced, each comprising nearly 5000 samples.

3.2 Create Training and Testing Image Sets

The datasets were divided into two subsets, with the division performed randomly to avoid bias. Eighty percent of the selected images from each subset were assigned to the training dataset, while the remaining 20% were allocated to the test dataset.

3.3 Pre-trained Models

For our proposed methods in facial expression recognition, employed in this study were pre-trained models as well as Sequential Convolutional Neural Network (CNN) models such as Inception-V3 and Sequential respectively. We conducted a comparative study of these models to determine which one performs better in recognizing emotional differences.

Pre-trained model means the previously trained model, that over all standard dataset on a relevant problem i.e. the problem that we are trying to solve right now.

Our proposed facial expression recognition system diagramed in the form of three stages (Image Pre-Processing, Facial Feature Extraction with Feature Classification using CNN) (in Fig. 1).

Inception-V3. A Convolutional Neural Network (CNN) model mainly used for ad image classification tasks of type Inception-V3 model is a hyper-hyper-optimized version of the original Inception-V1 which was first presented as GoogLeNet in 2014. The Inception-V3 convolutional was a Google team doing much more advanced and polished version of their predecessor [In inception 1].

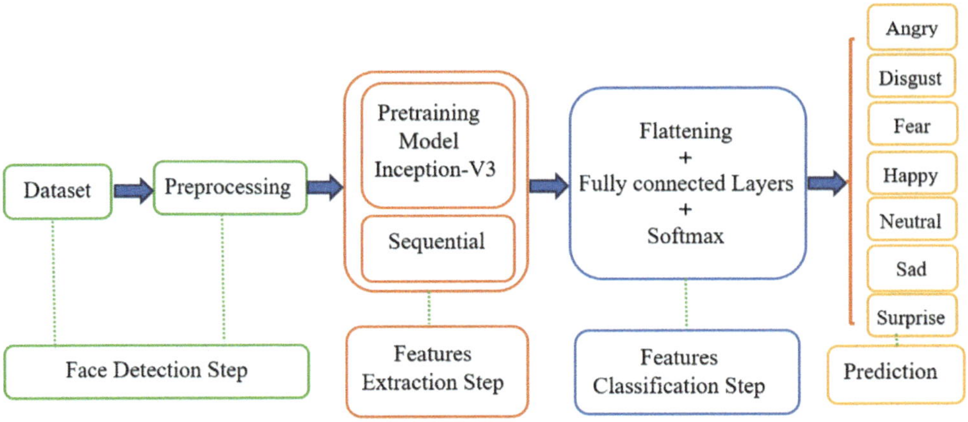

Fig. 1 The process architecture for a transfer learning model (Inception-V3) and Sequential model applied to emotion recognition

Table 1 Comparison of results with approaches in metric measurements

Models	Sensitivity (%)	Specificity (%)	Accuracy (%)	$F1$-score (%)
Sequential	68	92	68	68
Inception-V3	71	86.2	71	71

In this strain for higher performance of the model, Inception-V3 model merges several methods to optimize the network. The formal version of this new model was made available in 2015 with 42 layers and a decreased error rate from earlier versions.

3.4 Train a Classifier with Additional Layers Using a Pre-trained Models (CNN)

With the procedure, we did the extract of trained image features from pre-trained Convolutional Neural Network (CNN) with the transfer learning for feature extraction. Specifically, feature extraction was done using an activation method on the fc1000 layer (i.e. before the classification layer). Then the features were extracted and used in the training/testing phase of classifier alongside the other layers.

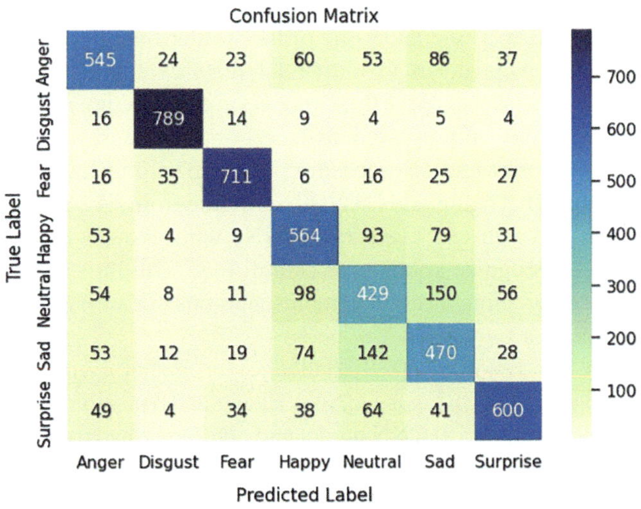

Fig. 2 Inception-V3 cross-validation confusion matrix for 7-Class FER 2013 emotional dataset

3.5 Evaluation Procedure

The evaluation of a method's performance in analyzing facial expression images often includes assessing specificity, sensitivity, accuracy, and the $F1$ score (Table 1).

4 Experimental Results and Discussion

A Tesla K80 GPU with 12 GB RAM from Google Colab was used in the suggested system. The method evaluation was mainly done with Python, a tool used for both classification and selecting features. 80% of the data was used for the training subset to train the network to classify, and then 20% left as a testing subset, evaluating if a face image is in-class with respect to one of our facial expression labels during training.

Results of Training a Convolutional Neural Net with Transfer Learning Network was evaluated by its average accuracy about performance. This section depicts confusion matrices in Figs. 2 and 3 that demonstrate the recognition accuracies for a total of seven facial expressions, as a convolutional weights with highest accuracy rate were used to generate training set. Inception-V3 71% and Sequential 68% for in the identification.

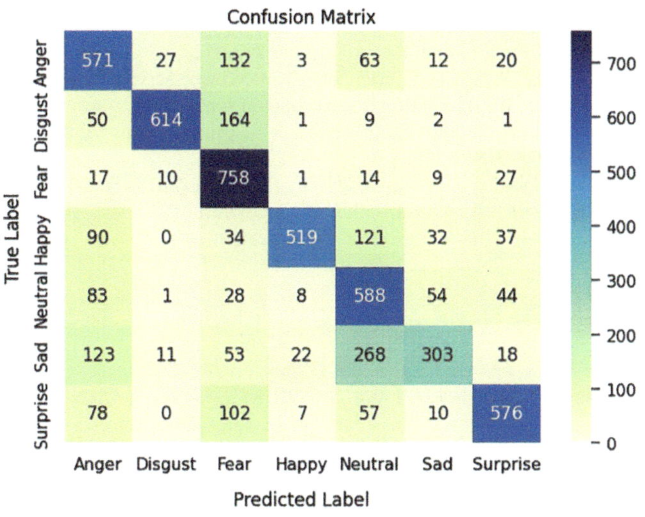

Fig. 3 Sequential cross-validation confusion matrix for 7-Class FER 2013 emotional dataset

The experimental results for Inception-V3 and Sequential models on the FER2013 dataset suggest several key performance distinctions between the two approaches. Let us break down these metrics:

Sensitivity: Inception-V3 outperforms the Sequential model with a sensitivity of 71%, compared to Sequential's 68%. Sensitivity, reflecting a model's ability to correctly identify positive cases, indicates that Inception-V3 is slightly more effective at detecting facial expressions on the FER2013 dataset. This suggests that Inception-V3 may generalize better, particularly in recognizing subtle expressions across diverse samples.

Specificity: Sequential is much more specific at 92% versus Inception-V3's 86.2% Specificity: how well the model can identify negative cases (i.e., correctly labeling non-expressions or incorrectly classified expressions).

Inheritance with a specificity of higher 94% (Sequential) against only 86.2% (Inception-V3), higher specificity in Sequential may suggest a strict classification approach, having less number of false positives but may also be failing to catch some nuanced expressions that Inception-V3 learns.

Accuracy: In terms of total accuracy, Inception-V3 does a better job than the Sequential model (71% versus 68%). This measures how well the model can classify positive and negative samples in all categories. This small accuracy boost has that heading on Inception-V3, hinting to how good it is at this. The more sophisticated structure permits it to pick up finer variations and is better at seeing differences in facial expressions.

$F1$ score: Inception-V3 is outperformed Inception-V3 on $F1$ with a score of 71% vs. the Sequentials 68%. This implies that while Inception-V3 classifies expression better Inception performs an excellent job in keeping false-positives low, and precise. Therefore it shows a consistent performance, even with both exotic and normal expressions in FER2013 dataset.

Discussion: Our $F1$ score results can be analyzed and compared against what Ashi Agarwal and Seba Susan presented in their study entitled Emotion Recognition from Masked Faces Using Inception-V3 [21] for the want of a better intuition regarding how good or bad Inception-V3 does in terms of recognizing and emotion. Study [21]: Inception-V3 on unmasked FER-2013 images E, Eg pro did $F1 = 0.68$ (for their research). They evaluate models performance in recognizing facial expressions when full visible, which mean its good non-occluded images representable model with confidence of feature extraction (as observed from performance trends).

Our work likewise outperforms in terms of $F1$ score (0.71) when applied Inception-V3 on FER-2013. This advantage might be due to technical refinements such as in preprocessings or hyperparameter adjustment, which gave a little boost to the recognition performance. Thus, this testing will also imply about potent Inception-V3 in emotion recognition tasks and show us the necessity of test training strategies, even when we are using the same dataset as FER-2013.

The results affirm that Inception-V3 is robust for both emotions in a wide range of facial expression datasets, indicating its effectiveness for multiple applications. The slight $F1$ score does not substantially improve our results highlight the capability of the model in adapting to FER-2013 variations and the vestiges of transfer learning and hyperparameters configured for facial expression analysis. This gives us important directions for future research in the space of developing further improved deep learning based emotion recognition.

5 Conclusion

Our findings transfer learning with Inception-V3 in particular, on FER-2013 dataset are a demonstration of the fact that much could be done by leveraging Transfer learning aspects. When fine-tuning, with well-chosen preprocessing and hyperparameters Inception-V3 is able to beat the state-of-the-art benchmarks for un-mangled facial expressions. It illustrates the excellent capability of Inception-V3 (great for handling large and complex datasets like FER-2013) being robust enough and predictable on unseen data. The results further highlight the capability of the model in emotion recognition problems, where high accuracy and generalization are desired (such as scenarios in which it should perform consistently between different facial expressions).

In the future, the efficient fine-tuning of Inception-V3 with other datasets can lead an intension to explore if Inception is capable of accommodating the different demographic groups in terms of face variations. Also developing the use of better domain-agnostic feature selection methods and/or multimodel ensembling methods for more accurate classification, or even in out-of-the-worldly situations (no image of a face), such as occlusion or facial masks. The existing study contributes to this line of research on transfer learning in face recognition, toward the goal of improving the robustness of emotion recognition models for use in application domains such as Human–Computer interaction, security, or mental health diagnostics.

Acknowledgements We would like to express our heartfelt thanks to all people who supported us during the successful running of this research. Beyond, we would also like to extend our thanks here to our funder which provided the necessary resources and climate to carry out this study. Moreover, we gratefully acknowledge the datasets authors and previous research that guided our approach. Their contributions have been instrumental in defining the current state-of-the-art for research in facial expression recognition and deep learning efforts.

References

1. Ekman P, Keltner D (1970) Universal facial expressions of emotion. Calif Ment Health Res Dig 8(4):151–158
2. Sanchez-DelaCruz E, Pozos-Parra P (2018) Machine learning-based classification for diagnosis of neurodegenerative diseases. Instituto Tecnologico y de Estudios Superiores de Occidente, Tlaquepaque, Jalisco
3. Schmidhuber J (2015) Deep learning in neural networks: an overview. Neural Netw 61:85–117. https://doi.org/10.1016/j.neunet.2014.09.003
4. Ji S, Xu W, Yang M, Yu K (2013) 3D convolutional neural networks for human action recognition. IEEE Trans Pattern Anal Mach Intell 35(1):221–231. https://doi.org/10.1109/TPAMI.2012.59
5. Satauri I (2023) New GIS approach using machine learning algorithm for early floods detection. Moroccan J Quant Qual Res 6(1):39–58. https://doi.org/10.48379/IMIST.PRSM/mjqr-v5i1.38974
6. Oguntokun JA, Adewusi AO (2023) Residential rental applications screening: a comparative performance of feedforward and recursive neural networks architectures. Moroc J Quant Qual Res 5(3). https://doi.org/10.48379/IMIST.PRSM/mjqr-v5i3.42602
7. Zeng Z, Pantic M, Roisman GI, Huang TS (2008) A survey of affect recognition methods: audio, visual, and spontaneous expressions. IEEE Trans Pattern Anal Mach Intell 31(1):39–58. https://doi.org/10.1109/TPAMI.2008.52
8. Bettadapura V (2012) Face expression recognition and analysis: the state of the art. Tech Report, arXiv:1203.6722
9. Rao J, Su X (2004) A survey of automated web service composition methods. In: International workshop on semantic web services and web process composition, pp 43–54. Springer. https://doi.org/10.1007/978-3-540-30581-1_5
10. Giannopoulos P, Perikos I, Hatzilygeroudis I (2018) Deep learning approaches for facial emotion recognition: a case study on FER-2013. In: Advances in hybridization of intelligent methods. Springer, Cham, pp 1–16. https://doi.org/10.1007/978-3-319-66790-4_1.
11. Torrey L, Shavlik J (2010) Transfer learning. In: Handbook of research on machine learning applications and trends: algorithms, methods, and techniques. IGI Global, pp 242–264
12. Hussain M, Bird JJ, Faria DR (2018) A study on CNN transfer learning for image classification. In: UK workshop on computational intelligence, pp 191–202
13. Muttu Y, Virani HG (2015) Effective face detection feature extraction neural network based approaches for facial expression recognition. In: IEEE international conference on information processing (ICIP), pp 102–107
14. Mousavi N, Siqueira H, Barros P, Fernandes B, Wermter S (2016) Understanding how deep neural networks learn face expressions. In: IEEE international joint conference on neural networks (IJCNN). https://doi.org/10.1109/IJCNN.2016.7727203
15. Al-Sumaidaee SA (2015) Facial expression recognition using local Gabor gradient code-horizontal diagonal descriptor. In: School of Electrical and Electronic Engineering, Newcastle University, England, UK. https://doi.org/10.1049/cp.2015.1766
16. Santiago HC, Ren T, Cavalcanti GDC (2016) Facial expression recognition based on motion estimation. In: 2016 International joint conference on neural networks (IJCNN), Vancouver, BC, Canada. Electronic ISSN: 2161-4407. https://doi.org/10.1109/IJCNN.2016.7727391
17. Li J, Lam EY (2015) Facial expression recognition using deep neural networks. In: Imaging systems and techniques (IST), IEEE international conference on, pp 1–6. https://doi.org/10.1109/IST.2015.7294547
18. Yu Z, Zhang C, Image-based static facial expression recognition with multiple deep network learning. In: ICMI proceedings. https://doi.org/10.1145/2818346.2830595
19. Ebrahimi Kahou S et al (2013) Combining modality specific deep neural networks for emotion recognition in video. In: The 15th ACM international conference on multimodal interaction. ACM. https://doi.org/10.1145/2522848.2531745
20. Hamester D, Barros P, Wermter S (2015) Face expression recognition with a 2-channel convolutional neural network. In: International joint conference on neural networks (IJCNN), pp 1787–1794. https://doi.org/10.1109/IJCNN.2015.7280539
21. Agarwal A, Susan S (2023) Emotion recognition from masked faces using Inception-v3. In: 5th international conference on recent advances in information technology (RAIT), pp 1–6. IEEE, Dhanbad. https://doi.org/10.1109/RAIT57693.2023.10126777

Main Topics of the AI Applications for Water Research: Terms and Concepts of the Recent Trends

Hicham Boutracheh, Yassine Mouhssine, Rachid El Ansari, Nezha Mejjad, Mohammed El Bouhadioui, and Aniss Moumen

Abstract

Economic development and population growth are putting considerable pressure on drinking water supplies, and the acceleration of climate change is making the situation more difficult. Given the complexity of water-related issues and the overabundance of data to be processed, researchers are increasingly turning to artificial intelligence to improve their understanding of phenomena and integrate this wealth of data as a lever for development. The present Scopus-based bibliometric study uses an incremental heuristic approach to explore the structure and trends of this research direction. We find that remote sensing, photometry, measurement of biological parameters, and physicochemical analysis of materials have emerged as strong trends for environmental and sustainability applications. The role of AI is essential due to the multitude of parameters to be analysed and the large quantities of information to be processed. Some AI concepts seem to be among the most common: 'artificial neural networks', 'random forest', 'long-term memory', 'Short-Term Memory', 'Extreme Learning Machine', 'Fuzzy Inference System', etc. The interest in observing and monitoring water quality, groundwater, evapotranspiration, and monitoring water levels seems to benefit agriculture and water resource management issues. Regarding international collaboration, there is a strong concentration on the most prolific countries (China, United States, India), remaining the main players in this field of research. Finally, the study revealed emerging topics of interest among the most dynamic Topics. This concerns, for example, the interactions of admixtures with water in technical mixtures, the dimensioning and calculation of stresses and compressions, and the dynamics of waves and large masses of water. This study thus offers indications and development prospects for research teams, particularly in countries on the margins of the current bibliometric distribution of the field. It is also very instructive to enrich these results with expert information, using other types of information (economic, ecological, etc.). This could improve the analysis consistency and the relevance of the findings.

Keywords

Water research · WRM · Artificial intelligence · ML · Bibliometrics · Research priority · Research trends · Prominent topics · Scopus · Scival

1 Introduction

Water is a fundamental resource for human survival and plays a central role in sustaining key sectors such as agriculture, health, energy, and industry [1]. However, climate change, rapid urbanisation, economic growth, and demographic pressures are intensifying water management challenges, making it not only an environmental but also a geopolitical concern [2, 3]. In some regions, water scarcity has already been linked to social unrest, migration, and conflict, highlighting its role as a potential catalyst for future geopolitical tensions [4, 5].

Simultaneously, artificial intelligence (AI) has emerged as a transformative force in data science and decision-making [6–8]. AI offers unprecedented capabilities to analyse

H. Boutracheh (✉) · Y. Mouhssine · R. El Ansari · A. Moumen
Laboratory of Engineering Sciences, National School of Applied Sciences, Ibn Tofaïl University, Kenitra, Morocco
e-mail: hicham.boutracheh@uit.ac.ma

H. Boutracheh
IMIST, CNRST, National Center for Scientific and Technical Research, Rabat, Morocco

N. Mejjad
National Centre for Nuclear Energy, Science and Technology, Kenitra, Morocco

M. El Bouhadioui
National School of Mines, Agdal-Rabat, Morocco

A. Moumen et al. (eds.), *Artificial Intelligence and Data Analytics for Innovative Applications in Engineering, Sustainability and Technology*, Sustainable Artificial Intelligence-Powered Applications, https://doi.org/10.1007/978-3-031-90318-2_6

complex datasets at scale, enabling researchers and practitioners to discover patterns and insights beyond human cognitive reach [9, 10]. Because of its ability to handle uncertainty and simulate multiple scenarios, AI is particularly well suited to solving complex environmental challenges [11]. Research stands to benefit significantly from these developments, as AI-based tools open new avenues for deeper analysis, hypothesis testing, and predictive modelling [12].

The intersection between AI and Water Research is still in its formative stages, but it holds immense potential. The application of AI in water-related fields, such as water quality monitoring, predictive flood modelling, and efficient resource management, can generate new knowledge and practical solutions [13, 14]. Identifying emerging research trends at this level is essential to understand the changing landscape and inform policy and decision-making processes.

This study draws on bibliometric analysis [15–17] to explore how AI is integrated into water research. Bibliometrics provides a valuable framework to quantify, map, and track the intellectual structure of a research field, revealing trends, influential work, and future directions [18–22]. By systematically analysing scientific publications on the intersection between AI and Water, this study aims to fill a gap in the existing literature, providing a comprehensive view of the knowledge landscape and highlighting areas where AI can address critical water-related challenges. While previous studies have addressed specific aspects of AI in environmental research [23–28] few have comprehensively explored the field from a bibliometric perspective, which is the focus of this work [29].

2 Methodological Approach

The two research fields, water and artificial intelligence (AI), have different perspectives in terms of anteriority, volumes, and scopes. This is reflected in the amount range of publications in Scopus: ~ $5.1 * 10^6$ for the Water Corpus and $2.5 * 10^6$ for the AI Corpus.

The intersection of these two major fields gives rise to a corpus (AI-Water) of around $66 * 10^3$ publications. This is a tiny proportion of the original corpus: 1.3% of the Water Corpus and 2.7% of the AI One. But, for the 2019–2024 period, the volume of the two constitutive corpora (AI and Water) became very similar and the intersection proportion grew and became mainly the same (~3%) for both original corpora. The remarkable evolution of the volume and proportions of this intersection corpus (AI-Water) clearly shows the mutual interest between the two fields and confirms that water research is taking hold of AI and tends to benefit from its advances.

This observation reinforces the rationale of our bibliometric study. It imposes the duty to try to understand this remarkable evolution of research activity at the interface of advances in Water and AI.

In the continuous development of our previous bibliometric approaches [30, 31], concerning water issues, we propose a heuristic method, favouring an exploratory attitude, guided by the main results which emerge from the successive stages (Fig. 1).

This incremental and intuitive approach is based on the fundamental laws of bibliometrics [32–36], which allows for exploring large volumes of data by reducing the size of the corpora and entities to be analysed.

We propose to conduct this bibliometric study on a corpus extracted from the bibliographic database Scopus and analyse it using the Scival [37, 38] analytical platform, provided by 'Elsevier'.

Figure 1 illustrates the methodological approach and presents the different stages of the study and the various parameters considered in the analysis.

3 The Core Activity of the AI-Water Domain

3.1 AI-Water Research, a Recent Strong Trend

At the end of September 2024, the AI-Water corpus on Scopus contained 66,150 publications, while + 90% of which were published after 2009 and + 67% (i.e. 44,663 Pub.) between 2019 and September 2024.

The evolution curve of the whole corpus (Fig. 2) also suggests that this field of research is quite recent, which predicts an interesting dynamism, of the actors and subjects, to be observed.

Indeed, at the end of September 2024, the 44,630 publications, of the period considered, generated an average of 10.8 citations, with an FWCI[1] of 1.5 (i.e. 50% higher than the overall average). Note, however, a modest level of international collaboration of 26.3%.

Furthermore, 21% of this corpus is in the top 10% of most cited publications worldwide and 24.3% of its publications are in the top 10% of journals by SNIP, 51.8% in the Top 25% (Q1 by SNIP).

These remarkable performances confirm the interest and dynamism of this field of research and indicate that it is probably at the beginning of its maturity phase.

[1] **FWCI** (Field Weighted Citation Impact) is a normalized metric of the average impact of a corpus (set of publications). The FWCI considers the context of citations and the type, year and field of each publication in the corpus. The average value of FWCI is 1. A value of 1.5 means that the corpus is 50% more performing than the average.

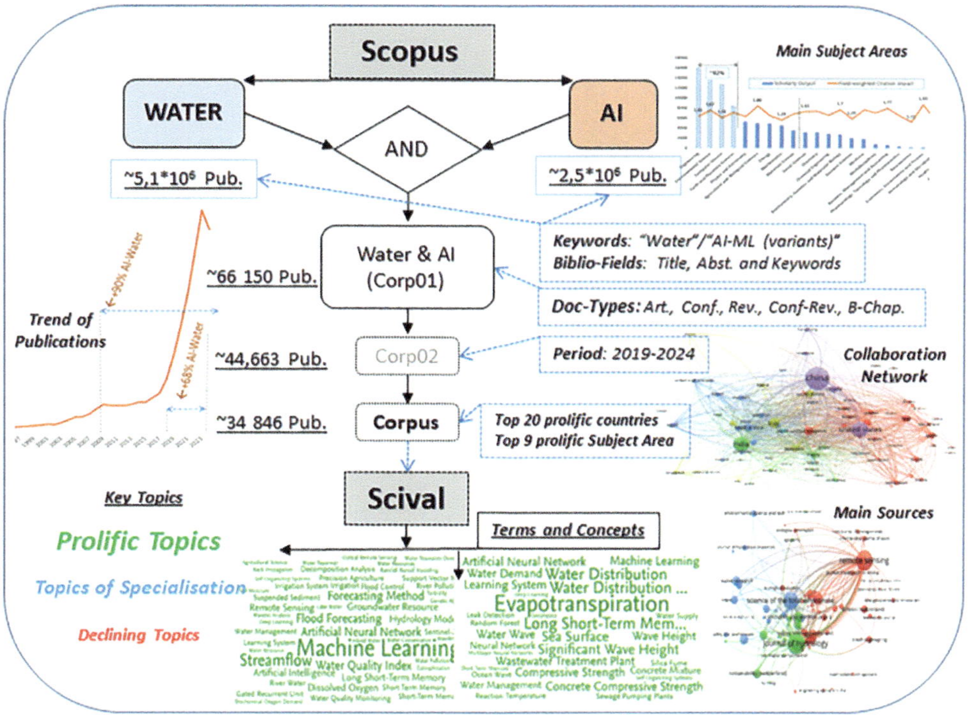

Fig. 1 Methodological approach

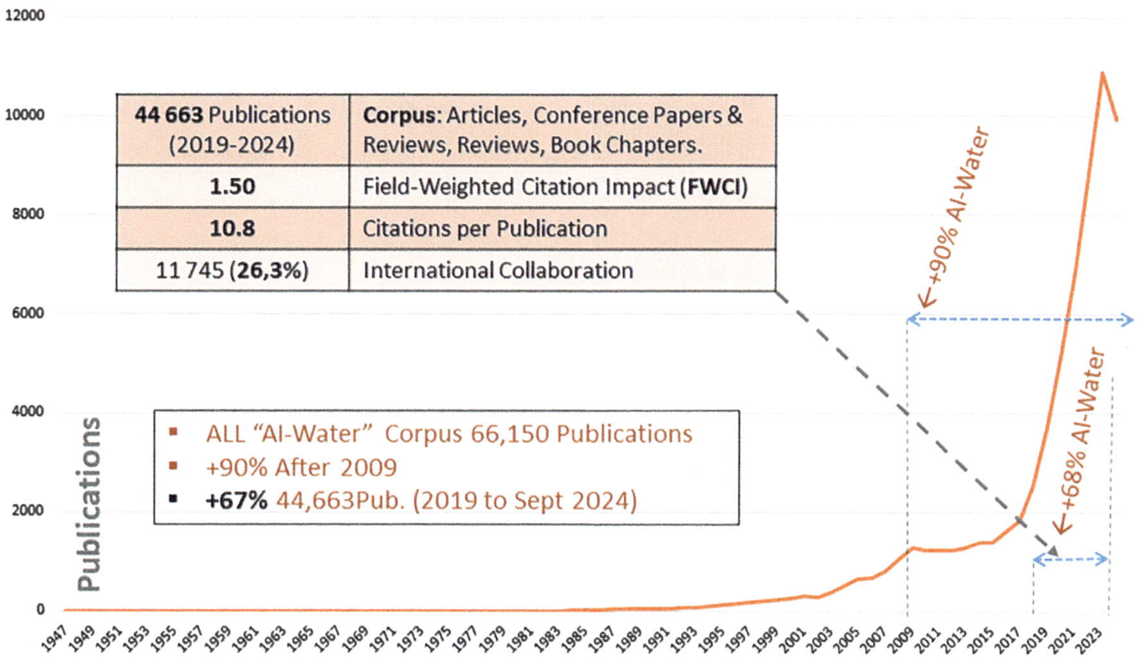

Fig. 2 Evolution and trend of publications in the AI-water field

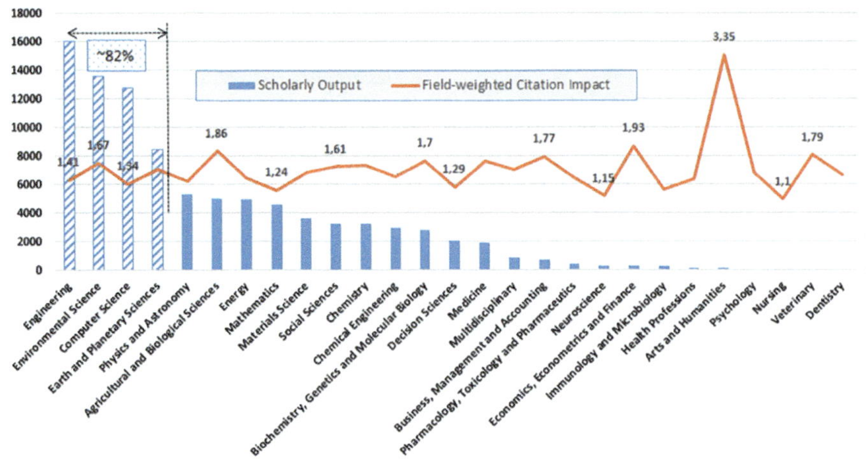

Fig. 3 Main subject areas in the AI-water field

3.2 Disciplinary Structure of the AI-Water Field

The AI-Water corpus presents a concentrated distribution of disciplinary fields. The disciplinary proportions in the corpus (Fig. 3) support a macroscopic analysis of the significant disciplines (~share > = 8%). We can thus distinguish three major disciplinary groups:

- G1: The first 4 disciplinary fields with contributions above 19% in the global corpus. Together, they make up ~ 82% of the total, with FWCIs between 1.34 and 1.67. This is the heart of the distribution: 'Engineering', 'Environmental Science', 'Computer Science', and 'Earth and Planetary Sciences';
- G2: The 5 disciplinary fields contribute between 8 and 12% to the corpus studied (1.24 < FWCI < 1.86). Together, they make up ~ 44% of the total corpus. Nevertheless, because of the disciplinary overlap, they only add ~ 11% more to the first group: 'Physics and Astronomy', 'Agricultural and Biological Sciences', 'Energy', 'Mathematics', and 'Materials Science'.
- G3: consists of the other disciplines that contribute barely 7% more to the first two groups. Let us mention in particular the 5 most prolific fields of this G3, contributing between 5 and 7% to the general corpus, with (1.29 < FWCI < 1.7): 'Social Sciences', 'Chemistry', 'Chemical Engineering', 'Biochemistry, Genetics and Molecular Biology', and 'Decision Sciences').

The head disciplinary fields group G1 shows a structuring trend of the AI applications addressed to the environment and climate phenomena. This is supported by the second group G2, which confirms the importance of mathematics, thermodynamics, and atmospheric studies, for agriculture and biology applications. The question of water resource management is fundamental. It is addressed in particular from the energy and materials sciences point of view.

Then, the main subject areas of group G3 confirm the first trends, particularly in chemical and biological engineering and decision sciences (Mathematics). The G3 also reminds us of the centrality of social sciences in a disciplinary field at the crossroads of Water and AI, especially concerning the Subcategory 'Geography, Planning and Development'.

3.3 Countries' Contribution and Collaboration Networks

Table 1 describes the front countries of the corpus distribution. There are only 100 countries that participate in the overall corpus with + 20 publications, and just 20 countries that contribute at + 2% to the corpus. These latter's accumulate ~ 84% of the world's achievement, with a maximum performance of 32% for China. The rest of the world brings only the remaining 16% of the world's corpus.

In addition to being prolific, these 20 countries show good qualitative results. Indeed, they all have FWCIs above the global threshold of '1', and only Brazil (1.3), Japan (1.46), and China (1.48) have FWCIs below the corpus average (1.5). Similarly, they all have proportions of publications above 10% in the Top 10% journal percentile by SNIP,[2] and only six have a rate below the corpus average (24.3%), with a minimum of 15.4% for India.

In terms of international collaboration, these 20 countries have relatively high rates, with an average of 58.8%, and only

[2] The **SNIP** (Source Normalized Impact per Paper) is a metric of the citation impact of Scopus journals. It normalizes citation practices across fields and neutralizes the effect of volume and publication year. Therefore, SNIP helps compare journals from different fields.

Table 1 Prolific countries in the AI-water field

Countries	Publications	% Inter. Collab	FWCI	% in Top 10% Journals by SNIP	Corpus %			
China	*14310*	*24,4*	*1,48*	*30,1*	*32%*	84% of the Overall Corpus	61%	57%
USA	*6588*	*48,3*	*1,94*	*34,6*	*15%*			
India	*5920*	*24,9*	*1,83*	*15,4*	*13%*			
Iran	*2564*	*56,6*	*2,15*	*27,2*	*6%*			
UK	1810	77,6	2,12	37,5	4%			
S-Korea	1626	44,3	1,75	30,7	4%			
Canada	1530	64,7	1,78	35,7	3%			
Germany	1491	65,8	2,02	30	3%			
S-Arabia	1445	80,6	2,68	26,5	3%			
Australia	1320	72,1	2,37	41,8	3%			
Italy	1253	55,5	1,96	29,2	3%			
Malaysia	1070	69,7	2,23	24,9	2%			
Japan	1009	50,4	1,46	23,9	2%			
Spain	999	61,1	1,65	32,1	2%			
Turkey	981	41	1,69	20,9	2%			
Brazil	889	39	1,3	20,4	2%			
France	887	68,5	1,65	32,1	2%			
Egypt	762	77,3	2,62	26,7	2%			
Pakistan	707	84,2	2,73	18,8	2%			
Iraq	674	69,1	2,29	22,7	2%			

China (24.4%) and India (24.9%) have rates below the overall corpus average (26.3%).

Furthermore, we know that it is difficult for a country to perform in volume (publications) and to maintain, at the same time, a high level of citations (high FWCI). We also know that publication in renowned journals (high SNIP) is a parameter that rather indicates a certain robustness and a willingness to excel, which does not systematically affect visibility (FWCI) [39]. On the other hand, international collaboration has a proven effect on citation performance (e.g. FWCI) [40, 41]. The above observations thus allow us to detect countries that exhibit their robustness (e.g. + 30% of publications in the Top 10% of journals by SNIP), countries that are structurally oriented towards international collaboration, or that possibly depend on it, (e.g., the UK, Canada, Saudi Arabia, Australia, etc.) and, finally, countries that manage to capture maximum visibility thanks to their strategies (e.g. FWCI > 1.9). These reflexions must be moderated by the volume effect, particularly in the case of the top 3 (China, the United States, and India) leading the race in this AI-Water field.

The countries that constitute the core of this research field also structure international collaboration. The network represented by (Fig. 4), built using Vosviewer (V1.6.20) [42], shows the co-publications in the field of AI-Water, over the period 2019–2024. It highlights the main actors and the intensity of the relations between the groups. It also shows the structuring role of the prolific countries and the networks established. For example, the China–USA couple, which includes Canada and Hong Kong in its close network, constitutes the essential bridge in international collaboration in this field. Other clusters display a regional logic, such as the European network (in red), led by the UK, Germany, Italy, and Spain, and the network of Middle East and Central Asia (in blue) mainly carried by the Saudi Arabia-Egypt couple. India and Iran also lead two clusters (green and yellow) and provide a link between the different networks.

4 Bibliometric Deepening by Selective Reduction

The shape of the publication curve (Fig. 2), the concentrations observed in countries (Table 2) and disciplines (Fig. 3) are largely consistent with the empirical laws that describe general bibliometric distributions [43–45]. Then, as recommended by some previous works [46], the macroscopic exploration of the basic corpus (44,630 Pubs.) suggests a selective reduction of the main entities of the corpus. This would improve the consistency of the data and go deeper to increase the relevance and reduce the 'cost' of the analytical exploration [47]. We thus limit the corpus to the 20 most prolific countries and the first nine Subject Areas of the distribution. The result represents the core distribution and summarises the dynamics of the AI-water research field. The resulting corpus is about 34,846 publications, meaning + 78% of the initial corpus.

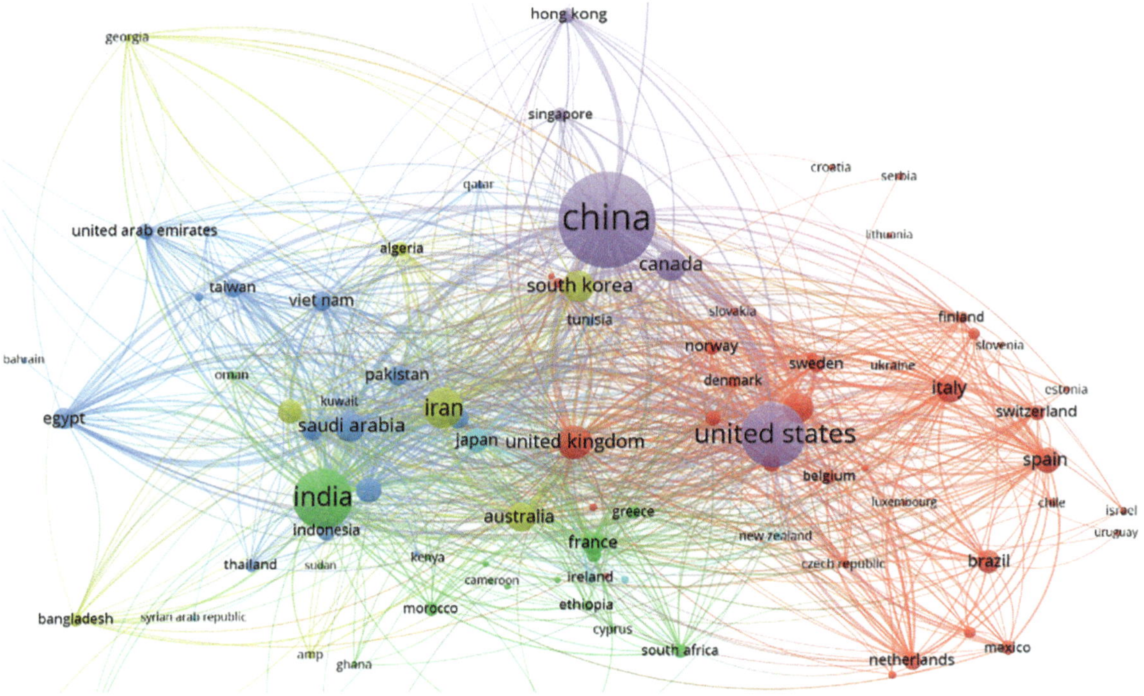

Fig. 4 Main collaborating countries (+50 publications)

Table 2 Main subject areas in the reduced corpus

Subject Areas	Publi-cations	%	2019	2020	2021	2022	2023	2024
Engineering	13302	38%	1027	**1431**	1810	2642	3339	3053
Environmental Science	11864	34%	817	**1193**	1712	2189	2939	3014
Computer Science	10097	29%	957	**1188**	1533	1968	2557	1894
Earth & Planetary Sc.	7270	21%	557	**834**	1163	1453	1663	1600
Agricultural & Biological Sc.	4421	13%	363	492	**674**	819	1049	1024
Physics and Astronomy	4372	12%	341	540	**680**	867	1066	878
Energy	3914	11%	289	448	534	**804**	**1008**	**831**
Mathematics	3582	10%	350	394	487	**666**	**957**	**728**
Materials Science	3105	9%	225	358	419	605	**761**	**737**
Social Sciences	2484	7%	192	279	356	486	**657**	514
Chemistry	2289	7%	149	242	327	445	540	**586**
Chemical Engineering	2033	6%	126	180	279	369	494	**585**
Biochemistry, Genetics and Molecular Biology	1993	6%	134	220	278	409	495	457
Decision Sciences	1498	4%	124	154	234	308	429	249

4.1 Effect of Reduction on Disciplinary Stability

The new corpus (34,846 Pub.) generates an average of 11.6 citations per publication, with an FWCI of 1.59 (i.e. 59% higher than the world average), with a better level of international collaboration than the previous corpus (29% versus 26.3%). In addition, 22.9% of this corpus is in the top 10% of most cited publications worldwide and 27.3% of its publications are in the top 10% of journals by SNIP, 54.9% in the Top 25% (Q1 by SNIP).

The resulting corpus performs better and maintains a general balance to represent the AI-water field. Therefore, we will conduct additional verification at the level of the disciplines covered.

Table 2 represents the 14 prolific disciplines of the resulting corpus (+ 1400 Publications, i.e. + 4% of the corpus). Note that the order and proportions have remained almost the same, with a general upward trend and a slight rank change between 'Physics and Astronomy', and 'Agriculture and Biological Sciences'. The first three disciplines 'Engineering', 'Environmental Sciences' and 'Computer Science' accumulate +

77% of the corpus, with a firm increase over the period studied. The other, less prolific Subject Areas also show a remarkable increase in volume.

These disciplines, including 'Social Science', are interdependent and confirm the multidisciplinarity that increasingly marks research activity [48, 49].

To do short, the first 5 disciplines total ~ 91% of the new corpus. This confirms the normal appearance of a bibliometric distribution [50, 51] and the representativeness of the first disciplines due to their interaction with the rest of the disciplines in this corpus. Indeed, the preserved disciplinary structure confirms the quality of the reduction carried out on the initial corpus and thus supports the relevance of the incremental bibliometric approach [52]. This will allow us to specify the preliminary analysis (Fig. 3) while maintaining satisfying confidence and representativeness.

4.2 Sources Consistency in the Reduced Corpus

Figure 5 shows the bibliographic coupling network between the first 100 sources (Journals, Conferences, etc.) that contain at least 50 publications and have received a minimum of 100 citations. Analysis of this list, using Vosviewer, suggests an exploratory approach to the main research orientations of the field [53–56].

Hence, a strong orientation to use observation and measurement techniques, particularly from remote sensing, photometry, and physicochemical properties of materials. These tools make it possible to generate masses of rich and valuable data, which implies sophisticated processing, using advances in AI and engineering approaches, from the fields of Networks and Systems, Geo-information, etc.

The main aims are oriented towards a better understanding of phenomena related to hydrology in general, such as the environment, energy, oceans, chemosphere, pollution, agriculture, construction, sustainability, etc.

5 Topics and Prominence in Water-AI Research

The post-reduction checks (Sect. 4) reassure us about the stability and improvement of knowledge that the retained corpus (AI-Water) could reveal. We are ready to delve into this research area, particularly by exploring the Prominent Research Questions (Topics).

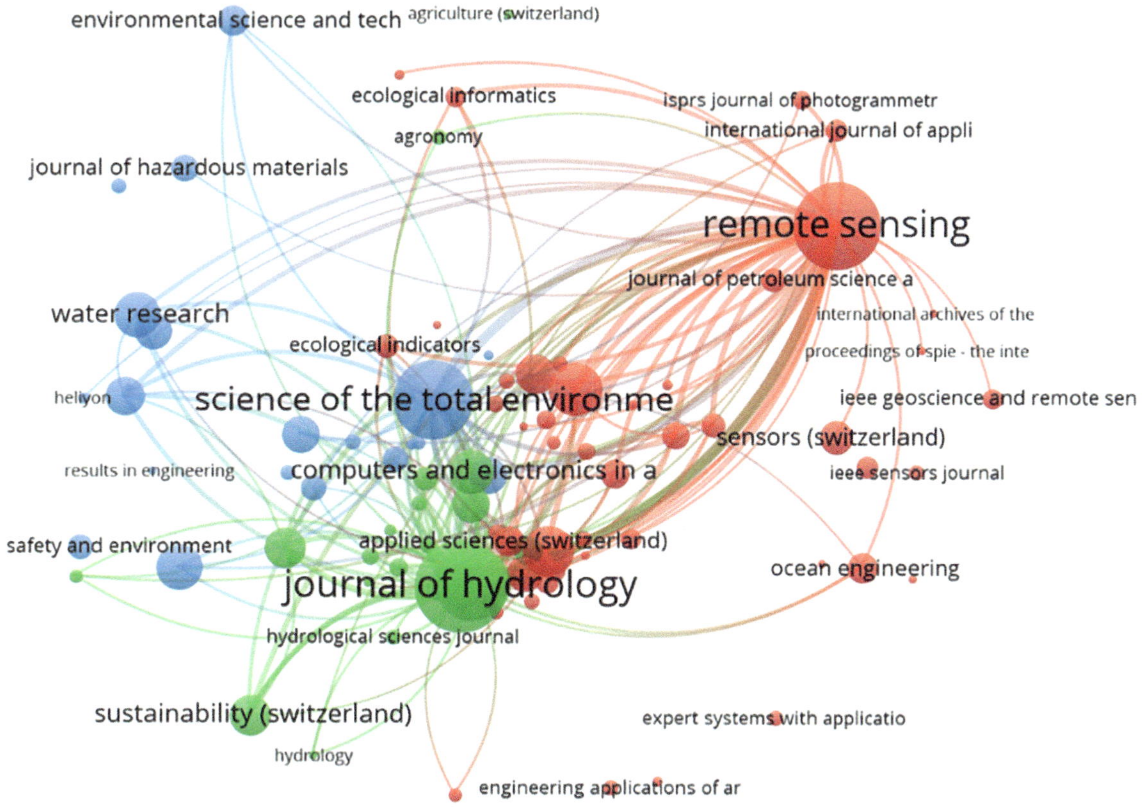

Fig. 5 Main sources (+ 50 publications and + 100 citations)

Indeed, Scival proposes a classification of research subjects, based on a direct clustering citation method. This results in a built 'Prominent Topics', or simply Topics.

A Topic is a unique, dynamic, and homogeneous set of publications, linked according to certain parameters (citations, views, and sources' performance) that reveal the '*momentum*' of this research question [57, 58]. An index of '*Prominence*' gives a numerical value for this momentum (a percentile classification) [59].

Scival supports a corpus analysis by calculating its contribution to a list of around 94,000 Topics, and 1500 Topic Clusters. This allows us to go deepen the analysis of the domain's research trends and main opportunities [60, 61].

For our case, the AI-Water corpus selected contributes to 6845 Topics (belonging to 1001 Topic Clusters). It should be noted that the new corpus still retains very satisfactory representativeness. Indeed, this retained corpus covers + 78% of the initial number of Topics (and + 86% of Topic Clusters).

For the analysis, we will use some parameters that allow us to evaluate the relative importance of each of the Topics for the studied domain. Thus, for each key topic, we have the following metrics:

- **Pubs.**: the number of corpus's publications shared with the Topic concerned. This gives an idea of the relative importance of the Topic in the corpus studied.
- **Topic Share** (%): Inversely, the contribution of the studied corpus in the Topic indicates the specialisation of this corpus and its relative importance for the Topic, compared to the whole database (Scopus).
- **Growth** (%): The volume evolution (%) of the corpus studied concerning the current Topic.
- **Prom-Perc.**: The **Prominence Percentile** (%) gives a general idea of the topic's 'momentum' and current opportunities for the research community.

5.1 Prolific Key Topics of the AI-Water Field

Table 3 represents Topics participating higher than 0.5% in the AI-Water corpus. All these topics have a high momentum (Prom-Perc. > 94.7%). Furthermore, the AI-Water corpus is structurally constitutive of the Topics: 'Artificial Neural Network; Support Vector Machine; Flood Control' [62, 63], 'Artificial Neural Network; Dissolved Oxygen; Water Pollution' [64, 65], 'Artificial Neural Network; Support Vector Machine; Groundwater Resource' [66, 67], and 'Neural Network; Chemical Oxygen Demand; Wastewater Treatment Plant' [68, 69], with participation of 41.3%, 64.2%, 58.8% and 46.1% respectively and FWCI values of 2.18, 2.41, 1.96 and 1.81.

Thus, from a volume point of view, the interest is focused on the use of AI tools, in particular the 'Artificial Neural Network (ANN) and the 'Support Vector Machine' (SVM), to study problems of control, supervision, and prediction of phenomena related to floods, water pollution, underground resources, sewage treatment plants, etc. The assessment of dissolved Oxygen in water, its chemical demand, and its interest in predicting biological phenomena seems to be a central issue at this stage.

On the other hand, 'Remote Sensing; Ocean Color; Coastal Water', 'Evapotranspiration; China; Climate Change' [70], 'Evapotranspiration; Penman–Monteith Equation; Crop

Table 3 Main topics by number of publications (+ 0.5% of the AI-water corpus)

Top Topics by Publications	Pubs.	Topic Share (%)	Growth (%)	FWCI	Prom-Perc.
Artificial Neural Network; Support Vector Machine; Flood Control	1237	41,3	12,8	2,18	99,58
Artificial Neural Network; Dissolved Oxygen; Water Pollution	977	64,2	-6,2	2,41	99,07
Remote Sensing; Ocean Color; Coastal Water	613	19,9	243	1,73	99,32
Irrigation System; Wireless Sensor Network; Internet of Things	469	9,2	93,8	2,24	99,729
Deep Learning; Convolutional Neural Network; Object Detection	402	1,3	128,5	1,34	99,986
Artificial Neural Network; Support Vector Machine; Groundwater Resource	366	58,8	6,7	1,96	97,05
Compressive Strength; Artificial Neural Network; Machine Learning	320	21,4	54	3,08	99,423
Evapotranspiration; Penman-Monteith Equation; Crop Coefficient	310	22,9	197,2	2,48	97,958
Remote Sensing; Surface Water; Climate Change	281	18,1	61	2,17	97,979
Neural Network; Chemical Oxygen Demand; Wastewater Treatment Plant	255	46,1	56,8	1,81	96,391
Deep Learning; Image Reconstruction; Image Color Processing	224	11,4	194,2	1,8	98,365
Nanofluid; Thermal Conductivity; Heat Convection	210	4,2	59,3	2,75	99,785
Evapotranspiration; China; Climate Change	185	3,9	235,2	1,92	99,766
Artificial Neural Network; Water Wave; Sea Level	182	33,6	31,7	1,79	95,263
Water Distribution System; Distribution Network; Hydraulics	176	29,4	123,4	1,33	94,728

Coefficient' [71, 72] and 'Deep Learning; Image Reconstruction; Image Color Processing' [73] are the Topics that show the highest growth rates over the period 2019–2023, with respective values of 243, 235.2, 197.2, and 194.2%.

These dynamic Topics show a firm tendency to study the Evapotranspiration of cultures and Coastal Waters. Remote sensing is the main observation technique used to assess the impacts of climate change. AI possibilities (i.e. deep learning) are mobilised to better exploit the large amounts of data collected, particularly for image and colour processing. AI thus implements scientific knowledge on these phenomena (crop coefficient, Penman–Monteith equation, etc.) to develop monitoring and predicting tools.

5.2 Key Topics of Specialisation in the AI-Water Field

Table 4 represents among the first ones in terms of volume, the Topics where the contribution of the AI-Water corpus is greater than 18%. Despite the strong contribution in the first three Topics (between 49 and 64%) and their respectable performances in FWCI (between 1.23 and 2.41), the growth observed is negative to very low (− 20 to 6.7%). These Topics deal with water pollution, flocculation in treatment plants, and groundwater [74–76].

A second group of Topics, with corpus participation between 40 and 46% and publications between 91 and 255, shows growths between 56.8 and 194% and FWCIs of 1.31–1.87. This group highlights the usage of ANN for water demand and management issues [77, 78], observation and measurement of water temperatures and surfaces [79], and wastewater treatment at the station level. These issues are addressed from the chemical (oxygen demand) and geophysical [80] (water masses and surfaces) points of view.

A third group, whose participation in the Topics is between 18 and 29% and growth between 123 and 1100%, brings issues related to evapotranspiration, systems and networks of distribution [81, 82], and management of drinking and agricultural water and monitoring of pollution [83]. Of particular note is the spectacular growth of 1100% of the Topic 'Remote Sensing; Farmland; Water Management'.

In this group, AI is useful for processing data and images, provided by remote sensing and UV/VIS Spectroscopy techniques [84], to reconstruct images and evaluate physicochemical dynamics, such as chemical oxygen demand [85].

Note finally, the performance of 5.76 (FWCI) of the Topic 'Remote Sensing; Farmland; Water Management' which deals with an advanced technique of enhanced oil recovery using carbon dioxide (CO2 EOR) [86, 87].

Table 4 Main topics by the AI-water contribution (+18% of the topic share)

Topic	Pubs.	Topic Share (%)	Growth (%)	FWCI	Prom-Perc
Artificial Neural Network; Dissolved Oxygen; Water Pollution	977	64	-6,2	2,41	99,07
Artificial Neural Network; Support Vector Machine; Groundwater Resource	366	59	6,7	1,96	97,05
Water Treatment Plant; Neural Network; Flocculation	50	49	-20	1,23	76,813
Neural Network; **Chemical Oxygen Demand; Wastewater Treatment Plant**	255	46	56,8	1,81	96,391
Surface Water; Sea Surface Temperature; Geophysics	118	46	81,1	1,54	85,497
Water Demand; Artificial Neural Network; Climate Change	168	42	62,1	1,31	89,85
Artificial Neural Network; Support Vector Machine; Flood Control	1237	41	12,8	2,18	99,58
Artificial Neural Network; Support Vector Machine; Water Management	91	40	194	1,87	87,512
Cyanobacteria; Phytoplankton; Harmful Algal Blooms	89	37	38,4	1,47	87,246
Artificial Neural Network; Water Wave; Sea Level	182	34	31,7	1,79	95,263
Neural Network; Reverse Osmosis; Desalination	42	33	-5,9	1,59	88,29
Water Distribution System; Distribution Network; Hydraulics	176	29	123,4	1,33	94,728
Coal Mine; Laser Induced Fluorescence; Flood	37	25	16,7	0,7	63,638
Evapotranspiration; Penman-Monteith Equation; Crop Coefficient	310	23	197,2	2,48	97,958
Compressive Strength; Artificial Neural Network; Machine Learning	320	21	54	3,08	99,423
Remote Sensing; Farmland; Water Management	36	20	1100	1,06	79,867
Remote Sensing; Ocean Color; Coastal Water	613	20	243	1,73	99,32
Flood Plain; River Basin; Wetland Management	28	20	-	1,9	79,934
Remote Sensing; Synthetic Aperture Radar; Image Classification	59	19	107,6	2,1	90,043
Compressive Strength; Artificial Neural Network; Rheology	23	19	-	2,79	84,476
Carbon Dioxide; Oil Well; Enhanced Oil Recovery	12	19	75	5,76	53,845
Chemical Oxygen Demand; UV/VIS Spectroscopy; Water Pollution	40	19	238	0,69	78,513
Water Distribution System; Failure Analysis; Potable Water	89	18	171,7	1,35	89,591
Remote Sensing; Surface Water; Climate Change	281	18	61	2,17	97,979

5.3 Declining Key Topics of the AI-Water Field

Table 5 represents, among the first in volume and those with topic share > 3%, the declining topics (growth (%) < 0).

Some Topics, such as 'Nuclear Fuel; Neural Network; Nuclear Power Plant' [88] and 'Hydraulic Conductivity; Infiltrometer; Soil Water' [89], have a sharp decline (resp. − 54.9% and − 30.1%), accentuated by a weak performance in FWCI (resp. 0.98 and 1.14). This group shows a strong trend of disengagement from this research area compared to these Topics.

Others, such as 'Infill Drilling; Rate of Penetration; Machine Learning' [90] and 'Neural Network; Deep Learning; Physics' [91], have a strong decline (resp. − 63.6% and − 21.4%) but remarkable performances in FWCI (resp. 2.99 and 496). This contrast could indicate a certain reorientation towards new research questions that shift from the main field of AI-Water currently studied.

Finally, some Topics, such as 'Greenhouse Gas; Nitrogen Oxide; Wastewater Treatment' [92] and 'Remote Sensing; Satellite Altimetry; Climate Change' [93], with a low decline (resp. − 4.1 and − 7%) have remarkable FWCIs (resp. 3.44 and 2.06). This would probably be a cyclical effect or the emergence of new trends that have not yet been well-established in specialist circles. In all these cases, it is necessary to deepen the explorations and verify all hypotheses and probable interpretations.

In sum, the analysis of the most representative topics corroborates the above exploration of the main Subject Areas (Sect. 4.1).

First, a high concentration of publications (469–1237 Pubs.) for the four topics 'Artificial Neural Network; Support Vector Machine; Flood Control', 'Artificial Neural Network; Dissolved Oxygen; Water Pollution', 'Remote Sensing; Ocean Color; Coastal Water' and 'Irrigation System; Wireless Sensor Network; Internet of Things', with the highest rate of growth for the third one (243% over the 2019–2024 period).

The 'Topic share' ranking proposes another set of growing Topics: 'Neural Network; Chemical Oxygen Demand; Wastewater Treatment Plant', 'Surface Water; Sea Surface Temperature; Geophysics', 'Water Demand; Artificial Neural Network; Climate Change', 'Artificial Neural Network; Support Vector Machine; Water Management', 'Water Distribution System; Distribution Network; Hydraulics' and 'Evapotranspiration; Penman–Monteith Equation; Crop Coefficient'. The AI-Water corpus is strongly represented in these Topics (23–46% of Topic Share) and shows a high Growth rate (56.8–197.2%).

On the other hand, 'Remote Sensing; Farmland; Water Management' is showing the highest Growth rate (1100%), with a Topic Share higher than 18%. While the Topic 'Carbon Dioxide; Oil Well; Enhanced Oil Recovery' performs the highest value of FWCI, with a growth rate of 75%.

Finally, declining Topics also allow us to capture inverse trends and try to detect the meaning behind some controversial signals presented by certain Topics (High FWCI and negative Growth Trend!).

6 Concepts Analysis of the Main AI-Water Topics

The topics explored previously are built on citation networks that reveal semantic coherence and thematic dependency. Terminological analysis would provide detailed knowledge of the trends and concepts addressed by the main Topics [30, 31, 59].

The algorithms behind Scival [94] allow us to attribute terms (or keyphrases) to each publication of the corpus studied. These terms (or Concepts) have a weight, between a minimum of 0 and a maximum of 1, which reflects their relative importance in the corpus studied.

Then, we will explore the main terms of the key topics and evaluate their relative importance and the major trends that emerge from this analysis.

6.1 Main Concepts of the Entire AI-Water Corpus

For the general corpus, the word cloud in Fig. 6 Main concepts of the whole corpus (first 50 terms)shows the most relevant terms (size ~ 0 to 1) that are growing (green colour). These first 50 key terms are present in ~ 82% of the publications in the AI-Water corpus, which gives them good representativeness. In addition, 62% of these terms show growth greater than 200%, over the period 2019–2024, with a peak of 722% for 'Random Forest' [64, 95] and a minimum of 78% for 'Backpropagation' [96].

Furthermore, concepts and algorithms of AI and 'Machine Learning' dominate this list of first terms. Indeed, except for 'Remote Sensing', the most relevant terms are rather related to these methods, such as 'Learning System', 'Artificial/Convolutional/Deep/Neural Network', 'Deep Learning', 'Long Short-Term Memory', 'Support Vector Machine', 'Sentinel-2', 'Deep Neural Network', etc.

The ranking by growth rate highlights additional concepts, more related to water resources, agriculture and image processing issues: 'Water Quality Index' [97], 'Image Enhancement', 'Water Body', 'Sea Surface', 'Precision Agriculture' [98], 'Convolutional Neural Network', 'Satellite Imagery', 'Water Resource', 'Evapotranspiration', 'Streamflow', 'Flood Control', 'Groundwater' [99], 'Soil Moisture', etc.

Table 5 Main declining topics in the AI-water corpus (Pubs. > 18; topic share > 3%)

Topic	Pubs.	Topic Share (%)	Growth (%)	FWCI	Prom-Perc
Artificial Neural Network; Dissolved Oxygen; Water Pollution	977	64,19	-6,2	2,41	99,07
Water Treatment Plant; Neural Network; Flocculation	50	49,02	-20	1,23	76,813
Neural Network; Reverse Osmosis; Desalination	42	32,81	-5,9	1,59	88,29
Gamma Radiation; Neural Network; Two Phase Flow	37	17,21	-31,4	1,6	86,559
Distribution System; Cyber Physical Systems; Network Security	21	17,07	-54,3	1,89	80,411
Remote Sensing; Synthetic Aperture Radar; Radar Imaging	30	16,22	-4,8	0,9	84,434
Nuclear Fuel; Neural Network; Nuclear Power Plant	69	11,86	-54,9	0,98	92,053
Water Distribution System; Genetic Algorithm; Potable Water	32	11,72	-46,1	1,34	85,368
Monitoring System; Early Warning System; Internet of Things	42	10,99	-0,2	0,99	83,682
Spillway; Numerical Model; Discharge Coefficient	34	8,85	-43,2	1,73	86,758
Rain Gage; TRMM; Soil Moisture	141	5,62	-9,9	1,66	98,858
Remote Sensing; Satellite Altimetry; Climate Change	38	5,47	-7	2,06	94,519
Synthetic Aperture Radar; Hydrogeology; Subsidence	21	5,19	-47,9	1,67	90,033
Infill Drilling; Rate of Penetration; Machine Learning	24	4,03	-63,6	2,99	92,173
Greenhouse Gas; Nitrogen Oxide; Wastewater Treatment	20	3,68	-4,1	3,44	96,074
Synthetic Aperture Radar; Remote Sensing; Vegetation	32	3,58	-42,7	1,47	94,932
Principal Component Analysis; Surface Water; Environmental Monitoring	27	3,35	-40,6	1,51	94,281
Neural Network; Deep Learning; Physics	106	3,18	-21,4	4,96	99,699
Fault Detection; Air Conditioning; Energy Engineering	33	3,18	-6,6	1,35	98,458
Hydraulic Conductivity; Infiltrometer; Soil Water	19	2,86	-30,1	1,14	92,059
Remote Sensing; Image Analysis; Vegetation	21	2,72	-31,2	1,57	89,531
Remote Sensing; Satellite Imagery; Image Processing	26	2,69	-6,3	1,24	95,282
Two-Phase Flow; Pressure Gradient; Computational Fluid Dynamics	23	2,55	-3,8	1,49	91,687

Fig. 6 Main concepts of the whole corpus (first 50 terms)

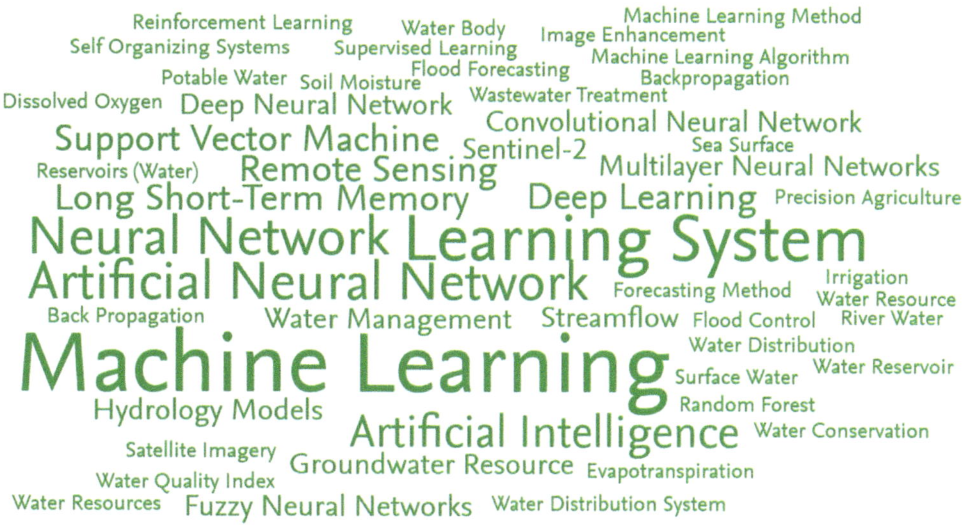

This observation confirms the previous analysis and further specifies the general trend of the AI-Water corpus.

6.2 Key Concepts of Prolific Topics

We now explore the main terms (Concepts) of the 6 most prolific key topics (between 366 and 1237 publications). These are 'Artificial Neural Network; Support Vector Machine; Flood Control', 'Artificial Neural Network; Dissolved Oxygen; Water Pollution', 'Remote Sensing; Ocean Color; Coastal Water', 'Irrigation System; Wireless Sensor Network; Internet of Things', 'Deep Learning; Convolutional Neural Network; Object Detection' and 'Artificial Neural Network; Support Vector Machine; Groundwater Resource.

The corpus constituted by these 6 prolific Topics contains 4064 publications, i.e. 11.7% of the general corpus. This set generates on average 15.4 citations per publication, with an FWCI of 2.07. It also shows a better rate of international

Fig. 7 Top 50 terms of the top 6 prolific topics of the AI-water corpus

Table 6 Emerging concepts from the top 6 prolific topics

Terms Concepts	Relevance (0 to 1)	Pub. Growth (%) 2019-2023	Pubs.
Short Term Memory	0,28	2050	140
Water Quality Monitoring	0,25	837,5	237
River Pollution	0,24	322,2	120
Suspended Sediment	0,24	45,5	102
Gated Recurrent Unit	0,23	1500	110
Short-Term Memory-2	0,22	825	123
Irrigation System	0,22	278,6	166
Decomposition Analysis	0,22	66,7	89
Lake Water	0,21	387,5	116
Biochemical Oxygen Demand	0,21	100	95
Rainfall-Runoff Modeling	0,21	88,9	89
Eutrophication	0,2	2900	116
Water Pollution	0,2	381,3	243
Inland Water	0,19	620	127
Wavelet Analysis	0,19	66,7	90
Optical Remote Sensing	0,18	1400	70
Agricultural Science	0,18	426,7	257
Turbidity	0,18	300	187
Water Resources Development	0,18	284,6	385
Genetic Algorithm	0,18	66,7	122
Variational Mode Decomposition	0,17	533,3	74
Wavelet	0,17	70,8	194

collaboration than the general corpus (30.7%). In addition, 29.2% of this corpus is in the top 10% of the most cited publications in the world and 22% of its publications are in the top 10% of journals according to SNIP and 49.1% in the Top 25% (Q1-SNIP).

The word cloud (Fig. 7) of the subset of the 6 prolific Topics (4064 Pubs.) confirms the trends of the general corpus, particularly concerning the most used AI methods to address water management issues in their different forms, facing the scientific challenges of acquisition, processing, reconstruction, and interpretation of data and images collected on physical and chemical phenomena in manifestation. This cloud also brings new complementary concepts, as shown in Table 6.

These concepts bring more interest to the issues of pollution and water quality in environments such as lakes, rivers or for irrigation [100–102]. The AI methods and algorithms, such as 'Short-Term Memory' [62], 'Gated Recurrent Unit' [103], 'Wavelet Analysis' [104], 'Variational Mode Decomposition' [105], etc. are used to process observation data

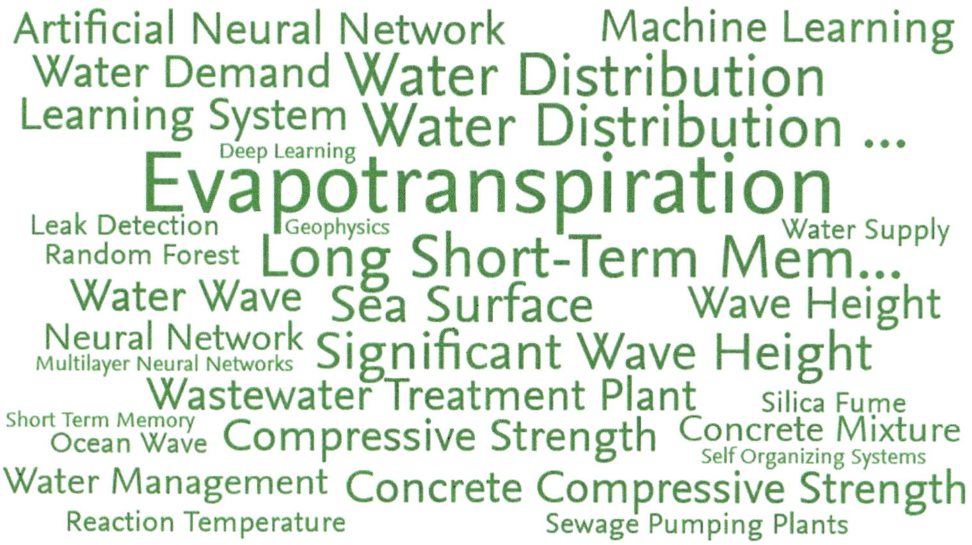

Fig. 8 Main concepts of the nine topics where the AI-water share is the highest

on phenomena, such as 'Eutrophication' [106], 'Turbidity' [107], 'Suspended Sediment' [108], etc.

Thus, imaging techniques, such as 'Optical Remote Sensing' [109, 110], and analytical techniques measuring biochemical or physical parameters [111, 112] ('Eutrophication', 'Biochemical Oxygen Demand', 'Suspended Sediment', etc.) generate large amounts of data and images and provide rich inputs for AI tools to enhance understanding and prediction of water phenomena.

6.3 Key Concepts of Topics of High Specialisation

Among the 500 most prolific topics, there are nine topics for which the AI-water corpus verifies the following conditions: participation greater than 21% and growth (2019–2023) greater than 30%. These are the Topics: 'Neural Network; Chemical Oxygen Demand; Wastewater Treatment Plant', 'Surface Water; Sea Surface Temperature; Geophysics', 'Water Demand; Artificial Neural Network; Climate Change', 'Artificial Neural Network; Support Vector Machine; Flood Control', 'Artificial Neural Network; Support Vector Machine; Water Management', 'Cyanobacteria; Phytoplankton; Harmful Algal Blooms', 'Artificial Neural Network; Water Wave; Sea Level', 'Water Distribution System; Distribution Network; Hydraulics', 'Evapotranspiration; Penman–Monteith Equation; Crop Coefficient' and 'Compressive Strength; Artificial Neural Network; Machine Learning'.

The corpus constituted by these nine topics contains 1709 publications, equivalent to 5% of the overall corpus. This set generates an average of 17 citations per publication, with an FWCI of 2.03 and a rate of international collaboration of 32.6%. In addition, 34.9% of this corpus is in the top 10% of the world's most cited publications, and 31.2% of its publications are in the top 10% of journals according to SNIP and 54.3% in the Top 25% (Q1).

The concepts related to AI and machine learning are among the most frequent terms in the subset of the nine topics [113–115].

On the other hand, the most relevant terms are related to some priorities (Fig. 8) like Evapotranspiration [116], Monitoring of Water Surfaces and Management of Drinking Water or Wastewater [117, 118]. In addition, the focus on these nine topics brings new concepts to the previous analyses (Table 7).

In addition to a reminder of some AI concepts (Short-Term Memory, Extreme Learning Machine, Fuzzy Inference System), at least four trends are identified in Table 7. The first concerns water issues and the interactions of adjuvants with water in the mixtures used to dimension and calculate the stresses and compressions involved [119–124]. This is translated by the concepts: Concrete/Compressive Strength; Aggregate; Mixture; High Performance; Cement/, Silica Fume, etc. This first trend is probably linked to the others, such as the one related to the wastewater field: Wastewater/Water Treatment Plant, Reaction Temperature, Sewage Pumping Plants, etc. [125–127] and the one associated with the management of drinking water demand: Water Demand/Supply, Forecasting, Leak Detection, etc. [128–131].

A last trend concerns the observation and study of geophysical phenomena: Wave Height, Water and Ocean Waves, etc. [132–135]. This field, of great interest for environmental and sustainability issues, benefits from AI possibilities to better exploit complex observational and imaging data, which are now available.

Table 7 Emerging concepts from the nine topics where the AI-water share is the highest

Terms Concepts	Relevance (0 to 1)	Pub. Growth (%) 2019-2023	Pubs.
Significant Wave Height	0,51	600	93
Wastewater Treatment Plant	0,49	153,3	130
Compressive Strength	0,47	361,5	225
Concrete Compressive Strength	0,43	966,7	103
Water Demand	0,43	283,3	84
Wave Height	0,42	900	96
Demand Forecasting	0,42	18,2	68
Water Wave	0,4	440	91
Concrete Aggregate	0,4	1000	67
Concrete Mixture	0,37	500	83
Water Treatment Plant	0,37	81,8	68
High Performance Concrete	0,37	325	58
High-Performance Concrete	0,33	325	56
Reaction Temperature	0,31	360	71
Sewage Pumping Plants	0,31	62,5	40
Waste Water Treatment Plant	0,29	80	64
Ocean Wave	0,27	650	50
Short-Term Memory	0,26	1300	51
Silica Fume	0,25	1200	43
Leak Detection	0,24	1300	37
Water Supply	0,22	600	175
Extreme Learning Machine	0,22	20	52
Aggregate Concrete	0,2	1800	53
Geophysics	0,18	280	64
Fuzzy Inference System	0,18	12,5	57
Cement Concrete	0,18	400	35

7 Summary and Conclusions

The intersection of AI and Water research is emerging as a dynamic field, with promising potential to address pressing global water challenges. This bibliometric study reveals significant trends and evolving opportunities in the AI-Water domain through an incremental heuristic approach, providing insight into how AI could shape water-related research, applications, and science policies.

The bibliometric analysis highlights the rapid growth of this interdisciplinary field. Although the AI-water corpus accounts for a relatively small proportion of the total AI and water research output (around 3% in recent years), the convergence between the two fields has intensified, with increasing volumes and proportional balance during the last few years. This evolution indicates the growing relevance of AI tools, such as ANN, SVM, LSTM, ELM, etc., to solve complex problems like flood control, water pollution management, water quality, and underground resource monitoring.

A critical finding of the study is the identification of thematic areas with both high growth rates and substantial AI integration. Indeed, the topics of Remote Sensing, Optics, Wireless Sensors, IoT, etc., demonstrate high integration of AI's capacity to advance agricultural water practices and drinking water supply through data-driven forecasting, crop coefficients, quality index, and evapotranspiration predictive models. Similarly, AI-powered image and colour processing tools are increasingly applied in climate studies to monitor coastal waters and analyse ocean colour, illustrating the broader environmental implications of AI.

On the other hand, some topics, such as groundwater management and water pollution treatment, show declining growth or minimal increase despite respectable FWCI scores. This discrepancy suggests either a saturation in research focus or a transition towards newer, more specialised sub-fields. This underscores the need for continuous adaptation and innovative research strategies to sustain progress in areas where AI has already established its utility.

The study also captures inverse trends, with high FWCI values but declining output in areas like water treatment flocculation. These contradictory signals could indicate shifts in research priorities or emerging technological constraints that warrant further investigation.

For instance, while wastewater treatment remains a relevant issue, newer AI applications in geophysical monitoring, such as water waves and sea surface temperature studies, are gaining momentum.

The word cloud analysis reinforces these observations, revealing that AI concepts, including convolutional neural networks, deep learning models, and random forests, are among the fastest-growing terms. This signals a strong methodological shift in the field, where AI-based algorithms are central to addressing critical water-related phenomena like streamflow forecasting, eutrophication, and sediment suspension. Moreover, the dominance of remote sensing, optic imaging, and biochemical data in this research reflects the field's increasing reliance on large datasets and sophisticated analytics to develop predictive and monitoring tools.

In terms of future opportunities, some key trends stand out: AI-driven management of drinking and agricultural water through predictive analytics and leak detection systems; advanced wastewater treatment methods; integration of AI into geophysical observations to monitor phenomena like wave height and ocean dynamics; Enhanced material science applications, such as the study of compressive strength in water-cement mixtures for sustainable construction, etc.

In conclusion, the integration of AI into water research is not only expanding the methodological toolkit available to scientists but also creating new avenues for tackling water management challenges. The convergence of these fields offers policymakers and practitioners actionable insights into sustainable water practices. Future research should focus on reinforcing interdisciplinary collaboration, developing innovative AI frameworks tailored to water issues, and addressing emerging gaps to sustain the positive growth trajectory observed in recent years. This evolving landscape offers an unprecedented opportunity to harness AI's potential for smarter, more efficient water management systems.

References

1. Gleick P, White GF (1993) Water in crisis : a guide to the world's fresh water resources, p 473
2. UNESCO WWAP (2019) The United Nations World Water Development Report 2019: Leaving No One Behind. UNESCO. Accessed 14 Oct 2024. [Online]. Available: https://unesdoc.unesco.org/ark:/48223/pf0000367306
3. UNESCO and World Water Assessment Programme (2023) The United Nations World Water Development Report 2023: partnerships and cooperation for water. Accessed 28 Oct 2023. [Online]. Available: https://unesdoc.unesco.org/ark:/48223/pf0000384655
4. Schwartz DM, Deligiannis T, Homer-Dixon T (2018) The environment and violent conflict. Environ Conflict 273–294. https://doi.org/10.4324/9780429500794-13
5. Unfried K, Kis-Katos K, Poser T (2022) Water scarcity and social conflict. J Environ Econ Manage 113:102633. https://doi.org/10.1016/J.JEEM.2022.102633
6. Russell SJ et al (2024) Artificial intelligence : a modern approach. Pearson 3:1151. Accessed 14 Oct 2024. [Online]. Available: https://thuvienso.hoasen.edu.vn/handle/123456789/8967
7. Provost F, Fawcett T (2013) Data science and its relationship to big data and data-driven decision making. Big Data 1(1):51–59. https://doi.org/10.1089/BIG.2013.1508/ASSET/IMAGES/LARGE/FIGURE1.JPEG
8. Gruetzemacher R, Whittlestone J (2022) The transformative potential of artificial intelligence. Futures 135:102884. https://doi.org/10.1016/J.FUTURES.2021.102884
9. Domingos P (2024) The master algorithm : how the quest for the ultimate learning machine will remake our world, p 329. Accessed 14 Oct 2024. [Online]. Available: https://books.google.com/books/about/The_Master_Algorithm.html?hl=fr&id=pjRkCQAAQBAJ
10. Reed SK (2019) Building bridges between AI and cognitive psychology. AI Mag 40(2):17–28. https://doi.org/10.1609/AIMAG.V40I2.2853
11. Rolnick D et al (2023) Tackling climate change with machine learning. ACM Comput Surv 55(2). https://doi.org/10.1145/3485128
12. Jordan MI, Mitchell TM (1979) Machine learning: trends, perspectives, and prospects. Science 349(6245):255–260. https://doi.org/10.1126/SCIENCE.AAA8415
13. Ngoc TTH, Khanh PT, Pramanik S (2024) AI-driven solution selection: prediction of water quality using machine learning. In: Using traditional design methods to enhance AI-driven decision making, pp 166–180. https://doi.org/10.4018/979-8-3693-0639-0.CH007
14. Aydin HE, Iban MC (2023) Predicting and analyzing flood susceptibility using boosting-based ensemble machine learning algorithms with SHapley Additive exPlanations. Nat Hazards 116(3):2957–2991. https://doi.org/10.1007/S11069-022-05793-Y
15. Donthu N, Kumar S, Mukherjee D, Pandey N, Lim WM (2021) How to conduct a bibliometric analysis: an overview and guidelines. J Bus Res 133:285–296. https://doi.org/10.1016/j.jbusres.2021.04.070
16. Bento C, Martins B, Calado P (2013) Predicting the future impact of academic publications. In: Lecture notes in computer science (including subseries lecture notes in artificial intelligence and lecture notes in bioinformatics), vol 8154 LNAI, pp 366–377. https://doi.org/10.1007/978-3-642-40669-0_32
17. Gogoglou A, Manolopoulos Y (2017) Predicting the evolution of scientific output. In: Lecture notes in computer science (including subseries lecture notes in artificial intelligence and lecture notes in bioinformatics), vol 10448 LNAI, pp 244–254. https://doi.org/10.1007/978-3-319-67074-4_24
18. Wang M-H, Yu T-C, Ho Y-S (2010) A bibliometric analysis of the performance of water research. Scientometrics 84(3):813–820. https://doi.org/10.1007/s11192-009-0112-0
19. Zhang L, Li S, Loáiciga HA, Zhuang Y, Du Y (2015) Opportunities and challenges of interbasin water transfers: a literature review with bibliometric analysis. Scientometrics 105(1):279–294. https://doi.org/10.1007/s11192-015-1656-9
20. Dai Y, Song Y, Gao H, Wang S, Yuan Y (2015) Bibliometric analysis of research progress in membrane water treatment technology from 1985 to 2013. Scientometrics 105(1):577–591. https://doi.org/10.1007/s11192-015-1669-4
21. Wang Y, Xiang C, Zhao P, Mao G, Du H (2016) A bibliometric analysis for the research on river water quality assessment and simulation during 2000–2014. Scientometrics 108(3):1333–1346. https://doi.org/10.1007/s11192-016-2014-2
22. Vasconcelos RN et al (2023) Bibliometric analysis of surface water detection and mapping using remote sensing in South America. Scientometrics 128(3):1667–1688. https://doi.org/10.1007/s11192-022-04570-9
23. Shikuku V (2023) Artificial intelligence applications in water treatment and water resource management. In: Artificial intelligence

applications in water treatment and water resource management, pp 1–270. https://doi.org/10.4018/978-1-6684-6791-6
24. Shen C (2018) A transdisciplinary review of deep learning research and its relevance for water resources scientists. Water Resour Res 54(11):8558–8593. https://doi.org/10.1029/2018WR022643
25. El Ansari R et al (2023) A review of Machine learning models and parameters for groundwater issues. In: Proceedings of the 6th international conference on networking, intelligent systems and security, pp 1–7. https://doi.org/10.1145/3607720.3607777
26. Seifeddine Zekrifa DM, Kulkarni M, Bhagyalakshmi A, Devireddy N, Gupta S, Boopathi S (2023) Integrating machine learning and AI for improved hydrological modeling and water resource management. In: Artificial intelligence applications in water treatment and water resource management, pp 46–70. https://doi.org/10.4018/978-1-6684-6791-6.CH003
27. Niazkar M et al (2024) Applications of XGBoost in water resources engineering: a systematic literature review (Dec 2018–May 2023). Environ Model Softw 174. https://doi.org/10.1016/J.ENVSOFT.2024.105971
28. Patel A et al (2023) Review of artificial intelligence and internet of things technologies in land and water management research during 1991–2021: a bibliometric analysis. Eng Appl Artif Intell 123. https://doi.org/10.1016/J.ENGAPPAI.2023.106335
29. Boutracheh H, Mejjad N, El Bouhadioui M, Moumen A (2024) Water research in the age of AI: a bibliometric heuristic analysis for trends and opportunities. In: Lecture notes in geoinformation and cartography, pp 3–45. https://doi.org/10.1007/978-3-031-63038-5_1
30. Boutracheh H, El Bouhaddioui M, Moumen A (2024) Current research priorities on fog harvesting as a clean water resource: a bibliometric approach. E3S Web Conf 489:05002. https://doi.org/10.1051/E3SCONF/202448905002
31. Boutracheh H, Mejjad N, El Bouhadioui M, Moumen A (2024) Water research in the age of AI: a bibliometric heuristic analysis for trends and opportunities. In: GIS, applied computing and data science for water management. Springer, Cham, pp 3–45. https://doi.org/10.1007/978-3-031-63038-5_1
32. Hood WW, Wilson CS (2001) The literature of bibliometrics, scientometrics, and informetrics. Kluwer Academic Publishers
33. Wallin JA (2005) Bibliometric methods: pitfalls and possibilities. https://doi.org/10.1111/j.1742-7843.2005.pto_139.x
34. Gigerenzer G, Todd PM (2000) Simple heuristics that make us smart. In: Behavioral and brain sciences, vol 23. Cambridge University, pp 727–780
35. Larivière V, Sugimoto CR (2018) Mesurer la science. Les Presses de l'Université de Montréal
36. Rostaing H (1996) La bibliometrie et ses techniques, Outils et méthodes. Éditions Sciences de la Société AND CRRM
37. Cardoso L, Silva R, de Almeida GGF, Santos LL (2020) A bibliometric model to analyze country research performance: scival topic prominence approach in tourism, leisure and hospitality. Sustainability (Switzerland) 12(23):1–27. https://doi.org/10.3390/su12239897
38. Mousavizadeh M, Chakoli AN, Pournaghi R (2020) Qualitative analysis of architectural dimensions and framework of SciVal in terms of analyzing, processing and information management of research. Iran J Inf Process Manag 35(3):755–784
39. Zanotto ED, Carvalho V (2021) Article age-and field-normalized tools to evaluate scientific impact and momentum. Scientometrics 126(4):2865–2883. https://doi.org/10.1007/S11192-021-03877-3/METRICS
40. Ibáñez A, Bielza C, Larrañaga P (2013) Relationship among research collaboration, number of documents and number of citations: a case study in Spanish computer science production in 2000–2009. Scientometrics 95(2):689–716. https://doi.org/10.1007/S11192-012-0883-6/METRICS
41. Aman V (2016) How collaboration impacts citation flows within the German science system. Scientometrics 109(3):2195–2216. https://doi.org/10.1007/S11192-016-2092-1/METRICS
42. McAllister JT, Lennertz L, Atencio Mojica Z (2022) Mapping a discipline: a guide to using VOSviewer for bibliometric and visual analysis. Sci Technol Libr (New York, NY) 41(3):319–348. https://doi.org/10.1080/0194262X.2021.1991547
43. Price DJDS (1965) The scientific foundations of science policy. Nature 206(4981):233–238. https://doi.org/10.1038/206233a0
44. de Solla Price D (1978) Cumulative advantage urn games explained: a reply to kantor. J Am Soc Inf Sci 29(4):204–206. https://doi.org/10.1002/asi.4630290410
45. de Solla Price D (1975) Society's needs in scientific and technical information. Ann N Y Acad Sci 261(1):126–136. https://doi.org/10.1111/j.1749-6632.1975.tb43309.x
46. Boutracheh H, El Ansari R, Mejjad N, Moumen A (2023) Application of bibliometrics as a data mining technique for research prioritization. In: Proceedings of the 6th international conference on networking, intelligent systems and security, pp 1–9. https://doi.org/10.1145/3607720.3607779
47. Bornmann L (2020) Bibliometrics-based decision trees (Bbdts) based on bibliometrics-based heuristics (bbhs): visualized guidelines for the use of bibliometrics in research evaluation. Quant Sci Stud 1(1):171–182. https://doi.org/10.1162/qss_a_00012
48. Braun T, Schubert A (2007) The growth of research on inter-and multidisciplinarity in science and social science papers, 1975–2006. Scientometrics 73(3):345–351. https://doi.org/10.1007/s11192-007-1933-3
49. Kim H, Hong I, Jung WS (2019) Measuring national capability over big science's multidisciplinarity: a case study of nuclear fusion research. PLoS One 14(2). https://doi.org/10.1371/journal.pone.0211963
50. Price DDS (1976) A general theory of bibliometric and other cumulative advantage processes. J Am Soc Inf Sci 27(5):292–306. https://doi.org/10.1002/asi.4630270505
51. Newman MEJ (2006) Power laws, Pareto distributions and Zipf's law. Contemp Phys 46(5):323–351. [Online]. Available: http://linkage.rockefeller.edu/wli/zipf/
52. Tversky A, Kahneman D (1974) Judgment under uncertainty: heuristics and biases. Science 185(4157):1124–1131. https://doi.org/10.1126/SCIENCE.185.4157.1124
53. Mingaleva Z, Chernova O, Mitrofanova IV (2023) Bibliometric analysis of research trends in water management aimed at increasing the sustainability of the socio-economic development of a region. Water (Switzerland) 15(20). https://doi.org/10.3390/w15203688
54. Nandiyanto ABD, Fiandini M, Al Husaeni DN (2024) Research trends from the Scopus database using keyword water hyacinth and ecosystem: a bibliometric literature review. ASEAN J Sci Eng 4(1):33–48. https://doi.org/10.17509/ajse.v4i1.60149
55. Huang Z, Sun R, Wang H, Wu X (2024) Trends and innovations in surface water monitoring via satellite altimetry: a 34-year bibliometric review. Remote Sens (Basel) 16(16). https://doi.org/10.3390/rs16162886
56. Mejjad N, Moumen A, Boutracheh H, Hilal I, Qurtobi M, El Bouhaddioui M (2024) A bibliometric analysis and classification of research on water resources management based on 17SDGs and ANZSRC indicators. In: Lecture notes in geoinformation and cartography, pp 47–61. https://doi.org/10.1007/978-3-031-63038-5_2
57. Klavans R, Boyack KW (2017) Research portfolio analysis and topic prominence. J Informetr 11(4):1158–1174. https://doi.org/10.1016/j.joi.2017.10.002
58. Small H, Boyack KW, Klavans R (2014) Identifying emerging topics in science and technology. Res Policy 43(8):1450–1467. https://doi.org/10.1016/j.respol.2014.02.005

59. Mokhnacheva YV, Tsvetkova VA (2021) Development of research topics based on the terminological approach (for example, immunology and microbiology according to scopus–SciVal data). Sci Tech Inf Process 48(2):139–145. https://doi.org/10.3103/S0147688221020106
60. Wen L, Lu Y, Li H, Long S, Li J (2020) Detecting of research front topic in artificial intelligence based on SciVal. In: ACM international conference proceeding series, pp 145–149. https://doi.org/10.1145/3421766.3421799
61. Waltman L, Van Eck NJ (2012) A new methodology for constructing a publication-level classification system of science. J Am Soc Inform Sci Technol 63(12):2378–2392. https://doi.org/10.1002/ASI.22748/ABSTRACT
62. Wu J, Wang Z, Hu Y, Tao S, Dong J (2023) Runoff forecasting using convolutional neural networks and optimized bi-directional long short-term memory. Water Resour Manage 37(2):937–953. https://doi.org/10.1007/S11269-022-03414-8
63. Xu D, Hu X, Wang W, Chau K, Zang H, Wang J (2024) A new hybrid model for monthly runoff prediction using ELMAN neural network based on decomposition-integration structure with local error correction method. Expert Syst Appl 238. https://doi.org/10.1016/J.ESWA.2023.121719
64. Uddin MG, Nash S, Rahman A, Olbert AI (2023) Performance analysis of the water quality index model for predicting water state using machine learning techniques. Process Saf Environ Prot 169:808–828. https://doi.org/10.1016/J.PSEP.2022.11.073
65. Zhu M et al (2022) A review of the application of machine learning in water quality evaluation. Eco-Environ Health 1(2):107–116. https://doi.org/10.1016/J.EEHL.2022.06.001
66. Wunsch A, Liesch T, Broda S (2021) Groundwater level forecasting with artificial neural networks: a comparison of long short-term memory (LSTM), convolutional neural networks (CNNs), and non linear autoregressive networks with exogenous input (NARX). Hydrol Earth Syst Sci 25(3):1671–1687. https://doi.org/10.5194/HESS-25-1671-2021
67. Pham QB et al (2022) Groundwater level prediction using machine learning algorithms in a drought-prone area. Neural Comput Appl 34(13):10751–10773. https://doi.org/10.1007/S00521-022-07009-7
68. Singh NK et al (2023) Artificial intelligence and machine learning-based monitoring and design of biological wastewater treatment systems. Bioresour Technol 369. https://doi.org/10.1016/J.BIORTECH.2022.128486
69. Zhao L, Dai T, Qiao Z, Sun P, Hao J, Yang Y (2020) Application of artificial intelligence to wastewater treatment: a bibliometric analysis and systematic review of technology, economy, management, and wastewater reuse. Process Saf Environ Prot 133:169–182. https://doi.org/10.1016/J.PSEP.2019.11.014
70. Yang H, Kong J, Hu H, Du Y, Gao M, Chen F (2022) A review of remote sensing for water quality retrieval: progress and challenges. Remote Sens (Basel) 14(8). https://doi.org/10.3390/RS14081770
71. Tikhamarine Y, Malik A, Souag-Gamane D, Kisi O (2020) Artificial intelligence models versus empirical equations for modeling monthly reference evapotranspiration. Environ Sci Pollut Res 27(24):30001–30019. https://doi.org/10.1007/S11356-020-08792-3
72. Zhang Y, Zhao Z, Zheng J (2020) CatBoost: a new approach for estimating daily reference crop evapotranspiration in arid and semi-arid regions of Northern China. J Hydrol (Amst) 588. https://doi.org/10.1016/J.JHYDROL.2020.125087
73. Mograne MA, Jamet C, Loisel H, Vantrepotte V, Mériaux X, Cauvin A (2019) Evaluation of five atmospheric correction algorithms over French optically-complex waters for the sentinel-3a OLCI ocean color sensor. Remote Sens (Basel) 11(6). https://doi.org/10.3390/RS11060668
74. Li L, Rong S, Wang R, Yu S (2021) Recent advances in artificial intelligence and machine learning for nonlinear relationship analysis and process control in drinking water treatment: a review. Chem Eng J 405. https://doi.org/10.1016/J.CEJ.2020.126673
75. Ghasemi M, Hasani Zonoozi M, Rezania N, Saadatpour M (2022) Predicting coagulation–flocculation process for turbidity removal from water using graphene oxide: a comparative study on ANN, SVR, ANFIS, and RSM models. Environ Sci Pollut Ress 29(48):72839–72852. https://doi.org/10.1007/S11356-022-20989-2
76. Shi Z, Chow CWK, Fabris R, Liu J, Sawade E, Jin B (2022) Determination of coagulant dosages for process control using online UV-vis spectra of raw water. J Water Process Eng 45. https://doi.org/10.1016/J.JWPE.2021.102526
77. Du B, Zhou Q, Guo J, Guo S, Wang L (2021) Deep learning with long short-term memory neural networks combining wavelet transform and principal component analysis for daily urban water demand forecasting. Expert Syst Appl 171. https://doi.org/10.1016/J.ESWA.2021.114571
78. Pesantez JE, Berglund EZ, Kaza N (2020) Smart meters data for modeling and forecasting water demand at the user-level. Environ Model Softw 125. https://doi.org/10.1016/J.ENVSOFT.2020.104633
79. Xiao C, Chen N, Hu C, Wang K, Gong J, Chen Z (2019) Short and mid-term sea surface temperature prediction using time-series satellite data and LSTM-AdaBoost combination approach. Remote Sens Environ 233. https://doi.org/10.1016/J.RSE.2019.111358
80. Sun Y, Yao X, Bi X, Huang X, Zhao X, Qiao B (2021) Time-series graph network for sea surface temperature prediction. Big Data Res 25. https://doi.org/10.1016/J.BDR.2021.100237
81. Wang N, Xie L, Zuo Y, Wang S (2023) Determination of total phosphorus concentration in water by using visible-near-infrared spectroscopy with machine learning algorithm. Environ Sci Pollut Res 30(20):58243–58252. https://doi.org/10.1007/S11356-023-26611-3
82. Waczak J et al (2024) Characterizing water composition with an autonomous robotic team employing comprehensive in situ sensing, hyperspectral imaging, machine learning, and conformal prediction. In: Remote Sens (Basel) 16. https://doi.org/10.3390/RS16060996
83. Zhang J, Meng F, Fu P, Jing T, Xu J, Yang X (2024) Tracking changes in chlorophyll-a concentration and turbidity in Nansi Lake using Sentinel-2 imagery: a novel machine learning approach. Ecol Inform 81. https://doi.org/10.1016/J.ECOINF.2024.102597
84. Cardia M, Chessa S, Franceschi M, Gambineri F, Micheli A (2023) Estimation of COD from UV-Vis spectrometer exploiting machine learning in leather industries wastewater. In: World congress on civil, structural, and environmental engineering. https://doi.org/10.11159/ICEPTP23.160
85. Zhou C, Zhang J (2023) Simultaneous measurement of chemical oxygen demand and turbidity in water based on broad optical spectra using backpropagation neural network. Chemom Intell Lab Syst 237. https://doi.org/10.1016/J.CHEMOLAB.2023.104830
86. Acheampong SA, Ampomah W, Khaniani H, Will R, Sarkodie-Kyeremeh J (2022) Quantitative interpretation of time-lapse seismic data at Farnsworth field unit: rock physics modeling, and calibration of simulated time-lapse velocity responses. Greenh Gases Sci Technol 12(6):671–697. https://doi.org/10.1002/GHG.2184
87. Koray AM, Bui D, Ampomah W, Kubi EA, Klumpenhower J (2023) Application of machine learning optimization workflow to improve oil recovery. In: Society of petroleum engineers—SPE Oklahoma City Oil and Gas Symposium, OKOG 2023. https://doi.org/10.2118/213095-MS

88. Kaminski M, Diab A (2024) Time-series forecasting of a typical pwr undergoing large break LOCA. Sci Technol Nucl Install. https://doi.org/10.1155/2024/6162232
89. Pal P, Tripathi S, Kumar C (2022) Single probe imitation of multi-depth capacitive soil moisture sensor using bidirectional recurrent neural network. IEEE Trans Instrum Meas 71. https://doi.org/10.1109/TIM.2022.3156179
90. Wood DA (2024) Real-time monitoring and optimization of drilling performance using artificial intelligence techniques: a review. In: Sustainable natural gas drilling: technologies and case studies for the energy transition, pp 169–210. https://doi.org/10.1016/B978-0-443-13422-7.00017-9
91. Donnelly J, Daneshkhah A, Abolfathi S (2024) Physics-informed neural networks as surrogate models of hydrodynamic simulators. Sci Total Environ 912. https://doi.org/10.1016/J.SCITOTENV.2023.168814
92. Huang L, Li H, Li Y (2024) Greenhouse gas accounting methodologies for wastewater treatment plants: a review. J Clean Prod 448. https://doi.org/10.1016/J.JCLEPRO.2024.141424
93. Hao Z et al (2024) GRDL: a new global reservoir area-storage-depth data set derived through deep learning-based bathymetry reconstruction. Water Resour Res 60(1). https://doi.org/10.1029/2023WR035781
94. Elsevier, "SciVal Topics|Elsevier," Elsevier. Accessed 16 Oct 2024. [Online]. Available: https://www.elsevier.com/products/scival/overview/topics#0-using-scival-topics
95. Giri S et al (2023) Revealing the sources of arsenic in private well water using Random Forest Classification and Regression. Sci Total Environ 857. https://doi.org/10.1016/j.scitotenv.2022.159360
96. Chen L, Wu T, Wang Z, Lin X, Cai Y (2023) A novel hybrid BPNN model based on adaptive evolutionary Artificial Bee Colony Algorithm for water quality index prediction. Ecol Indic 146. https://doi.org/10.1016/J.ECOLIND.2023.109882
97. Uddin MG, Nash S, Rahman A, Olbert AI (2023) A novel approach for estimating and predicting uncertainty in water quality index model using machine learning approaches. Water Res 229. https://doi.org/10.1016/J.WATRES.2022.119422
98. Zeng H, Dhiman G, Sharma A, Sharma A, Tselykh A (2023) An IoT and Blockchain-based approach for the smart water management system in agriculture. Expert Syst 40(4). https://doi.org/10.1111/EXSY.12892
99. Ibrahim H et al (2023) Evaluation and prediction of groundwater quality for irrigation using an integrated water quality indices, machine learning models and GIS approaches: a representative case study. Water (Switzerland) 15(4). https://doi.org/10.3390/W15040694
100. Najafzadeh M, Basirian S (2023) Evaluation of river water quality index using remote sensing and artificial intelligence models. Remote Sens (Basel) 15(9). https://doi.org/10.3390/RS15092359
101. Wen Z et al (2024) Remote estimates of suspended particulate matter in global lakes using machine learning models. Int Soil Water Conserv Res 12(1):200–216. https://doi.org/10.1016/J.ISWCR.2023.07.002
102. Al-Mashreki MH et al (2023) Integration of geochemical modeling, multivariate analysis, and irrigation indices for assessing groundwater quality in the Al-Jawf Basin, Yemen. Water (Switzerland) 15(8). https://doi.org/10.3390/W15081496
103. Min X, Hao B, Sheng Y, Huang Y, Qin J (2023) Transfer performance of gated recurrent unit model for runoff prediction based on the comprehensive spatiotemporal similarity of catchments. J Environ Manage 330. https://doi.org/10.1016/J.JENVMAN.2022.117182
104. Nagaraju TV, Sunil BM, Chaudhary B, Prasad CD, Gobinath R (2023) Prediction of ammonia contaminants in the aquaculture ponds using soft computing coupled with wavelet analysis. Environ Pollut 331. https://doi.org/10.1016/J.ENVPOL.2023.121924
105. Ahmadi F, Tohidi M, Sadrianzade M (2023) Streamflow prediction using a hybrid methodology based on variational mode decomposition (VMD) and machine learning approaches. Appl Water Sci 13(6). https://doi.org/10.1007/S13201-023-01943-0
106. Zhu L et al (2024) Robust remote sensing retrieval of key eutrophication indicators in coastal waters based on explainable machine learning. ISPRS J Photogramm Remote Sens 211:262–280. https://doi.org/10.1016/J.ISPRSJPRS.2024.04.007
107. Souza AP et al (2023) Integrating remote sensing and machine learning to detect turbidity anomalies in hydroelectric reservoirs. Sci Total Environ, 902. https://doi.org/10.1016/J.SCITOTENV.2023.165964
108. Fan J, Liu X, Li W (2023) Daily suspended sediment concentration forecast in the upper reach of Yellow River using a comprehensive integrated deep learning model. J Hydrol (Amst) 623. https://doi.org/10.1016/J.JHYDROL.2023.129732
109. Nasir N et al (2023) Deep learning detection of types of water-bodies using optical variables and ensembling. Intell Syst Appl 18. https://doi.org/10.1016/J.ISWA.2023.200222
110. Leggesse ES, Zimale FA, Sultan D, Enku T, Srinivasan R, Tilahun SA (2023) Predicting optical water quality indicators from remote sensing using machine learning algorithms in tropical highlands of Ethiopia. Hydrology 10(5). https://doi.org/10.3390/HYDROLOGY10050110
111. Abba SI, Benaafi M, Usman AG, Aljundi IH (2023) Sandstone groundwater salinization modelling using physicochemical variables in Southern Saudi Arabia: application of novel data intelligent algorithms. Ain Shams Eng J 14(3). https://doi.org/10.1016/J.ASEJ.2022.101894
112. Li H et al (2023) Prediction of the freshness of horse mackerel (Trachurus japonicus) using E-nose, E-tongue, and colorimeter based on biochemical indexes analyzed during frozen storage of whole fish. Food Chem 402. https://doi.org/10.1016/J.FOODCHEM.2022.134325
113. Davoodabadi Farahani S, Helforoush Z, Karimipour A, Sheremet MA (2024) Heat transfer and erosion control of nanofluid flow with tube inserts: Discrete phase model and adaptive neuro-fuzzy inference system approach. Appl Therm Eng 257. https://doi.org/10.1016/J.APPLTHERMALENG.2024.124181
114. Can M (2024) Invasive-weed-optimization-based extreme learning machine for prediction of lake water level using major atmospheric-oceanic climate scenarios. Sustainability 16(17):7825. https://doi.org/10.3390/SU16177825
115. Sharafi M, Rezaverdinejad V, Behmanesh J, Samadianfard S (2024) Development of long short-term memory along with differential optimization and neural networks for coagulant dosage prediction in water treatment plant. J Water Process Eng 65. https://doi.org/10.1016/J.JWPE.2024.105784
116. Mustapha M, Zineddine M, Majikumna UK, Alaoui AEH (2024) Forecasting reference evapotranspiration using LSTM and transformer. In: Lecture notes in networks and systems, vol 1098 LNNS, pp 267–276. https://doi.org/10.1007/978-3-031-68650-4_26
117. Zaidi Farouk MIH, Jamil Z, Abdul Latip MF (2023) Towards online surface water quality monitoring technology: a review. Environ Res 238. https://doi.org/10.1016/J.ENVRES.2023.117147
118. Sánchez-Guerrero GE, Viera González PM, Martinez Guerra E, Acuña Askar K, Cortes Martinez R (2023) Plasmonic sensor for house wastewater monitoring, p 68. https://doi.org/10.1117/12.2676086
119. Javid A, Toufigh V (2024) Utilizing ensemble machine learning and gray wolf optimization to predict the compressive strength of silica fume mixtures. Struct Concr. https://doi.org/10.1002/SUCO.202301135

120. Abbas YM, Alsaif A (2024) Influence of feature-to-feature interactions on chloride migration in type-I cement concrete: a robust modeling approach using extra random forest. Mater Today Commun 40. https://doi.org/10.1016/J.MTCOMM.2024.109419.
121. Yehia SA, Shahin RI, Fayed S (2024) Compressive behavior of eco-friendly concrete containing glass waste and recycled concrete aggregate using experimental investigation and machine learning techniques. Constr Build Mater 436. https://doi.org/10.1016/J.CONBUILDMAT.2024.137002
122. Fan M, Li Y, Shen J, Jin K, Shi J (2024) Multi-objective optimization design of recycled aggregate concrete mixture proportions based on machine learning and NSGA-II algorithm. Adv Inengineering Softw 192. https://doi.org/10.1016/J.ADVENGSOFT.2024.103631
123. Zhang L et al (2024) Understanding and predicting micro-characteristics of ultra-high performance concrete (UHPC) with green porous lightweight aggregates: insights from machine learning techniques. Constr Build Mater 446. https://doi.org/10.1016/J.CONBUILDMAT.2024.138021
124. Abdellatief M, Wong LS, Din NM, Mo KH, Ahmed AN, El-Shafie A (2024) Evaluating enhanced predictive modeling of foam concrete compressive strength using artificial intelligence algorithms. Mater Today Commun 40. https://doi.org/10.1016/J.MTCOMM.2024.110022
125. Sheng B, Liu S, Xiong K, Liu J, Zhu S, Zhang R (2024) Microbial community dynamics in different floc size aggregates during nitrogen removal process upgrading in a full-scale landfill leachate treatment plant. Bioresour Technol 413. https://doi.org/10.1016/J.BIORTECH.2024.131484
126. Li F, Li G, Lougou BG, Zhou Q, Jiang B, Shuai Y (2024) Upcycling biowaste into advanced carbon materials via low-temperature plasma hybrid system: applications, mechanisms, strategies and future prospects. Waste Manage 189:364–388. https://doi.org/10.1016/J.WASMAN.2024.08.036
127. Sathish T et al (2024) DeepNNet 15 for the prediction of biological waste to energy conversion and nutrient level detection in treated sewage water. Process Saf Environ Prot 189:636–647. https://doi.org/10.1016/J.PSEP.2024.06.119
128. Obaid A, Al Mubarak M (2024) Using data-base deep learning artificial intelligence in leak detection for sustainable water resources management. In: Studies in systems, decision and control, vol 537, pp 69–83. https://doi.org/10.1007/978-3-031-62106-2_6
129. McMillan L, Fayaz J, Varga L (2023) Flow forecasting for leakage burst prediction in water distribution systems using long short-term memory neural networks and Kalman filtering. Sustain Cities Soc 99. https://doi.org/10.1016/J.SCS.2023.104934
130. Ul Hassan I, Lone ZA, Swati S, Gamal A (2023) Forecasting weather and water management through machine learning. In: Innovations in machine learning and IoT for water management, pp 71–94. https://doi.org/10.4018/979-8-3693-1194-3.CH004
131. Li L, Chen H (2023) Artificial intelligence and internet of things-based leak detection method for the water supply network. In: International transactions on electrical energy systems, vol 2023. https://doi.org/10.1155/2023/3443047
132. Gaidai O, Sheng J, Cao Y, Zhu Y, Liu Z (2024) Evaluating areal windspeeds and wave heights by Gaidai risk evaluation method. Nat Hazards Rev 25(4). https://doi.org/10.1061/NHREFO.NHENG-2184
133. Wu K, Li XM (2024) Deep learning for retrieving omni-directional ocean wave spectra from spaceborne synthetic aperture radar. Remote Sens Environ 314. https://doi.org/10.1016/J.RSE.2024.114386
134. Chen L, Li B, Luo C, Lei X (2024) WaveNets: physics-informed neural networks for full-field recovery of rotational flow beneath large-amplitude periodic water waves. Eng Comput. https://doi.org/10.1007/S00366-024-01944-W
135. Ti Z, Kong Y (2024) Single-instant spatial wave height forecast using machine learning: an image-to-image translation approach based on generative adversarial networks. Appl Ocean Res 150. https://doi.org/10.1016/J.APOR.2024.104094

Big Data Analytics for Water Resource Management: A Review of Applications and Opportunities in Morocco

Elhassan Jamal, Youssef Rissouni, Hicham Jamil, Rachid El Ansari, Mohamed Amine Mitach, Aimad Tahi, Abdelali Taouss, Jamal Chao, and Aniss Moumen

Abstract

As the challenges in water resource management escalate, especially in water-scarce regions like Morocco, utilizing big data analytics has emerged as a promising approach for optimizing resource allocation and improving decision-making processes. This review synthesizes existing literature on using big data analytics in water resource management, specifically focusing on its applications and potential within the Moroccan context. By examining various studies and projects, this paper explores the benefits, challenges, and opportunities associated with integrating big data analytics into water management practices in Morocco. Through a comprehensive analysis, we identify key insights and gaps in current research, providing valuable recommendations for policymakers and researchers aiming to leverage big data analytics for sustainable water resource management in the country.

Keywords

Big data analytics · Water resource management · Sustainable water management · Data-driven decision-making · Morocco

E. Jamal (✉) · Y. Rissouni · H. Jamil · R. E. Ansari · A. Moumen
Laboratory of Engineering Sciences, National School of Applied Sciences, Ibn Tofaïl University, Kenitra, Morocco
e-mail: jamal.elhassan@gmail.com

J. Chao
Laboratory of Natural Resources and Sustainable Development, Ibn Tofaïl University, Kenitra, Morocco

M. A. Mitach
Sous Massa Hydraulic Bassin Agency, Agadir, Morocco

A. Tahi
Lenidit, Paris, France

A. Taouss
Geosphere, Temara, Morocco

1 Introduction

Morocco, like many arid and semi-arid regions, faces significant water resource management challenges. With a growing population, expanding urbanization, and the impacts of climate change, water demand is increasing, while water availability is becoming more uncertain [1]. This exacerbates water scarcity, inefficient water use, and inadequate infrastructure, posing substantial risks to socio-economic development and environmental sustainability [2].

In this context, the application of big data analysis is highly promising for tackling the complex issues of water resource management in Morocco. Big data analytics can process vast amounts of data from a variety of sources, including remote sensing, weather stations, sensors and socio-economic indicators. By exploiting advanced analysis techniques such as machine learning, data mining and predictive modeling, decision-makers can gain valuable insights into water availability, patterns of demand and the effectiveness of management strategies [3].

The aims of this article are twofold: firstly, to explore the importance of big data analysis in the context of water resource management in Morocco, and secondly, to provide an overview of the literature that exists on this topic. Through an exhaustive review of relevant studies and projects, including works [1], we aim to elucidate the potential applications, challenges and opportunities associated with integrating big data analytics into water management practices in Morocco.

By setting the context for the challenges of water resource management in Morocco and highlighting the importance of big data analytics in addressing these challenges, this introduction sets the scene for the following sections of the article. Through a systematic literature review, we will strive to contribute to a better understanding of how big data analytics can be harnessed to improve water resource management strategies and foster sustainable development in Morocco.

A. Moumen et al. (eds.), *Artificial Intelligence and Data Analytics for Innovative Applications in Engineering, Sustainability and Technology*, Sustainable Artificial Intelligence-Powered Applications, https://doi.org/10.1007/978-3-031-90318-2_7

2 Context of Water Resource Management in Morocco

Morocco, with its arid to semi-arid climate, faces major water resource management challenges. The country's fast-growing population and high demand for water have made it a priority to meet the growing demand for water. With a rapidly growing population and demand in sectors such as agriculture, industry and domestic use, pressure on water resources. Pressure on water resources is intensifying. In addition, the country is experiencing precipitation, exacerbated by the effects of climate change, leading to water shortages. Of climate change, leading to recurrent droughts and scarcity episodes of water.

One of the main challenges facing water management in Morocco is the imbalance between water supply and demand. Between water supply and demand. Despite initiatives aimed at developing water infrastructure and implementing conservation actions, water scarcity remains a persistent issue, particularly in rural and peri-urban areas. Moreover, water use practices are inefficient, outdated irrigation techniques and inadequate water governance exacerbate the problem.

Previous research on water resource management in Morocco has emphasized these challenges and suggested multiple strategies for overcoming them. Much of the research has centered around water allocation mechanisms, irrigation efficiency, integrated water resource management concepts and water pricing policies (Wang et al. By drawing on the results and study recommendations, policymakers and stakeholders gaining valuable insights into the challenges and opportunities facing water management in Morocco and helping inform more sustainable practices and policies in the future.

3 Fundamental Concepts of Big Data Analytics

Big data analytics extracts actionable insights from large and complex datasets, often characterized by volume, velocity, and variety. It involves systematically analyzing data to uncover hidden patterns, correlations, and trends that can inform decision-making and drive innovation across various domains.

Critical concepts in big data analytics can be summarized as follows (Fig. 1).

Data Collection and Acquisition: This step plays the first and an essential role in the initial phase; data are collected from different sources, which can include sensors, social media, mobile devices, IoT devices, etc. Structured data, from databases and spreadsheets, and unstructured data from text, images, and multimedia sources may be collected.

Data Storage and Management: Considering the huge volumes of data, sufficient storage and management tools are critical. Commonly, big data technologies like HDFS (Hadoop Distributed File System) and NoSQL databases (MongoDB, Cassandra) will structure the data in a scalable and efficiently accessible manner.

Data Processing and Analysis: Big data analytics cover various methods and techniques that process and analyze data. It includes descriptive analytics for the summation and visualization of data, diagnostic analytics for the detection of patterns and anomalies, predictive to estimate future behavior and prescriptive to recommend optimal actions.

Machine Learning and Artificial Intelligence: Machine learning algorithms are the backbone of big data analytics, allowing computers to learn from data and make predictions or decisions without explicit programming. Methods such as supervised, used to create predictive models, classified as supervised, unsupervised, and reinforcement learning.

Data visualization and interpretation: Effective communication of the insights is as critical for big data analytics. Other data visualization techniques like charts graphs. And dashboards convey results visually, in a chart-like manner, to make it easier for stakeholders to digest and use the data.

It covers all the methods and tools to derive value from big data analytics, in other words, of large and heterogeneous data. Using these basic ideas and methods organizations are able to have greater insights, innovate and make data driven decisions to meet objectives.

4 Applications of Big Data Analytics in Water Resource Management

The Applications of Big Data Analytics in Water Resource Management Yet, in the past few years, big data analytics has entered the water resource management arena as a new tool to tackle challenging issues in novel ways. Numerous studies and projects have proven the power of big data analytics over the board of water management, such as water supply, distribution, quality monitoring, and demand forecasting. Water quality monitoring and pollution control are some of the key applications of big data analytics. As

Fig. 1 Big data analytic concepts

Table 1 Key findings on big data utilization in Morocco

Key finding	Description
Enhanced water availability monitoring	Use of remote sensing and ML to monitor resources in real-time
Improved infrastructure management	Analysis of smart meter data for identifying leaks and improving distribution efficiency
Future demand forecasting	Use of climate and socioeconomic data to model water demand trends

an example, sensor data, satellite images, from monitoring resources to senor track record and microclimate data to get parameters of water quality like pH, turbidity, and concentrations of pollutants in real time. Through the analysis of big data, these studies for early detection of water contamination events and providing timely incursions to avoid environmental hazards [3, 4].

Moreover, it has also been boosting the efficiency of water distribution and optimizing the maintenance of infrastructure through big data analytics. By integrating data with smart meters, flow sensors, and GIS mapping systems, utilities can see who is connected to which pipes and how much water they are using.

Despite the significant potential of big data analytics in water resource management, several challenges need to be addressed, particularly in Morocco. Limited data availability, inadequate infrastructure, and technological barriers hinder the widespread adoption of big data analytics in water management practices. Moreover, data privacy, security, and interoperability issues require careful consideration to ensure the ethical and responsible use of data-driven technologies [5, 6].

In summary, Table 1 shows that while big data analytics offers promising opportunities for enhancing water resource management, its successful implementation in Morocco hinges on addressing these challenges and harnessing the full potential of data-driven technologies to achieve sustainable and equitable water management practices.

5 Methodology

The methodology in Fig. 2 employed in this literature review involved a systematic approach to identify, select, and analyze relevant studies on utilizing big data for water resource management in Morocco. The following criteria were established for selecting studies included in the review.

Relevance to the topic: Studies focused on applying big data analytics in water resource management, specifically in Morocco.

Quality of research: To ensure the reliability and validity of the findings, preference was given to peer-reviewed articles published in reputable journals and conference proceedings.

Date of publication: Studies considered for inclusion were limited to those published within the last decade to capture recent advancements and developments in the field.

Diversity of perspectives: Efforts were made to include studies from diverse disciplinary backgrounds, including but not limited to hydrology, environmental science, data science, and engineering.

The search process involved comprehensive literature searches across multiple academic databases, including PubMed, Scopus, Web of Science, and Google Scholar. Keywords and search terms related to big data analytics, water resource management, and Morocco were used in various combinations to identify relevant studies. Additionally, citation chaining and reference list scanning were employed to identify additional sources not captured through the initial database searches.

After retrieving relevant studies, a systematic screening process was conducted to assess their eligibility for inclusion based on the predefined criteria. Titles and abstracts of identified articles were initially screened to determine their relevance to the research topic. Subsequently, full-text screening was performed for potentially relevant articles to assess their suitability for inclusion further.

Finally, the selected studies were critically reviewed and analyzed to extract critical findings, methodologies, and insights related to using big data in water resource management in Morocco. Data synthesis techniques, such as thematic analysis and comparative analysis, were employed to identify common themes, trends, and gaps in literature.

This systematic methodology ensured a comprehensive and rigorous approach to identifying and synthesizing existing research on the topic, thereby providing valuable insights into the current state-of-the-art in considerable data utilization for water resource management in Morocco.

6 Results and Discussions

The literature review revealed several key findings regarding using big data analytics for water resource management in Morocco.

Firstly, studies have demonstrated the potential of big data analytics in improving water availability and quality monitoring. Remote sensing techniques coupled with machine learning algorithms have enabled real-time monitoring of water resources, facilitating early detection of pollution incidents and optimization of water allocation strategies [7, 8].

Secondly, big data analytics has shown promise in enhancing water distribution efficiency and infrastructure management. Utilities can identify leakages, predict

Fig. 2 Methodology of literature search

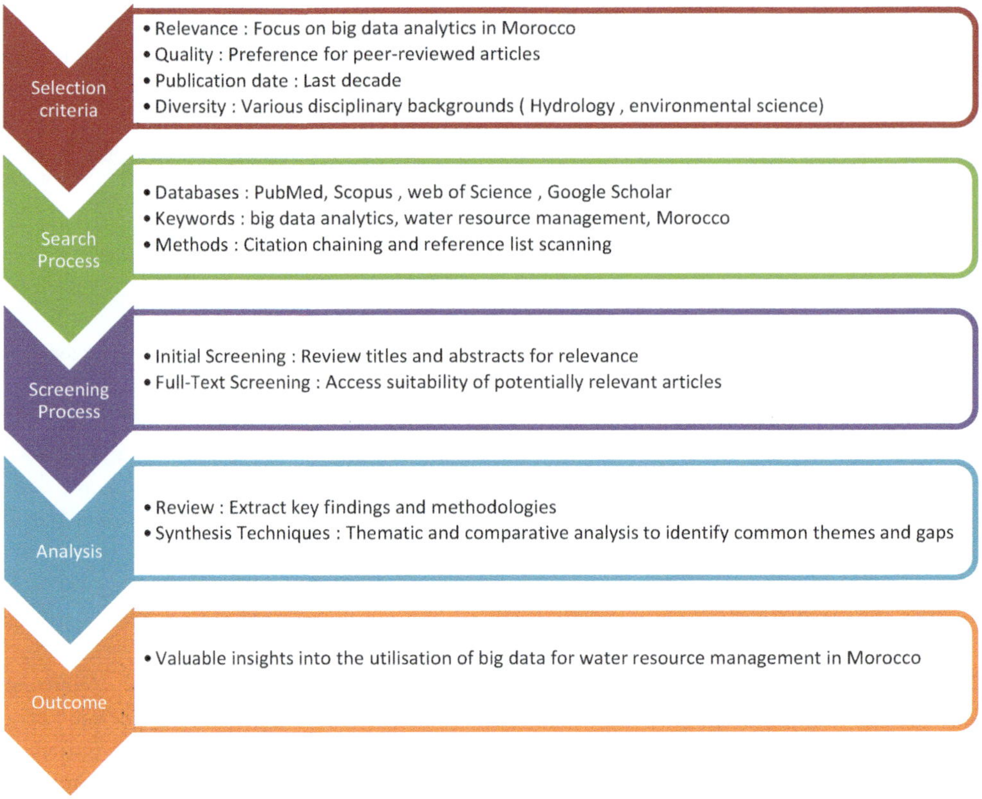

system failures, and optimize water distribution networks by analyzing smart meters and sensors data to minimize losses and improve service reliability [9, 10].

Moreover, big data analytics has the potential to revolutionize water demand forecasting and resource allocation in Morocco. By integrating socio-economic data, climate projections, and demographic trends, predictive models can provide valuable insights into future water demand patterns, enabling policymakers to implement Selection criteria * Relevance: Focus on big data analytics in Morocco.

- Quality: Preference for peer-reviewed articles * Publication date: Last decade * Diversity: Various disciplinary backgrounds (Hydrology, environmental science) Search Process * Databases: PubMed, Scopus, web of Science, Google Scholar * Keywords: big data analytics, water resource management, Morocco * Methods: Citation chaining and reference list scanning Screening Process * Initial Screening: Review titles and abstracts for relevance * Full-Text Screening: Access suitability of potentially relevant articles Analysis * Review: Extract key findings and methodologies * Synthesis Techniques: Thematic and comparative analysis to identify common themes and gaps Outcome * Valuable insights into the utilization of big data for water resource management in Morocco targeted conservation measures and prioritize investments in water infrastructure[6, 11].

Despite these advancements, several gaps in literature were identified. Firstly, there needs to be more comprehensive studies that integrate multiple data sources and disciplines to provide holistic solutions to water management challenges in Morocco. Interdisciplinary research collaborations involving hydrologists, data scientists, policymakers, and stakeholders are essential to address this gap and develop integrated approaches to water resource management. Furthermore, there is a need for greater emphasis on data governance, privacy, and security considerations in the context of big data analytics for water management.

By filling up the noted gaps and problems, researchers can help to create data-driven policies supporting sustainable and resilient water management practices in Morocco and surrounding countries (Table 2).

Table 2 Comparison of big data applications in water management

Application	Data type	Analysis method	Outcome/ Benefit
Water quality monitoring	Sensor, satellite	Predictive analytics	Early detection of pollution
Water distribution	Smart meters	Machine learning	Leak detection, efficiency optimization
Demand forecasting	Historical data	Predictive modeling	Anticipating future water needs

Table 3 Challenges and solutions in water management using big data

Challenge	Description	Proposed solution
Data availability	Limited data on water resources in Morocco	Investment in data infrastructure
Technological barriers	High cost of implementing big data tools	Government and private sector collaboration
Data privacy	Ensuring secure use of sensitive data	Ethical guidelines and regulatory frameworks

7 Big Data Utilization for Water Resource Management in Morocco

The use of big data in water resource management has slowly but steadily started to gain recognition in Morocco. While there are many studies that explore the role of big data analytics in water management, there are very few studies that have dealt specifically with Morocco.

One notable example of work done in this area is integrating remote sensing data and hydrology modeling techniques to assess water availability and to predict drought conditions in Morocco. This work has proven that satellite imagery and climate data can be effectively utilized to evaluate the water resource allocation and drought mitigation strategies [12, 13].

Moreover, they have designed decision support systems (DSSs) for water allocation and water use efficiency by applying big-data analytics in agriculture, which accounts for the greatest share of water consumption in Morocco. These DSS's utilize data on soil moisture, crop water needs, and meteorological parameters to provide timely recommendations for irrigation scheduling and crop management practices [12, 14].

Despite these advances in Morocco, the literature on big data use for water resource management is characterized by substantial gaps and is scarce, such as by developing comprehensive data-driven models that pull multiple data sources in different socio-economic, water quality, and groundwater monitoring contexts for holistic decision-making in water resources management.

Also, this opportunity should involve conducting cross-disciplinary research by pooling scientific input from various fields to interactively come up with creative solutions to Morocco's water-related challenges.

Data privacy and governance issues should be adequately tackled in line with the data governance, ensuring transparency, ethical use, and regulatory compliance for big data use in the arid Mediterranean context. There have been notable advances in Big Data for water use in agricultural settings over the past few years, it has been observed that the science is still in its preliminary steps. By addressing the identified gaps and challenges, researchers can contribute to developing data-driven strategies that promote sustainable and resilient water management practices in Morocco and beyond (Table 3).

To bridge the existing gaps and address the complex water resource management issues in Morocco, we propose a comprehensive big data analytics architecture designed to leverage diverse data sources and support multi-layered decision-making. This architecture is structured to enhance water forecasting, optimize allocation, and ensure sustainable water management across sectors.

This table concisely justifies each selected tool, highlighting the primary advantage over alternative options.

Layer	Chosen tool	Purpose	Advantages over alternatives
Streaming layer	Kafka	Real-time data streaming	High throughput and fault tolerance are better suited for massive data ingestion than RabbitMQ
	Apache spark	Real-time data processing	Unified batch and real-time processing are better for complex analytics than Flink or Storm
Data Lake	MongoDB	NoSQL data storage	Flexible schema, adaptable to evolving datasets, more accessible for unstructured data than Cassandra
	Amazon S3	Scalable object storage	Extensive integration options, high durability, well-established in big data ecosystem compared to Google Cloud Storage or Azure Blob
Batch Processing	Hadoop	Large-scale batch processing	Open-source, high control, and cost-effective for long-term storage, unlike Snowflake or BigQuery
Machine learning	Spark MLlib	Distributed machine learning	Natively integrated with spark, optimized for large-scale data, and a more unified approach than TensorFlow or Scikit-Learn
Serving layer	PostgreSQL	Relational database for batch view	It supports complex queries and data types and is better suited for analytical queries than MySQL
	Redis	In-memory data store for real-time view	Low-latency access supports more data structures and persistence than Memcached
Data visualization	Power BI/Grafana	Data visualization and reporting	It is cost-effective, integrates well with the Microsoft ecosystem, is easy to use, and is more affordable than Tableau for many organizations

By incorporating a multilayered approach, our big data architecture addresses the critical gaps in data integration, predictive capabilities, and stakeholder engagement in Morocco's water resource management. This architecture enhances resilience in water management through real-time insights, risk mitigation strategies, and optimized resource use tailored to Morocco's unique socio-environmental context. Such an approach will support immediate water resource challenges and foster sustainable, long-term water management practices in the region (Fig. 3).

8 Conclusion

In conclusion, this article has provided a comprehensive overview of using big data analytics for water resource management in Morocco. Several key findings can be drawn through a systematic review of the existing literature.

Firstly, big data analytics offers significant potential for improving various aspects of water resource management in Morocco, including water availability monitoring, infrastructure optimization, and demand forecasting. Big data techniques' efficacy for real-time decision-making that allows more sustainable water management has been proven by various studies.

Secondly, although huge strides have been made concerning big data analytics for improving water management, several unresolved issues and opportunities pertain to the implementation. These include data privacy and security challenges, limited availability of data, and some technological challenges that need to be addressed so that the benefits of big data can be fully availed within the realm of water resource management in Morocco.

Accordingly, several recommendations can be made to water resource management policymakers and researchers in Morocco based on the findings of this review. First, there is an urgent need for an increase in investment in data infrastructure and capacity building to optimize data collection, storage, and analysis capacities. Collaboration of government, research institutions, and private sector could utilize big data-analytics effectively in water management.

Moreover, policymakers should give priority to the establishment of a set of regulatory frameworks and ethical guidelines dealing with how data is being used so that the privacy rights of individuals are being respected. Capacity-building initiatives, training programs, and knowledge-sharing platforms may help in the adoption of big data analytics in water management professions in Morocco.

In future investigations targeted at developing the big data analytics area in water resource management, these gaps and constraints encountered should be addressed. Interdis-

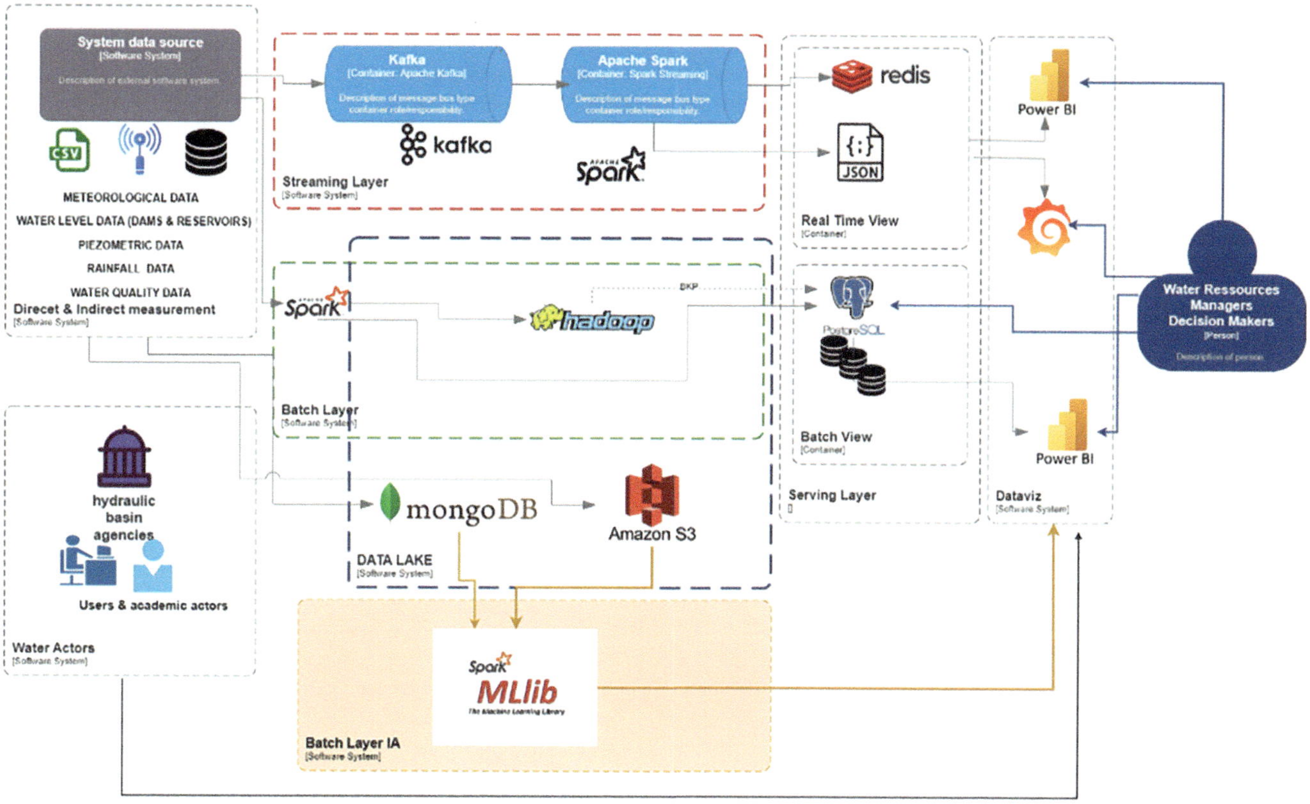

Fig. 3 Proposed big data architecture

ciplinary research collaboration, innovative data-driven solutions, and stakeholder involvement are the three pillars toward realizing the promise of big data analytics for sustainable water management in Morocco.

All in all, it would be through big data analytics and integrated water-resources management that Morocco would fulfill its water challenges and pave the way to a sustainable and resilient tomorrow.

References

1. Jamal E, Moumen A, Rissouni Y, Chao J, Tahi A (2021) Big data analytic and IoT for water resources, janv, pp 433–439. https://doi.org/10.5220/0010736000003101
2. Elhassan J, Aniss M, Jamal C (2020) Big Data analytic architecture for water resources management: a systematic review. In: Proceedings of the 4th edition of international conference on Geo-IT and water resources 2020, Geo-IT and water resources 2020, in GEOIT4W-2020. Al-Hoceima, Morocco: Association for Computing Machinery, mars 2020, pp 1–5. https://doi.org/10.1145/3399205.3399225
3. Chalh R, Bakkoury Z, Ouazar D, Hasnaoui MD, Mouatasim AE (2019) Regression analysis for estimating watershed potentialities and environmental indicators cover/use surface in Loukkos. Tangerois and Mediterranean Coastal Basin in Morocco 8(5):10
4. Abdullah M, Ibrahim M, Zulkifli H (2017) Big data technology implementation in managing water related disaster: Nahrim's experience, août
5. Kang GK, Gao JZ, Xie G (2017) Data-driven water quality analysis and prediction: a survey. ResearchGate. https://doi.org/10.1109/BigDataService.2017.40
6. Rissouni Y et al (2024) Fit-for-purpose cadaster architecture for Moroccan land-use planner: proposal and perception. E3S Web Conf 489. https://doi.org/10.1051/e3sconf/202448904017
7. Lrhoul H, Assaoui NE, Turki H (2021) Mapping of water research in Morocco: a scientometric analysis. Mater Today Proc févr. https://doi.org/10.1016/j.matpr.2020.12.1222
8. Nair JP, Vijaya MS (2021) Predictive models for river water quality using machine learning and big data techniques—a survey. In: 2021 International conference on artificial intelligence and smart systems (ICAIS), pp 1747–1753. https://doi.org/10.1109/ICAIS50930.2021.9395832
9. Kao CC, Lin YS, Wu GD, Huang CJ (2017) A comprehensive study on the internet of underwater things: applications, challenges, and channel models. Sensors 17(7), Art. no 7. https://doi.org/10.3390/s17071477
10. Abdillah AF et al (2017) Design and development of low cost coral monitoring system for shallow water based on internet of underwater things. J Telecommun, Electron Comput Eng (JTEC) 9(2–5), Art. no 2–5
11. Wang S, Xue H, Jiang X, Zhou Y, Duan X, Bai M (2018) Construction of information management system of steppe-watershed multiple water resources based on big data. In: Proceedings of the 2018 international conference on e-business and applications,

in ICEBA 2018. Da Nang, Viet Nam: Association for Computing Machinery. pp 78-81. https://doi.org/10.1145/3194188.3194204

12. Krishnan B, Nimhan A, Atole AM, Vikhe S (2018) Research for efficient water resource visualization using big data analytics. Consulté le: 31 octobre 2024. [En ligne]. Disponible sur: https://www.semanticscholar.org/paper/Research-For-Efficient-Water-Resource-Visualization-Krishnan-Nimhan/3a39793e9efe4aed2470913653c309596307d339
13. Uddin V, Hussain Rizvi SS, Hashmani M, Jameel Syed M (2020) A study of deterioration in classification models in real-time big data environment|request PDF. ResearchGate. https://doi.org/10.1007/978-3-030-33582-3_8
14. Alitane A et al (2022) Towards a decision-making approach of sustainable water resources management based on hydrological modeling: a case study in central Morocco. Sustainability 14(17), Art. no 17. https://doi.org/10.3390/su141710848

Capital Concentration and Value Creation in Listed Firms on the Casablanca Stock Exchange

Khalil EL Kouiri and Abdillah Kadouri

Abstract

Examining the correlation between concentrated ownership and business value is the major goal of this study. According to our research, the generation of value as measured by Tobin's Q is unaffected by an increase in the largest shareholder's. Indeed, the parameter linked with this attribute is not is not significant in statistical terms. This outcome, therefore, confirms the research hypothesis.

Keywords

Business value · Shareholder's · Concentration · Tobin

1 Introduction

A corporate governance structure is an organization through which the firm implements the managerial decision-making process about corporate affairs through its board of managers of the company. The benefit of both internal and external stakeholders must be incorporated. A high association exists between effective corporate governance and the financial results of an organization. Consequently, numerous corporate governance mechanisms are in place to assess performance, particularly financial performance. Ownership concentration (OC) is one such mechanism. The active participation of significant shareholders in overseeing managers can influence future performance. Notably, how the organization's present performance influences its forthcoming endeavors, such as corporate governance, demonstrates the dynamic relationship between the variables. A company's value will increase with good corporate governance. Thus, this article explores the connection among ownership concentration and firm value in Moroccan nonfinancial firm.

2 Literature Review

Your contribution Ownership structure", "shareholding structure", or even "geography of capital" concepts share the same meaning, which is the study of the distribution of power between one or more shareholders. In other words, it is a form of capital distribution according to the shareholding of voting rights and the percentage of financial interests generated [1]. It is considered a pillar in the study of corporate governance, which is not standard but differs according to geographical area, legal arsenal, nature of markets, etc. [2].

The degree of concentration of ownership in a company is an essential factor in the power distribution between managers and shareholders [3]. According to Jensen and Meckling [4], concentrated shareholding can alleviate the usual problems in corporate management, which are linked to agency conflict due to more effective control. Concentration gives the company a high degree of external management power at a minimum cost.

Other agency concerns may also occur between minority and majority shareholders.

The principal shareholder can lead to the latter expropriating the shares of minority shareholders, thereby jeopardizing the majority's interests. As a result, Pedersen and Thomsen [5] have argued that shareholder structure is a function of the regulations and institutions that prevail in market economies.

K. EL Kouiri (✉) · A. Kadouri
National School of Business and Management, Ibn Tofail University, Kenitra, Morocco
e-mail: khalil.elkouiri@uit.ac.ma

A. Kadouri
e-mail: abdillah.kadouri@uit.ac.ma

A. Moumen et al. (eds.), *Artificial Intelligence and Data Analytics for Innovative Applications in Engineering, Sustainability and Technology*, Sustainable Artificial Intelligence-Powered Applications, https://doi.org/10.1007/978-3-031-90318-2_8

2.1 Forms of Ownership Structure

Ownership structure can take many forms, depending on the firm's external context through laws, regulations, and market conditions. According to Lee [6], when the context surrounding the company is not advantageous for shareholders, in other words, when this context does not provide the necessary guarantees for shareholders to control management, they choose a concentrated structure. According to Lee, this choice is justified because the gain from concentration exceeds that from diversification. On the other hand, when these conditions benefit shareholders, the room for maneuver given to management is modest. As a result, concentrating shareholding is more costly in this case, which is why shareholders prefer a dispersed ownership structure.

Several authors have classified shareholding structures into several categories, including Pedersen and Thomsen [5], who have classified them into three main categories: concentrated, dispersed, and dominated.

- Concentrated shareholder structure

Concentrated shareholder structure is the most common ownership structure in many countries, in which the majority shareholder has more than 50% of the company's shares [2].

A shareholding is said to be concentrated when the majority of shares are held by one or a restricted group of shareholders known as "concentrated or controlling shareholders" [3]. This type of shareholding is prevalent in European and Asian countries. It is distinguished by a comparatively substantial concentration of ownership in the possession of one shareholder or company, especially a family-owned one. This observation has been confirmed by La porta [7] for continental European countries and [8] for Asia.

Shareholders are in a comfortable position to exert effective control through their presence on the Board of Directors and their voting rights to pressure management (on the management side and the resource allocation side) [3]. Similarly, shareholder concentration, according to Agrawal and Mandelker [9] and Shleifer and Vishny [10], is proof of the relevance of shareholder control over management. Unlike the dispersed structure, this type limits agency conflicts between shareholders or among executives and shareholders [2]. However, shareholders with a majority shareholding find it attractive to take control of managers since this will positively impact their returns. According to Charletry [11], shareholders are inclined to give as much as possible to improve the value of their shares and the company.

From the above, concentrated ownership is a crucial element in corporate management. For this reason, in firms with concentrated shareholding, the disciplinary power of the board of directors will be weak since it will be exercised directly by the majority shareholders. As a result, several researchers have indicated a negative relationship among the degree of board oversight (percentage of outside executives, independence between chairman and CEO, board size) and capital concentration [12–15].

Due to conflicts of interest among minority and majority shareholders, concentrated shareholding has several drawbacks despite its benefits. In other words, majority shareholders may involve the company in strategies that counter minority shareholders and the company's other internal partners (such as employees) or external partners (such as suppliers). These shareholders may decide, for example, to venture into investments that are too risky or impose special dividend payments that could jeopardize the interests of minority shareholders [3].

- The dispersed shareholder structures

In this structure, majority shareholders no longer hold over 20% of the company's shares [2]. Several researchers have based their foundations on the spread ownership structure in the literature [4]. According to La Porta [14], a shareholding structure dispersed among several shareholders is characterized by "a strong separation between shareholding and control". This constrains the strategies for expropriating minority stockholders by majority shareholders, who are held back by adopting a certain number of legal protection rules adopted by each country. This is the case, for example, in some Anglo-Saxon nations, that are highly oriented towards the stock market and have adopted drastic measures to protect owners' interests, such as the United Kingdom and the United States of America. A decline in family ownership accompanied the adoption of this type of structure in these countries and subsequently limited the power of family control in companies [16].

When monitoring management's opportunistic attitudes, dispersed shareholders need more power to exercise this control since the voting rights corresponding to the proportion of shares held are no longer significant, which gives rise to many agency conflicts. Of course, in a company with a widely dispersed shareholder structure, it is not in any shareholder's interest to invest in such control since they alone will have to bear the costs, not to mention that the other shareholders will benefit from the operation. We can, therefore, observe free-rider attitudes in a dispersed shareholder structure, which leaves a margin of freedom for managers and an opportunity to use their opportunism against the interests of shareholders [3]. This result is endorsed by Mayer [17], who pointed out that investing in control is not worthwhile for dispersed shareholders, as exit is less costly than intervention in control. This intervention may even harm the company's image on the stock market by sending out the wrong signal about the sustainability of the company's performance, even if the objective is a takeover strategy, for example, which may jeopardize the value of the shares. Also, this dispersion may minimize risk through diversification, and the resulting cost may exceed the agency cost of management controls [3].

Therefore, the ownership structure changes shape according to the level of value maximization. Indeed, replacing a concentrated shareholding structure with a diffuse one requires precisely determining the effects of the loss of control on the company's managers.

2.2 Shareholder Concentration and the Creation of Shareholder Value: Contrasting Relationships

Several theoretical and empirical studies have examined the capital concentration in a single individual's hands or group and its impact on value creation within companies. Some have found results that support a link between ownership concentration and value creation, while others have argued that there is no such link.

- **A positive relationship**

For Berle and Means [18], companies with a concentrated shareholder base can control managers well. On the other hand, for companies with diffuse shareholding (non-concentrated), control is taken out of the hands of shareholders, allowing managers to take advantage of this situation, which can hurt company performance. They found a linear and positive association among shareholder concentration and performance. According to Shleifer [10], a firm's achievement is inclined by the existence of its largest shareholder who has some authority over the board of directors.

Tomičić [19] worked on this relationship by studying 32 Croatian banking groups with ownership centralized in the possession of major (top 10 shareholders in most cases) at 89%. The authors discovered a strong correlation among performance and ownership concentration (presented by ROA and ROE). It is important to consider how other elements, including market share and innovation rate, affect performance.

Li [20] studied this link by working on a sample of 1241 Chinese companies. They discovered "that ownership concentration has a positive effect on firm performance as measured by market and accounting ratios" [21].

For the purpose to examine the relationship among and performance ownership concentration, Haldar and Nageswara Rao [22] worked on an unbalanced panel of businesses traded on the Stock Exchange of India (BSE-500) in 2011. They discovered a positive influence on result when the founders of the company control the great majority of the capital. Ganguli and Guha Deb [23], who worked on a sample of companies between 2009 and 2013, found the same result. Concentration also impacts company performance as represented by market and accounting indicators.

In the same vein, Abbas et al. [24] examined a selection of 100 Pakistani non financial firms, studying the link between performance (presented by ROE ROA indicators) and ownership concentration (presented by the total percentage of ownership). They found a close relationship between these variables in the positive direction, at a shareholding level of 10 and 50% of capital. Above this threshold, the authors noted a negative effect. This result is justified by the opportunistic behavior of majority shareholders, who may abuse their position to take advantage of personal gain at the outflow of the firm's overall welfare. Chandrapala [25] also confirmed the same result, but this time, two accounting variables were used: profits and book value.

As for Qatar, in 2012, Almudehki and Zeitun [26] worked on 29 nonfinancial firms registered on the Doha Stock Exchange. They discovered that concentration and ROE were positively correlated.

- **Negative relationship**

The existence of a dominant shareholder may also create "a conflict of interest between the majority and minority" shareholders [27], as there "is a negative correlation between ownership concentration and performance" [28]. To put it another way, when the manager of the firm also happens to be the majority shareholder, the manager may exploit this circumstance to obtain an advantage over minority owners, which could negatively impact the corporation's results [29]. Kirchmaier and Grant [30] studied this relationship in certain European countries.

Shareholder blocks can also hurt the company since the number of shareholders traded and listed is small, making the company illiquid [31].

Wang and Shailer [32] took as their objective the study of a sample of work "carried out in various countries have investigated the correlation between corporate performance" [33] and ownership concentration after correcting for endogeneity issues. These variables were determined to be negatively correlated by the authors.

Afgan [34] examined how ownership concentration impacts the investment performance of publicly listed Pakistani firms utilizing panel data. Ownership concentration adversely affects performance, as such as demonstrated by Tobin's Q.

- **Absence of relationship**

Other research have shown that performance is not always correlated with ownership concentration but may also be correlated with a diffuse shareholding structure, in contrast to the previously mentioned findings that demonstrate a "relationship between ownership concentration and performance" [35].

Thomsen and Pedersen [36] worked in the European context to determine the explanations and consequences of dominance of ownership regarding outcomes. As a result, they reported no link among ownership concentration and performance in their studies using ROE. However, they did indicate that knowledge of the causes of concentration is essential in choosing the optimal ownership structure. These causes can be either the firm's size (decreasing concentration) or the degree of profits (increasing concentration). There is also the nature of the financial market and its maturity regarding shareholder protection.

All the studies presented and carried out on this issue no longer share expected outcomes but somewhat contradictory results.

3 Data and Methodology

3.1 Data

This research employs comprehensive data encompassing all publicly traded firm's on the Casablanca stock exchange throughout 2016–2020. The data was gathered from the stock exchange website, the Moroccan capital market regulator, and the companies' websites under investigation. Following the typical approach, Entities inside the financial sector are omitted from the analysis and overseas companies that may own distinct ownership configurations. Following the removal of absent observations considering the independent and dependent variables, the final sample consists of 367 firms for analysis.

Several variables are mobilized, as follows:

Explanatory variables. Several explanatory variables were adopted. These include shareholding concentration, the proportion of "the family" that control the capital, and "shareholder pacts" (Table 1).

Explained variable. Generally, in most writings that have dealt with the calculation of Tobin's Q, the market value of the liability is equal to the book value of the liability as a result of the value of the liability not being taken into account in the calculation of the market value, Tobin's Q formula is, therefore, as follows [37]:

$$\text{Tobin's Q} = \frac{\text{Market value of shares}}{\text{book value of shareholder's equity}} \quad (1)$$

Table 1 Explanatory variables linked to ownership concentration

Explanatory variable	Indicators	Names
Shareholder concentration	% of capital held by first, second and third shareholders	Act
Family shareholding	The "family" nature of the holder of a block of shares of at least 10% in value	SeulAc family
Shareholders' agreement	Presence of a control or ownership agreement	Pact

Control variables. The control variables, as analyzed by Eelderink [38], make it possible to confirm whether these or other variables have an impact on any phenomenon. Size (Height) as "a control variable, measured by the logarithm of total assets" [39], tolerates a neutral effect of the various deviations between the chosen sample and the size so as not to impact the other variables in the model. Age is measured by the logarithm of number of years between the observation year and the firm's founding year. Financial leverage is calculated by dividing financial debts by total economic assets.

3.2 Estimation Model

Given the variables that influence value creation and that are controlled, the regression is carried out using the following equation:

$$\begin{aligned} QT_{it} = {} & \beta_i + \beta_1.Ac1_{it} + \beta_2.Ac2_{it} + \beta_3.Ac3_{it} \\ & + \beta_4.\text{AloneAcFamily}_{it} + \beta_5.\text{Pacte}_{it} \\ & + \beta_6.\text{financial leverage}_{it} + \beta_7.\text{size}_{it} + \beta_8.\text{Age}_{it} + e_{it} \end{aligned} \quad (2)$$

3.3 Estimation Method

In this regression analysis, ownership concentration is operationalized as the sum of shareholdings by the top three controlling block holders (at least 10% of voting shares) and by the presence of a single holder having at least 10% of the shares. The presence of a pact serves as a proxy for ownership and control separation.

4 Results and Discussion

4.1 Descriptive Statistics

The descriptive shareholding statistics for the top three shareholders shown in Table 2 that the first shareholder owns an average of 51.84% of the shares. For some corporations (Centrale Danone), this shareholding might approach 99%, while the second shareholder owns an average of 13.24%. However, the average holding of the third shareholder is only 6.66%, and it can even exceed 22% (Balima).

These results show no significant correlation above the 0.7 threshold, which, according to Pras et al. [40], is the minimum threshold at which we can say that there is a collinearity problem (Table 3).

There is a negative correlation between the first, second, and third shareholders. The greater the ownership of the second shareholder, the greater the possibility of the presence of the third shareholder (Table 4).

The higher the age, the more block holders there are, and the more there is a tendency and orientation towards shareholder concentration.

Table 2 Shareholding levels of the top three shareholders

Statistics	First shareholder	Second shareholder	Third shareholder
Nb. of observations	265	265	265
Minimum	9000	0.362	0.010
Maximum	99,680	33,340	21,720
1st quartile	36,839	7282	3419
Median	51,840	11,245	5940
3rd quartile	64,855	17,580	8875
Average	51,843	13,243	6663
Standard deviation (n)	21,011	8329	4802

Table 3 Bivariate correlation of the shareholding of the top three shareholders

Shareholder correlations				
		Act1	Act2	Act3
Act1	Pearson correlation	1	–, 322**	–, 522**
Act2	Pearson correlation	–, 322**	1	275**
Act3	Pearson correlation	–, 522**	275**	1

**. "Correlation is significant at the 0.01 level (two-tailed)"

Table 4 Bivariate correlation between ownership of top three shareholders and age

Correlations					
		Age	Act1	Act2	Act3
Age	Pearson correlation	1	156**	189**	135*
Act1	Pearson correlation	156**	1	–, 322**	–, 522**
Act2	Pearson correlation	189**	–, 322**	1	275**
Act3	Pearson correlation	135*	–, 522**	275**	1

** "Correlation is significant at the 0.01 level (two-tailed)"
* "Correlation is significant at the 0.05 level (two-tailed)"

4.2 Regression Results

Variations in Tobin's Q due to the existence of a pact, financial indebtedness, the size of the company, its age, and the sector in which it operates are controlled (Table 5).

The outcomes of the estimation of the elements of this equation reveal that in the BVC, the growth in the power of the greatest holder of blocks of shares does not always affect the creation of value assessed by Tobin's Q. This variable's parameter is, in fact, not statistically significant. In contrast to the holder of the second-largest block of shares, this is also true for the power of the holder of the third-largest block of shares. Tobin's Q's value is negatively impacted by the power of the second-largest controlling shareholder; for every 1% increase in this shareholder's share percentage, Tobin's Q's value decreases by 2.27%.

Estimation of the parameters of Eq. 2 yields additional results concerning the effect of capital concentration on value creation. While an increase in the power of the holder of the

Table 5 Effect of capital concentration on value creation (results of Eq. 2)

Variable	Regression
Act1	– 0.8551
Act2	– 2.2759**
Act3	– 3.5490
AloneAcFamily	– 1.2189**
Pact	– 0.0653
Financial-leverage	2.6740**
Size	0.0394
Age	1.7072
Constant	1.3837
R^2	0.4726

largest block of shares does not hurt value creation in the presence of a second and third block holder, it does harm value creation in the presence of a family with sole power. The study's findings support the entrenchment hypothesis, which holds that the most significant block of shareholders' desires to extract personal gains is caused by the lack of a countervailing force to that power. This opportunism is anticipated by BVC investors, who turn away from the shares of companies controlled by a family, resulting in their stock market discount.

The results of estimating Eq. 2 demonstrate the growth in value creation for companies listed on the BVC when their financial debt rises. This result implies that the increase in a company's indebtedness on this stock exchange constitutes a positive signal to investors about favorable growth prospects. Additionally, it means that value-adding initiatives are being funded by debt.

The negative and statistically significant sign of the "AloneAcFamily" variable demonstrates the negative effect of increasing the power of the holder of the largest block of shares on value creation. This result confirms the research hypothesis.

The negative effect of the increase in power of the holder of the largest block of shares when he is alone in power has been demonstrated in several studies. Indeed, among a sample of companies listed on the Seoul Stock Exchange during 2015–2017, Lee [41] indicates that increasing the power of block holders hurts Tobin's Q value. This result is also obtained by Dlugosz [42] with companies in the Standard and Poor's index. These authors found that an increase in the percentage of shares held by outside ownership negatively affects value creation as measured by Tobin's Q.

5 Conclusion

In this manuscript, we have examined and discussed the results of the tested hypotheses. We showed that increasing the power of holders of large blocks of shares no longer hurts the value creation process, even with a second or third shareholder, except for a single shareholder with family power. On the other hand, value creation increases with debt, which sends a positive signal to investors about the company's prospects. This study enhances our comprehension of corporate governance and finance literature by investigating the relationship between ownership concentration among major shareholders and business value in the contemporary Moroccan context.

Acknowledgements We are incredibly grateful to all Ibn Tofail and Ibn Zohr University staff participating in the study. Any errors and omissions remain our responsibility.

References

1. Barry TA (2010) Structure Actionnariale Des Banques, Risque Et Efficience. UNIVERSITE DE LIMOGES
2. Adm R, Sonza IB (2017) Patterns of efficiency in dispersed, dominant and concentrated ownership structures in Brazil. Mackenzie Manag Rev 18:232–259
3. Yammeesri J (2003) Corporate governance : ownership structure and firm performance—evidence from Thailand. University of Wollongong
4. Jensen MC, Meckling WH (1976) Theory of the firm: managerial behavior, agency costs and ownership structure. J Financ Econ 3:305–360
5. Pedersen T, Thomsen S (1997) European patterns of corporate ownership: a Twelve-Country study. J Int Bus Stud 28:759–778
6. Lee J (2005) Structure de propriété , stratégies de diversification et gouvernance des entreprises coréennes. Thèse de doctorat. Université Toulouse 1
7. La Porta R, Lopez-de-Silanes F, Shleifer A, Vishny RW (1998) Law and finance. J Polit Econ 106:1113–1155
8. Claessens S, Djankov S, Lang LHP (2000) The separation of ownership and control in East Asian Corporations
9. Agrawal A, Mandelker GN (1990) Large shareholders and the monitoring of managers. J Financ Quant Anal 25:143–161
10. Shleifer A, Vishny RW (1986) Large shareholders and corporate control. J Polit Econ 94:461–488
11. Charléty P (2001) Activisme Des Actionnaires : Le Cas Particulier Des Fonds De Pension. Bulletin COB, n° 354. Paris
12. Agrawal A, Knoeber CR (1996) Firm performance and mechanisms to control agency problems between managers and shareholders. J Financ Quant Anal 31:377–397
13. Charreaux G, Pitol-Belin J (1985) La théorie contractuelle des organisations : une application au conseil d'administration. In: Colloque de l'institut de Mathématiques Economiques de l'Université de Dijon. Dijon, pp 1–35
14. Fernández C, Gómez-Ansón S (2005) Does ownership structure affect firm performance? Evidence from a continental-type governance system. Corp Ownersh Control 3:75–89
15. Godard L (2001) La taille du conseil d'administration: déterminants et impact sur la performance. Rev Sci Gest 33:125–148
16. Franks J, Mayer C (2017) Evolution of ownership and control around the world: the changing face of capitalism, 1st ed. Elsevier B.V.
17. Mayer C (1998) Financial systems and corporate governance: a review of the international evidence. J Institutional Theor Econ 154:144–165
18. Mard Y, Marsat S, Roux F (2012) Structure de L'actionnariat et performance : le cas français. In: CIGE, pp 1–24
19. Tomičić I, Ćorić A, Čalopa MK (2012) Croatian banking sector research: relationship between ownership structure, concentration, owners' type and bank performance. J Inf Organ Sci 36:159–167
20. Li K, Lu L, Mittoo UR, Zhang Z (2015) Board independence, ownership concentration and corporate performance—Chinese evidence. Int Rev Financ Anal 41:162–175
21. Ynal A (2017) Ownership concentration: its determinants and the impact on firm performance. Evidence from MENA region
22. Haldar A, Nageswara Rao SVD, Tayde M (2010) Ownership structure and firm performance: evidence from India. In: Second Annual general business conference proceedings. Huntsville, Texas, pp 1–12
23. Ganguli SK, Guha Deb S (2016) Board composition, ownership structure and firm performance: New Indian evidence in a unique regulatory environment. SSRN Electron J 1–29

24. Abbas A, Naqvi HA, Mirza HH (2013) Impact of large ownership on firm performance: a case of non financial listed companies of Pakistan. World Appl Sci J 21:1141–1152
25. Chandrapala P (2013) The value relevance of earnings and book value: the importance of ownership concentration and firm size. J Compet 5:98–107
26. Almudehki N, Zeitun R (2012) Ownership structure and corporate performance: evidence from qatar. SSRN Electron J 1:1–21
27. Muntahanah S, Kusuma H, Harjito DA, Arifin Z (2021) The effect of family ownership and corporate governance on firm performance: a case study in Indonesia. J Asian Financ Econ Bus 8:697–706
28. Hu X, Yao G, Zhou T (2022) Does ownership structure affect the optimal capital structure? A PSTR model for China. Int J Financ Econ 27:2458–2480
29. Mard Y, Marsat S, Roux F (2014) Structure de l'actionnariat et performance financière de l'entreprise : le cas français. Financ Contrôle Strat 17:1–26
30. Kirchmaier T, Grant J (2005) Corporate ownership structure and performance in Europe. Eur Manag Rev 2:231–245
31. Becker B, Cronqvist H, Fahlenbrach R (2011) Estimating the effects of large shareholders using a geographic instrument. J Financ Quant Anal 46:907–942
32. Wang K, Shailer G (2015) Ownership concentration and firm performance in emerging markets: a meta-analysis. J Econ Surv 29:199–229
33. Nasiri M, Ramakrishnan S (2020) Earnings management, corporate governance and corporate performance among Malaysian listed companies. J Environ Treat Tech
34. Afgan N, Gugler K, Kunst R (2016) The effects of ownership concentration on performance of Pakistani listed companies. CBU Int Conf Proc 4:214–222
35. Holderness CG, Sheehan DP (1988) The role of majority shareholders in publicly held corporations. An exploratory analysis. J Financ Econ 20:317–346
36. Thomsen S, Pedersen T (2000) Ownership Structure and economic performance in the largest European companies. Strat Manag J 689–705. Palgrave Macmillan UK, London
37. Hayes A (2021) Q Ratio: Tobin's Q. https://www.investopedia.com/terms/q/qratio.asp. Accessed 20 May 2022
38. Eelderink GJ (2014) Effect of ownership structure on firm performance. University of Twente
39. Dube T (2018) An analysis of effects of ownership on capital structure and corporate performance of South African firms. University of Pretoria (South Africa)
40. Pras B, Evrard Y, Roux E (2009) Market: études et recherches en marketing, 4th edn. Dunod
41. Lee YK (2022) The effect of ownership structure on corporate payout policy and performance: evidence from Korea's exogenous dividends tax shock. Pacific-Basin Financ J 73:101763
42. Dlugosz J, Fahlenbrach R, Gompers P, Metrick A (2006) Large blocks of stock: prevalence, size, and measurement. J Corp Financ 12:594–618

Enhancing Models Portability Using Moodle Users' Traces

Nour Eddine El Fezazi, Ilyas Alloug, Ilham Oumaira, and Mohamed Daoudi

Abstract

The prediction of academic failure is one of the most pressing concerns for researchers in the field of education. Research in the field of E-learning has predominantly prioritized the precision and accuracy of machine learning models designed for specific courses. However, there is a dearth of studies exploring the transferability and generalization of prediction models from a source course to other courses. By portability of machine learning models, we generally mean the ability of a model to be used or deployed in different environments and platforms. Many factors can affect the portability of a model. The objective of this study is to investigate the portability of models obtained directly from the Moodle logs of 15 courses hosted in the Moodle platform of Ibn Tofail University. The approach used aims at verifying whether the number and level of use of activities provided by the Moodle logs, and whether the use of numerical or categorical attributes affects the portability of predictive failure models in educational contexts. We used the KNN classification algorithm on a dataset of courses of the same level to obtain a model and test their portability to other courses by evaluating accuracy and loss of accuracy. The results show that the portability of the predictive models and their application to different courses is possible in some cases but with an acceptable loss of accuracy.

Keywords

Portability of models · Failure prediction · Machine learning · KNN · Ibn Tofaïl University

N. E. E. Fezazi · I. Alloug · I. Oumaira
Ibn Tofail University, Kénitra, Morocco

M. Daoudi (✉)
Hassan II University, Casablanca, Morocco
e-mail: mohamed.daoudi@uit.ac.ma

A. Moumen et al. (eds.), *Artificial Intelligence and Data Analytics for Innovative Applications in Engineering, Sustainability and Technology*, Sustainable Artificial Intelligence-Powered Applications, https://doi.org/10.1007/978-3-031-90318-2_9

1 Introduction

In recent years, E-learning has become particularly important in the academic lives of Moroccan students. Unlike other forms, this mode of learning has several advantages, such as more flexible interaction in terms of time and space and the large amount of information circulated between the teacher and the student via learning management systems (LMS). Moodle is a powerful, open-source LMS that allows users to create robust, flexible, and engaging online learning experiences [1]. Moodle's logging of interactions with student learning resources provides a very important dataset for building models of student behavior and prediction of student performance.

Most approaches to analyze Moodle platform data offer predictive models adapted to a particular course. Yet the challenge is to create models for a particular course that can be useful when used in other courses [2]. This is what we call transferable or portable models [3].

In this work, we will study the degree of portability between courses of the models generated from the log data of the Moodle platform of Ibn Tofail University. The models were constructed using student interactions with Moodle logs, and the target class attribute is binary, indicating whether the student passed or failed the course (Pass/Fail). For this study, courses were categorized according to their utilization levels of activities, resources, and interactions in the Moodle course. The primary objective of this research is to answer the following research question: is it possible to apply a prediction model of student failure built from an online (source) course to online (target) courses with similar level of usage without losing prediction quality?

The structure of this research is as follows: In Sect. Related Work" we review the literature on this study. Section "Data Collection & Data Preprocessing" describes the extracted data. In "Results" section, we present and discuss the results of the research. Finally, we present the conclusions and future lines of this research.

2 Literature Review

Boyer [4] most studies focus on analyzing and predicting student behavior in E-learning platforms, with a set of EDM tools to quantify the student's performance [5] or detect student at risk [6].

However, most of the studies in this area have focused on obtaining the most accurate models for a specific class and course. In this regard, the EDM community tends to standardize the use of machine learning models in the field of E-learning by achieving generalizable and portable predictive models that can be applied to different classes and courses without losing their prediction accuracy, which remains an important challenge in the EDM field. In [7], they proposed a study that carried out the portability of predictive models between universities courses where two experiments executed the J48 classification algorithms over 24 courses. The study reveals that the portability's feasibility still depends on specific circumstances.

These previous works present different approaches for the generalization of predictive models of students' performance.

To outline and contextualize our work with the mentioned studies, our study aims to explore the concept of generalizability differently. According to [5], the study shows that the portability or transferability of predictive models depends strongly on high-level attributes that guarantee the interoperability between models in terms of prediction accuracy gain. While in this paper, the main efforts will be dedicated to using the portability of predictive models to prevent student failure in different courses by spotting and detecting the most accurate features that assure the interoperability of EDM predictive models.

We would like to mention that our work is mainly based on the work of Javier López-Zambrano [8], applying it to data from the Moodle platforms of Ibn Tofaïl University, while considering the problem of model portability linked to student failure.

3 Materials and Methods

3.1 Data Collection and Data Preprocessing

In this section, the description and preprocessing of the data used in this research will be presented. We reveal the collected data as well as the sources and methods of data extraction. Then, we announce the methods applied to transform the raw data collected from the Moodle log tables. We also describe the hands-on activities we conducted to answer our research question.

3.1.1 Data Collection

The data used in this research are collected from the Moodle platform for three institutions of the University Ibn Tofail: the National School of Applied Sciences, the National School of Business and Management, and the Faculty of Economics and Management. Data are collected from a diverse range of courses to enhance the models' generalizability. The same treatment is performed on courses selected according to criteria. A SQL query was developed to extract information on the courses and on the students' activity.

In the Moodle database, the system stores all user actions in log tables. The '*logstore_standard_log*' table records information such as the actions taken by a user for each course, quiz, poll, forum, etc. To find out if activities have been scheduled by a teacher in a specific course, the choice, quiz, modules, and course_modules tables will be used. The user table contains the learner's information. Figure 1 shows the physical data model and the relationships between the different tables used.

We used data from the log tables of 2408 Ibn Tofail University students from 15 different courses delivered after 2013. Table 1 summarizes the information from these courses. For each course, it shows the subject or title of the course (Subject), a code, the number of activities that are planned by the professor, the number of students (#Users), the number of actions performed for each course, and the level of Moodle usage (Low, Medium, or High). For ethical and confidential reasons related to the use of the data, we have received permission from the Moodle platform manager, and we have hidden the personal information of the teachers and students.

3.1.2 Data Preprocessing

In our study, a rigorous preprocessing process was applied (Fig. 2). This included data cleaning to eliminate missing information, encoding of categorical variables, and data normalization to standardize scales.

In this phase, we decided to skip the variable selection step. This decision is justified using the KNN (K-Nearest Neighbors) algorithm for our model. The advantage of this algorithm is that it bases predictions on the spatial proximity between observations, minimizing the effects of multicollinearity, in contrast to traditional techniques such as linear regression [9]. As a result, we retained all features, including those with significant correlations, without affecting the model's predictive performance.

In accordance with ethical and confidentiality standards, we requested the approval of the Moodle platform administrator and anonymized the personal information of

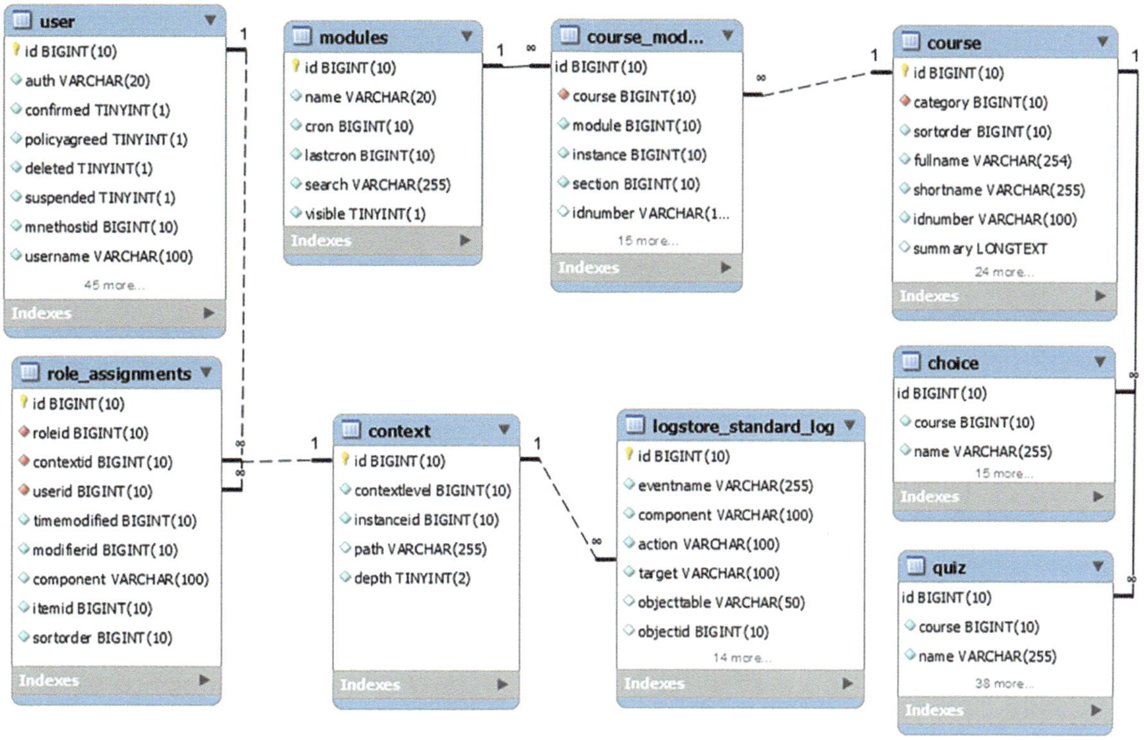

Fig. 1 Physical relational data model

Table 1 Information of all courses

Code	No. of activities	#Users	Total actions	Moodle usage
TDN	0	115	26,858	High
ME	0	182	5485	High
FPVR	0	78	97	Low
RI	0	35	797	Low
EMF	4	472	33,568	High
DC	0	102	1264	Low
PR	0	453	19,474	High
APP	15	227	127,597	High
SP	68	32	111	Low
EL	0	131	18,218	High
OP	0	68	414	Low
SD	0	370	24,172	High
IB	3	46	44	Low
AN	1	65	160	Low
EM	0	32	5350	High

teachers and students. The initial preprocessing consists of ensuring that learner data is correctly anonymized, in line with Moroccan ethical standards, confidentiality, and regulations. Our study mainly uses the technique of "Data Masking", which involves replacing student and teacher names with numerical codes to prevent individual identification. Similarly, course codes are used in place of titles to prevent course-teacher associations.

Fig. 2 Data preprocessing steps

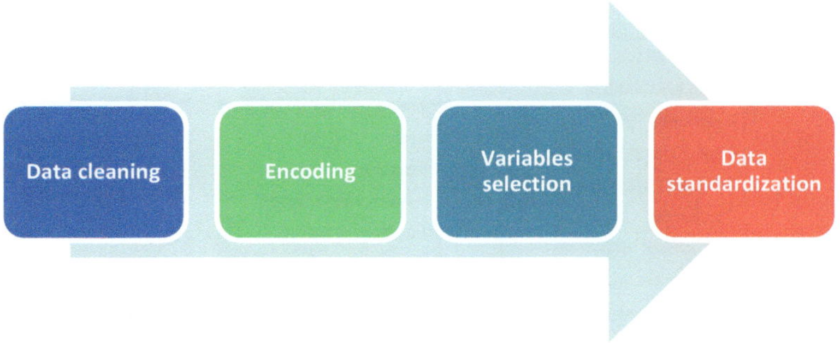

Fig. 3 Experimental methodology of our research

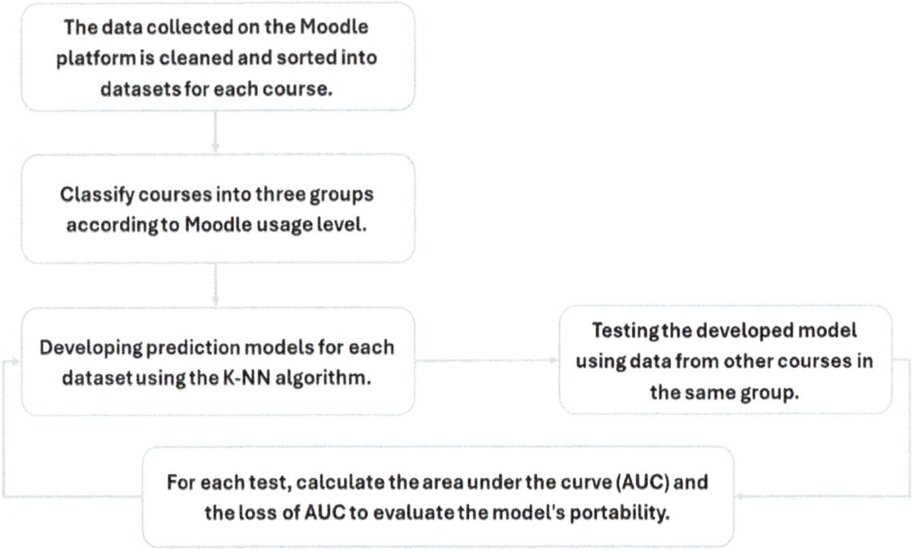

3.2 Methodology

Our research provides a structured methodology (see Fig. 3) for assessing the portability of failure prediction models in various educational contexts. Initially, the steps consist of data extraction and preparation, meticulously cleaning data from the Moodle platform and organizing it into datasets prepared in numerical and discretized formats.

The next steps focus on model development and course classification. We build predictive failure models for each dataset using the KNN algorithm and classify courses according to their level of use of the Moodle platform. From these categories, we select a model for further testing.

The methodology includes testing the selected model on datasets from other courses in the same group. The AUC has an important statistical property: the AUC of a classifier is equivalent to the probability that the classifier will rank a randomly chosen positive instance higher than a randomly chosen negative instance [10]. AUC loss is used to evaluate the model's portability.

This methodology provides an approach to evaluating the transferability of predictive failure models in different learning contexts.

Table 2 List of groups by Moodle usage

No.	Group	No. of subjects
1	High	8
2	Medium	1
3	Low	6

We grouped our 15 different courses (see Table 2) into three levels of use of Moodle activities in courses (see Table 1). Moodle provides us with resources (text and web page, link to files or websites, and label) and different types of activities (assignments, chat, choice, database, forum, glossary, lesson, quiz, survey, wiki, workshop, etc.) [7].

We have categorized the levels of use based on the quantity of activities used and the students' actions within the course:

- **Low level**: The course exhibits no activity and less than 10,000 actions performed by the students.
- **Medium level**: The course has between 10,000 and 20,000 actions performed by students.
- **High level**: The course has more than 20,000 actions performed by the students.

4 Results and Discussion

There are two categories into which the courses have been classified: high-level courses and low-level courses. The experimentation aims to predict student failure in different courses using KNN algorithm to measure models' portability on the same category of courses. The performance of the KNN algorithm was assessed using both numerical and discretized datasets.

Courses with a high level of use

Course code/AUC (numerical dataset)	APP	EMF	EM	AVG
APP	0.684	0.179	0.705	0.522
EMF	0.5	0.5	0.5	0.5
EM	0.536	0.689	0.333	0.519
Course code /AUC loss (numerical dataset)	APP	EMF	EM	AVG
APP	–	– 0.505	+ 0.021	0.242
EMF	0	–	0	0
EM	+ 0.203	+ 0.356	–	+ 0.280
Course code/AUC (discretized dataset)	APP	EMF	EM	AVG
APP	0.684	0.454	0.586	0.575
EMF	0.5	0.5	0.5	0.5
EM	0.408	0.501	0.333	0.414
Course code /AUC loss (discretized dataset)	APP	EMF	EM	AVG
APP	–	– 0.230	– 0.098	– 0.164
EMF	0	–	0	0
EM	+ 0.075	+ 0.168	–	– 1.22

Courses with low level of use

Course code/AUC (numerical dataset)	RI	DC	OP	AVG
RI	0.916	0.470	0.505	0.630
DC	0.5	0.5	0.5	0.5
OP	0.478	0.541	0.423	0.480
Course code/AUC loss (numerical dataset)	RI	DC	OP	AVG
RI	–	– 0.446	– 0.411	– 0.428
DC	0	–	0	0
OP	+ 0.055	+ 0.118	–	+ 0.086
Course code/AUC (discretized dataset)	RI	DC	OP	AVG
RI	0.916	0.470	0.472	0.619
DC	0.5	0.5	0.5	0.5
OP	0.472	0.541	0.423	0.478
Course code/AUC loss (discretized dataset)	RI	DC	OP	AVG
RI	–	– 0.446	– 0.444	–0.445
DC	0	–	0	0
OP	+ 0.049	+ 0.118	–	+ 0.083

4.1 High-Level Courses

Using numerical dataset or discretized dataset, the KNN algorithm performs better on the APP course, with an AUC of 0.684, followed by the EMF course (AUC = 0.5) and the EM course (AUC = 0.333). The average AUC for high-level courses using numerical or discretized data is 0.505.

Using numerical data, the performance of the models decreased significantly only for the model developed with the APP course data and applied on the EMF course, while for the other courses, the performance increased relatively by 4.4%. For the discretized data, we notice a loss of accuracy of 14% on average, which is still acceptable.

4.2 Low-Level Courses

For both types of data, the KNN algorithm achieves the best performance on the RI course, with an AUC of 0.916, followed by the DC course (AUC = 0.5) and the OP course (AUC = 0. 423). The average AUC for low-level courses using numerical or discretized data is 0.613.

For courses with a low level of activity, the performance of the models has decreased significantly for the courses. We observe a loss of accuracy of 34% on average when using numerical data and 36% on average when using discretized data. This raises the matter of the quantity of data.

In summary, the KNN algorithm performs better on low-level courses with an average AUC of 0.613, In contrast, for high-level courses, the KNN algorithm shows similar performance with both numerical and discretized datasets, with average AUC values of 0.505. The performance of the KNN algorithm on individual courses varies, with the best performance observed for the RI course in low-level courses (AUC = 0.916) and the APP course in high-level courses (AUC = 0.684).

Moreover, we found that the models developed with high-activity courses kept their performance when applied to the same level of activity courses. Concerning the portability of models developed on courses with low activity, we observed a significant loss in accuracy.

Further research should explore how different data preprocessing techniques, such as discretization, can affect the performance of the KNN algorithm and identify factors that contribute to the observed variations in performance across different courses.

5 Conclusion

This study focused on examining the portability of models created from student traces in Moodle log data. The findings revealed that predictive models developed from high-activity courses remained effective when applied to similar courses with similar levels of activity. However, when applying models from low-activity courses to other courses, a notable decrease in accuracy was observed, prompting considerations about data quantity and preprocessing techniques.

Future research should explore different data preprocessing methods and identify factors that influence performance variations between courses. Overall, this study contributes to enhancing the generalization of predictive models in E-learning.

Our future research aims to incorporate ontological structures into our data to enrich our understanding of the relationships underlying the data and improve the portability of the models.

Acknowledgements The authors express their gratitude to Mr. Charaf Moulay Hassan for his valuable guidance and constructive comments, which significantly contributed to improving the quality and scientific rigor of this paper.

References

1. Daoudi M, Lebkiri N, Ouali Y, Oumaira I (2022) Student involvement in mobile-learning: case of Ibn Tofail University. Stat Optim Inf Comput 10:59–74. https://doi.org/10.19139/soic-2310-5070-1217
2. Baker RS (2019) Challenges for the future of educational data mining: the baker learning analytics prizes. J Educ Data Min 11:1–17. https://doi.org/10.5281/zenodo.3554746
3. Boyer S, Veeramachaneni K (2015) Transfer learning for predictive models in massive open online courses. In: Artificial intelligence in education. Springer, Cham, pp 54–63
4. Marinho T, Costa EB, Dermeval D, Ferreira R, Braz LM, Bittencourt II, Luna HPL (2010) An ontology-based software framework to provide educational data mining. In: Proceedings of the 2010 ACM symposium on applied computing, pp 1433–1437. Association for Computing Machinery, New York, NY, USA
5. Yagci M (2022) Educational data mining: prediction of students' academic performance using machine learning algorithms. Smart Learn Environ 9. https://doi.org/10.1186/s40561-022-00192-z
6. Waheed H, Hassan S-U, Nawaz R, Aljohani NR, Chen G, Gasevic D (2023) Early prediction of learners at risk in self-paced education: a neural network approach. Expert Syst Appl 213:118868. https://doi.org/10.1016/j.eswa.2022.118868
7. López-Zambrano J, Lara JA, Romero C (2022) Correction to: improving the portability of predicting students' performance models by using ontologies. J Comput High Educ 34(1):1–19. https://doi.org/10.1007/s12528-021-09273-3). J Comput High Educ 34:20. https://doi.org/10.1007/s12528-021-09287-x
8. López-Zambrano J, Lara JA, Romero C (2020) Towards portability of models for predicting students' final performance in university courses starting from moodle logs. Appl Sci (Switzerland) 10. https://doi.org/10.3390/app10010354
9. Ramli NA, Ismail MT, Hooy C-W (2014) An analysis on two different data sets by using ensemble of k-Nearest Neighbor Classifiers. WSEAS Trans Math 13:780–789
10. Fawcett T (2006) Introduction to ROC analysis. Pattern Recogn Lett 27:861–874. https://doi.org/10.1016/j.patrec.2005.10.010

Smart OSC-MAC CT Mode in Wireless Sensor Networks Based on WI-LEM Technology

Mohamed Mbida

Abstract

Nowadays, wireless sensor networks (WSN) are one of the most important fields of research, especially in their operating cycle. Let's take the example of the MAC layer, which controls communications and media sharing. In this work, we manage the energetic balance of sensor nodes to maintain a longer lifetime and high transmission quality using Wireless Electrical Measurements (WI-LEM) technology and General Artificial Intelligence (GAI) cooperative transmission.

Keywords

Smart · MAC system · WI-LEM · Technology · Mobile node · AI · Duty cycle

1 Introduction

Among the most used protocols that ensure data transmission [1], the protocol On-Demand Scheduling Cooperative MAC (OSC-MAC), which schedules sending times on demand of the nodes in transmission state, with the technique of cooperation with predefined routes of nodes. The major concern is in the case of an out of service state of one or more nodes (energy exhaustion), which will result in a totally lost data transmission and the cooperative transmission technique becomes blocked. More precisely, OSC-MAC is based on a Transmitter Root Node (TRN) which serves as an intermediary to pass from one sub-network to another according to the predefined route, it is up to him to delegate one of these sub-nodes according to an exchange of planification's in progress if they are going to cooperate or not in a transmission of data and to update its table of information, this implies that the TRN will be in active mode longer than the others, so the risk of energy exhaustion leads to the isolation of a part of the network.

2 Classical OSC-MAC

Each mode consumes a portion of a node's energy. For example, idle listening means that nodes continue to listen to the channel when there are no incoming packets, and overhearing means that nodes decode packets intended for others. Collisions result in corrupted packets and involve MAC layer retransmissions and additional energy consumption, and a high risk of isolation of part or all of the network. Many researches have tried to reduce the idle listening time but this was not enough [2]. In our work, we focus on the three modes using the WI-LEM technique in CT mode to minimize network isolation risk.

3 Smart OSC-MAC WI-LEM Technology with CT Mode

In this work, we based on the Wireless Local Energy Meter system of monitoring measurement of each node located in a sub-network [3], which will be implemented in a TRN via a cognitive radio communication adapted between nodes, the use of this technology offers advantages at the level of the electrical cabinet and the installation of the network and its exploitation, however several research focused on the energy consumption of each node [4, 5]. As a continuation of this previous studies our work make this measures controlled by an intelligent system installed on a TRN, providing visibility of the network's energy status and planning duty cycles that better predict and reduce energy consumption in the mentioned modes.

M. Mbida (✉)
Department of Mathematic and Informatics, Emerging Technologies Laboratory (LAVETE), Faculty of Sciences and Technology, Hassan 1st University, Settat, Morocco
e-mail: m.mbida@uhp.ac.ma

A. Moumen et al. (eds.), *Artificial Intelligence and Data Analytics for Innovative Applications in Engineering, Sustainability and Technology*, Sustainable Artificial Intelligence-Powered Applications, https://doi.org/10.1007/978-3-031-90318-2_10

Indication: The SW OSC-MAC is an upgraded protocol from the classical OSC-MAC.

3.1 Intelligent Selection of Cooperative Nodes in a Transmission Mode

This section explains the measurement mechanism and selection of nodes for data transmission mode (CT). The system identifies transceiver nodes (Pseudo_code 1) with higher energy levels. The intelligent management system then selects the optimal route based on energy consumption and transmission time, incorporating it into each TRN's duty cycle. The measurement of the energy needed for transmission is done with the help of the WI-LEM Technology and the intelligent system which is based on an algorithm of calculation of the transmission energy (ETni) of M bits of each node, to make the selection of the nodes that will enter CT end-to-end mode (Pseudo_code 2), so that this last one will choose the route according to the summation of the optimal energies toward the final destination (Formula 1):

$$\text{OpEnergy} = \sum E\, ni/i = \{0 \ldots .ni\}. \tag{1}$$

```
Pseudo Code 1: Selection TRN RCSF Nodes
Input:
Lists of Network Nodes organized decreasing by
Energy Level: LNN
Energy Level: ELI
Distance between Transmitter and next hop: DTN
List Sub Networking: LSN
Output:
List of TRN nodes for Smart CT Mode: LSCM
Begin
For each li ∈ LSN do
If Eli >=(Elect*S) +(Empl*S*D2) && Eli>=Eli-1
Li-1<=li;
LSCM<=li+1;
Endif
Send LSCM ();
End
Pseudo Code 2: WI-LEM Energy Measurement's
for transmission of M bits
Input:
Emf-data: Amplifying energy
EFS: Free space energy
Output:
Necessary Energy of sending M bits: Ens
If DR is Distance route && do =sqrt (EFS/Emf)
Then Send (M bits data, DR)
Endif
If Dr >=do
Then Ens =Ent-(M*Etx+M*Emf*(DR^4));
Else
Ens=Ent-(M*Etx+M*Emf*(DR^2));
Endif
End
```

4 Duty Cycle Allocation for Transmission/Reception of Data Packets

To guarantee an optimal transmission and data loss rate, we have to plan a sending schedule for each TN executed by the intelligent supervision system and sent to each TRN node which will be the same in turn for its elected sub-nodes, another feature is the choice of packets to be transmitted of primary or secondary data in order not to overload a route with secondary information and leave priority data on standby. Another very important alternative is to keep an eye on the energy level of the simple nodes and TRN in case of a drop in energy level, and other nodes with more energy are delegated to maintain an extended life span of the network (Fig. 1).

Monitoring the energy levels of simple nodes and TRN is crucial; when energy drops, nodes with higher energy are delegated to prolong network lifespan.

5 Active Listening Status of the Transmission Channel and Collision Detection

Based on the intelligent management system of the sensor network, this mode will be almost non-existent in terms of energy consumption, because when a transmission request is triggered, the intelligent system informs the nodes that are going to enter CT mode of the optimal route in terms of energy and transmission quality without leaving a permanent listening of the medium. In the case of a collision, the intelligent network management system informs the transmitting node via TRN to indicate that it must retransmit within a preset time interval to synchronize with the others in a CT mode RDV schedule.

6 Experiments

6.1 Presentation Scenarios of Simulation

In this part, we have experimented under Contiki OS in LoT COOJA Simulator an execution of the CL-OSC/SW-OSC-MAC algorithms of the radio energy consumption during transmission state, and the total average energy consumed at

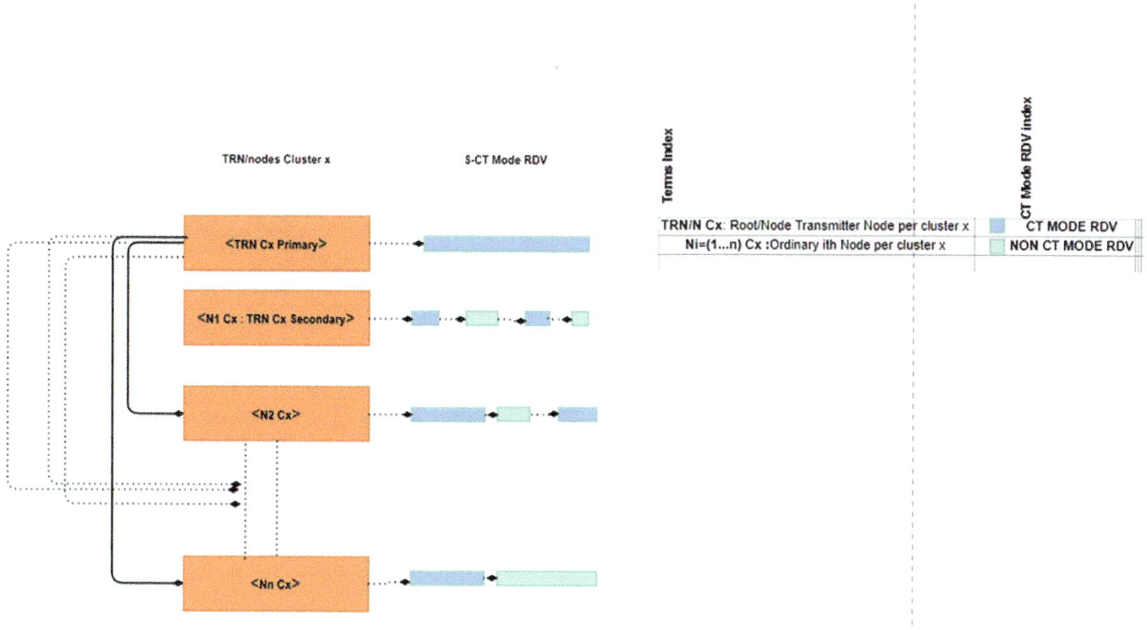

Fig. 1 Prototype of CT RDV in Transmission Mode

the battery level Figs. 2 and 3, both scenarios participated in the relaying of packets in end-to-end transmission mode for a 100-bit packet, according to specific parameters Tables 1 and 2.

Table 1 SW-OSC-MAC TRN algorithm parameters

Number of used motes sky	DTxR $_{\text{SW-OSC-MAC}}$ (m)	DIR $_{\text{SW-OSC-MAC}}$ (m)
5	43	23
10	85	48
30	150	68
60	320	104

Table 2 CL-OSC-MAC algorithm parameters

Number of used motes sky	DTxR CL-OSC-MAC (m)	DIR CL-OSC-MAC (m)
5	230	225
10	328	363
30	480	464
60	848	870

Note: We add progressively motes between end-to-end sender/receiver (TRN for SW-OSC-MAC and simple transmitter for CL-OSC-MAC).

Execution of CL-OSC/SW-OSC-MAC algorithms under Contiki OS using the Cooja Simulator, focusing on the radio energy consumption during the transmission state and calculating the total average energy consumed at the battery level throughout the process.

Indications

Dynamic Tx-range (meters unit): DTxR	TX-ratio success (percentage unit): TxRS	Average radio consumption energy/sky mote (percentage unit): ARCE
Dynamic int range (meters unit): DIR	RX-ratio success (percentage unit): RxRS	Average energy consumption battery (millivolts unit)/sky mote: AECB

6.2 Statistics and Analysis

6.2.1 Tx/Rx Ratio Success

According to the statistics obtained at the level of probability success in data transmission (Fig. 4), we notice that the rate of loss information takes too low values in the TxRS (SW)-OSC-MAC compared to the TxRS (CL)-OSC-MAC which presents a too high rate. The values obtained from the TxRS (SW)-OSC-MAC algorithm are the results of the intelligent management of the nodes in transmission mode which allows to change the transmission topologies (TRN/CT mode nodes) each time, and to test the QoS performance of the nodes that are going to enter a planned cooperative

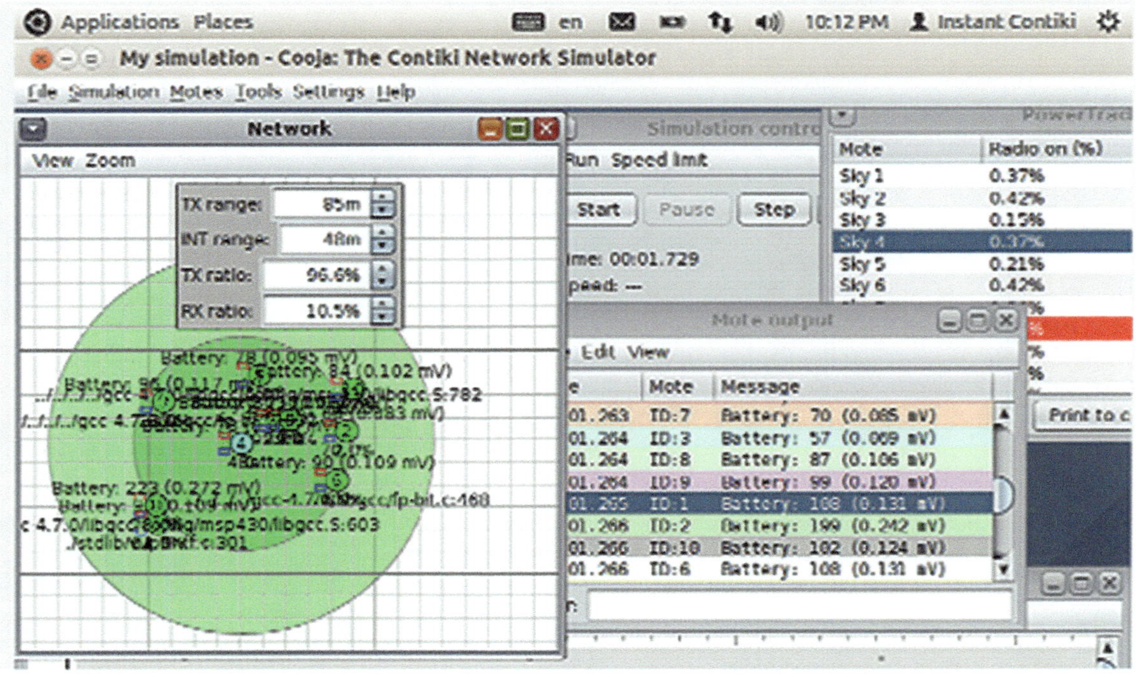

Fig. 2 Radio and battery level energy consumption in transmission mode for SW-OSC-MAC algorithm for 10 motes Sky

duty cycle, so that each node goes from the off state to the on state, and comes back to the off state as soon as it finishes the relaying of all the information outside its cluster. However, for the $RxRS_{(Sw)\text{-OSC-MAC}}$, we obtained 0% as a percentage of success in transmission mode. This can be explained by the total suspension of the reception mode at the level of the nodes that become transmitters in CT mode, and also in the $RXRS_{(CL)\text{OSC-MAC}}$, continues to activate the reception mode alternately with the sending of data, this leads to a loss of additional performance that will affect in turn the success of data transmission.

Additional Explanation for Fig. 4: This figure compares the data transmission success rates for different MAC algorithms. The **TxRS (SW)-OSC-MAC (blue)** has much lower data loss than the **TxRS (CL)-OSC-MAC (red)**, due to better node management during transmission. The **RxRS (SW)-OSC-MAC (green)** has a 0% success rate because nodes stop receiving while transmitting, while the **RxRS (CL)-OSC-MAC (purple)** alternates between sending and receiving, leading to performance losses.

6.2.2 Energy Consumption of Radio Transmission and Battery Level En Mode Transmission ARCE/AECB

In reference to the simulation of the energy consumption during the radio transmission between nodes (Fig. 5), the energy consumption of radio transmission of the (SW)-OSC-MAC algorithm (ARCE(SW)-OSC-MAC) takes a too low energy percentage compared to the ARCE values of the (CL)-OSC-MAC algorithm (ARCE(CL)-OSC-MAC), which is about the triple values of the ARCE (SW)-OSC-MAC. The results obtained from the latter cited algorithm are justified by the intelligent and planned activation according to a sufficient duration for the total transfer of data to the next hop outside the cluster; hence, the (CL)-OSC-MAC algorithm leaves its radio open during the end-to-end transmission (final destination), which causes a high loss of energy. However, this slightly affects the percentage of battery consumption allocated to the transmission in the (SW)-OSC-MAC algorithm, unlike the (CL)-OSC-MAC algorithm which takes large consumption values.

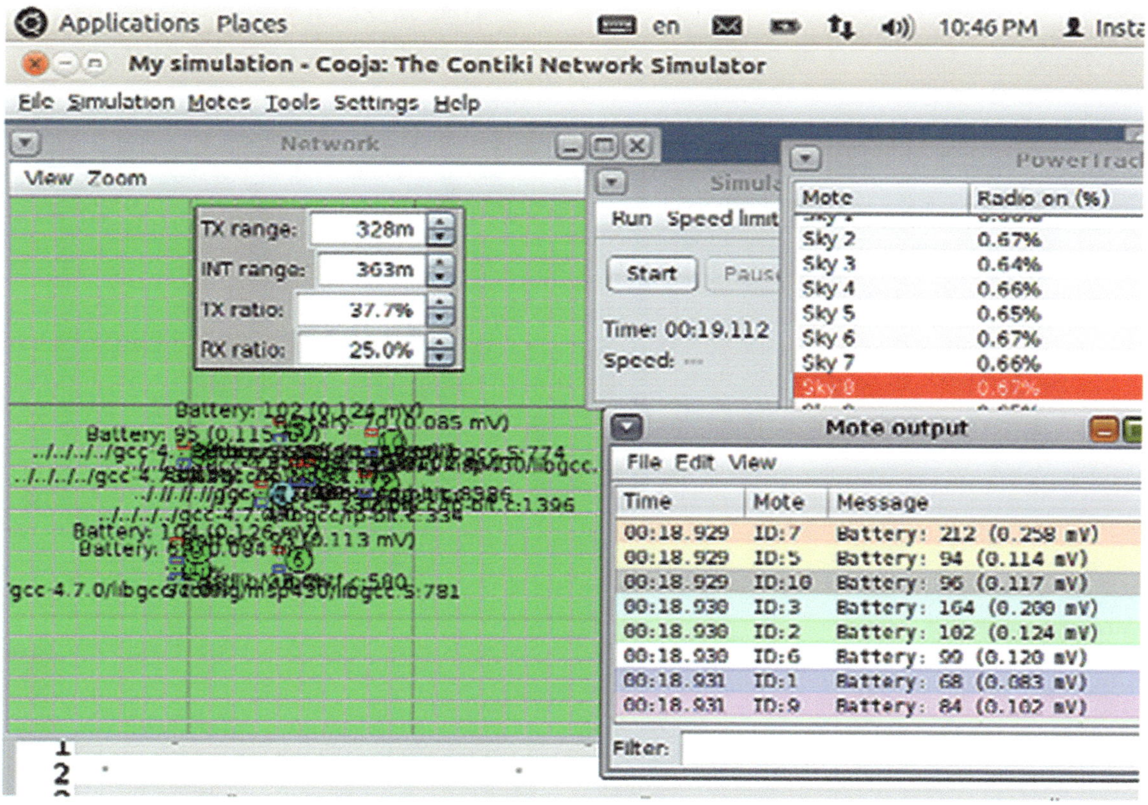

Fig. 3 Radio and battery level energy consumption in transmission mode for CL-OSC-MAC algorithm for 10 motes Sky

NB: The statistics of Figs. 4 and 5 take values in function of the number of nodes in sky motes according to an interval of [5…,60].

Additional Explanation for Fig. 5: This figure illustrates the energy consumption during radio transmission for two MAC algorithms. The ARCE(SW)-OSC-MAC (blue) demonstrates significantly lower energy usage compared to the ARCE(CL)-OSC-MAC (red), which consumes about three times more energy. This difference is attributed to the efficient management of node activation in the SW-OSC algorithm, which activates nodes only as needed for data transfer. In contrast, the CL-OSC algorithm keeps the radio on for the entire duration of the transmission, leading to excessive energy loss.

7 Conclusion

The intelligent SW-OSC-MAC algorithm implemented in this work is characterized by the construction of node routes that change the topology dynamically, which allow the elected nodes to enter the cooperation mode in end-to-end transmission mode based on energy measurements with WI-LEM technology of wireless sensor networks, and also to activate the radio channel communication only in a specific interval, and alternatively disable the reception function, which gave an additional energy saving that will serve to extend the network lifetime and prolong the network transmission performance. As a perspective to this project, we opt to integrate the SW-OSC-MAC algorithm a technique of classification of data by order of importance according to criteria that can be changed

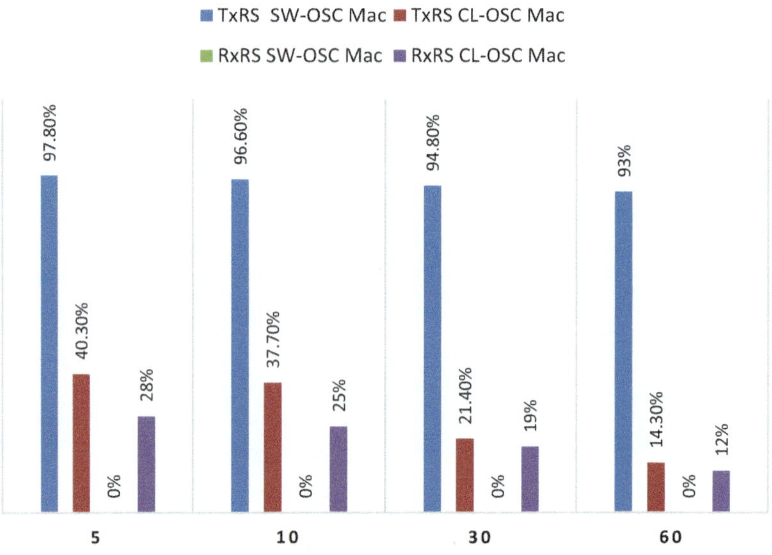

Fig. 4 TxRS (CL/SW)-OSC-MAC and RxRS (CL/SW)-OSC-MAC performances in WSN

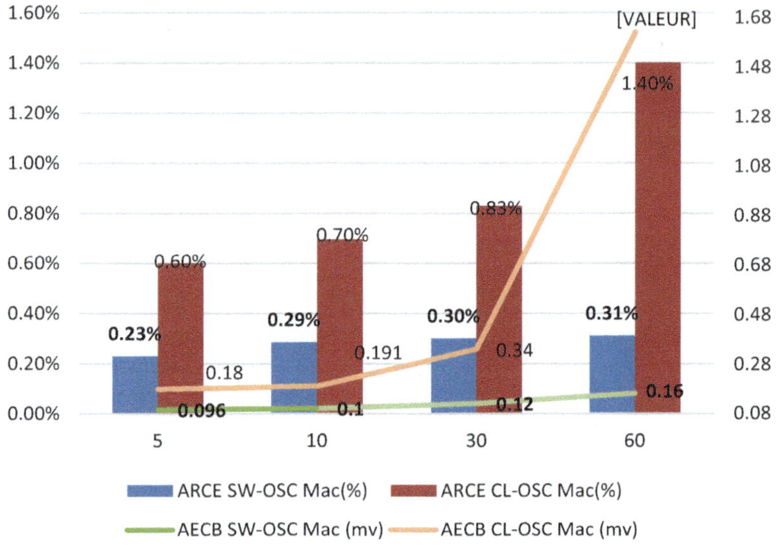

Fig. 5 ARCE(CL/SW)-OSC-MAC and AECB (CL/SW)-OSC-MAC performances in WSN

on demand, in order to direct the energy consumption toward the most interesting transmissions to be executed, and all simulations will be done in Contiki NG [6].

References

1. Lin J, Weitnauer MA (2018) Range extension cooperative MAC to attack energy hole in duty-cycled multi-hop WSNs. Wirel Netw 24:1419–1437. https://doi.org/10.1007/s11276-016-1408-7
2. Zhang DG, Zhou S, Tang YM (2018) A low duty cycle efficient MAC protocol based on self-adaption and predictive strategy. Mob Netw Appl 23(4):828–839
3. LEM International SA Route du Nant-d'Avril, 1521217 Meyrin Switzerland. https://www.lem.com/en/product-list
4. Tutunović M, Wuttidittachotti P (2019) Discovery of suitable node number for wireless sensor networks based on energy consumption using Cooja. In: 2019 21st international conference on advanced communication technology (ICACT). IEEE, pp 168–172
5. Li S, Kim KS, Zhang L, Huan X, Smith J (2022) Energy-efficient message bundling with delay and synchronization constraints in wireless sensor networks. Sensors 22(14):5276
6. Oikonomou G, Duquennoy S, Elsts A, Eriksson J, Tanaka Y, Tsiftes N (2022) The Contiki-NG open source operating system for next generation IoT devices. SoftwareX 18:101089. ISSN 2352-7110

A Mathematical Algorithmic Analysis of Water Quality Variability Using Kohonen's Self-organizing Maps

R. El Chaal and M. O. Aboutafail

Abstract

In this study, Kohonen's self-organizing maps (SOM) were applied to assess environmental challenges by mapping and categorizing physicochemical factors in the Inaouen watershed. Using a classification method based on the SOM artificial neural network, five distinct clusters were identified in the water quality of the region. Classes 2 and 3 showed low of sodium, potassium, magnesium, calcium, sulfates, and total dissolved solids. With respect to classes 1 and 4, they showed higher values of bicarbonates (HCO_3), total dissolved solids (TDS), total alkalinity ($CaCO_3$), magnesium (Mg), calcium (Ca), and electrical conductivity. Most of the parameters were found to be extremely high for Class 5 except the D.O. and NO_3, which indicates localized water quality issues in certain areas. The research highlights successfully using Kohonen's self-organizing map classification technique in evaluating the spatial distribution of water quality. The study of SOM offers great insight into the environment of the members of the Inaouen basin, thus improving the understanding of the very complex ecosystem. Thus, it helps the researchers to reach a better decision-making capacity to implement proper management in water resources.

Keywords

Classification methods · Kohonen mapping · SOM techniques · Statistical analysis

R. El Chaal (✉) · M. O. Aboutafail
ENSA of Kenitra, Engineering Sciences Laboratory, Data Analysis, Mathematical Modeling and Optimization Team, Ibn Tofail University, Kenitra, Morocco
e-mail: Rachid.elchaal@uit.ac.ma

1 Introduction

Self-organizing maps or Kohonen maps are a robust category of artificial neural network methodologies that are extensively used in diverse patterns, including but not limited to recognition, data visualization, and clustering. Self-organizing maps (SOMs) were developed in the 1980s by the Finnish researcher Teuvo Kohonen. SOMs are particularly useful for analyzing complex multidimensional data to reveal the underlying structure of the data.

Self-organizing maps (SOMs) are classified as unsupervised learning algorithms, in contrast to most artificial neural networks which are used for supervised learning tasks like classification or regression. That means they do not need any labeled training data; instead, they autonomously discover the structure of the input data through a self-organizing process. The SOM method employs competitive learning, in which the neurons in the SOM grid compete to represent different areas or clusters in the input data space. When samples are trained, the neuron associated with the most similar weight vector is chosen, which is the best matching unit (BMU) for the input data. The best matching unit's (BMU's) weights and the weights of nearby neurons are adjusted to fit the input data distribution. This creates a low-dimensional mapping of the input data space.

One of the most essential features of SOMs is the ability to preserve the topological properties of the input data space. That means similar input data examples get mapped to nearby neurons in the SOM grid, keeping the topology of the original space. As a result, because of their ability to provide novel visual embedding's, SOMs are often used for data visualization tasks, providing an intuitive understanding of the relationships and arrangement of complex data.

In this introduction, we will review the basics of self-organizing maps (SOMs), how they are used in many different domains, and the key components of the SOM approach.

A. Moumen et al. (eds.), *Artificial Intelligence and Data Analytics for Innovative Applications in Engineering, Sustainability and Technology*, Sustainable Artificial Intelligence-Powered Applications, https://doi.org/10.1007/978-3-031-90318-2_11

In this article, we will explore some of the benefits and limitations of SOMs and highlight their relevance in the current era of big data and machine learning. In environmental study (self-organizing maps (SOMs)), self-organizing maps (SOMs) are potent and unsupervised learning algorithm that works well to highlight patterns and wrinkles in space and time [1]. With a proven record of classification and visualization, SOMs have been applied in many environmental disciplines, such as soil, water, and air quality assessments. Shen et al., Zhou et al. [2], using a self-organizing map (SOM) and multivariate statistical methods, divided groundwater chemistries and qualities of a complex multilayered groundwater system in Northwest China Coal. Likewise, Bigdeli et al. used SOM algorithm integrating with technique in identifying geochemical anomalies (2022) [3] in Moalleman district, Northern Iran.

Santos et al. examined the hydrogeochemical spatialization and controlling factors for the Serra Geral aquifer system, southern Brazil, using Kohonen's SOM and k-means clustering of the hydrogeochemical data from the Serra Geral aquifer system (2020) [4]. Furthermore, Amiri et al. in this regard, [5] did a spatio-temporal assessment of groundwater quality for a coastal aquifer using Kohonen linear discriminant analysis (LDA) and self-organizing map (SOM) approaches in LDA/SOM for complementary information. This project will use a statistical method (SOM) to map and examine the spatial patterns of the distribution of water samples and their physicochemical' properties across the Inaouen watershed. Using SOM's hierarchical classification (SOM-CHA), we highlight specific land uses and spatial variability of the physicochemical variables of the study watershed.

2 Materials and Methods

2.1 Sampling Site Description and Geographic Context

The Oued Inaouene watershed is situated in northeastern Morocco between the Middle Atlas and the Pre-Rif, with an area of approximately 5109 km^2 (Fig. 1). The region, characterized by a Mediterranean climate with oceanic characteristics, is subject to significant seasonal variations and distinctive irregularities in precipitation due to its proximity to the ocean. This region is recognized by its moderately low total annual precipitation (600 mm on average), with a strongly continental monthly rainfall distribution. There are two bands of rainfall recorded; the lowest, below 500 m, receives approximately 800 mm of rain a year; the next band, between 500 and 1000 m, has annual rainfall between 800 and 1500 mm. The rainy season lasts from November through April, and

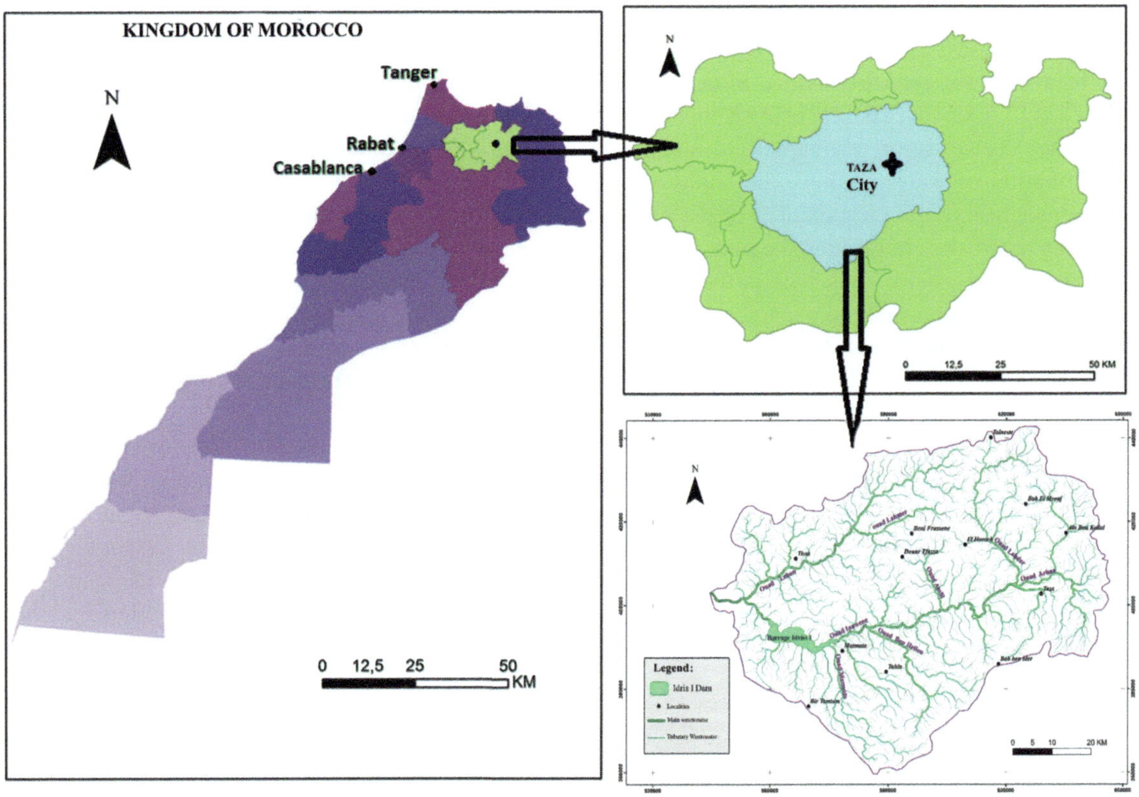

Fig. 1 Geographic position and digital terrain model (DTM) of the study site

the peak rainfall is in December and January. On the other hand, July and August are considered the driest months of the entire year. An important lithological dissimilarity differentiates the two river banks in the Oued Inaouene watershed. The right bank, related to the pre-Rif region, presents marly outcrops, which as they support a specific cultivated flora. Lateral Chorobal on the left bank constitutes the northern termination of the Middle Atlas beyond a series of outcrops stretching from the Paleozoic formations of Tazzeka to the Triassic outcrops [6].

2.2 Data Collection and Source

The dataset for this investigation comprises 16 physicochemical variables. This study employs a dataset consisting of 16 physicochemical parameters derived from 100 surface water samples taken in the Inaouen watershed between 2014 and 2015. The collection, transportation, and storage of water samples were executed by our team in accordance with the protocol established by the National Office of Drinking Water [7]. Some analyses were performed on-site, while the others were executed at the CURI laboratory in Fez by our team.

The dataset collects a robust framework of the physicochemical parameters, each yielding considerable information content on the attributes of the surface waters under analysis. The parameters encompass:

- pH: which measures how acidic or basic the water is or the concentration of hydrogen ions.
  Bicarbonate (HCO_3): Indicates the relative abundance of bicarbonate ions in the water and, therefore, the likely contribution to alkalinity.
- Dissolved Oxygen (Oxy. Diss): Measures the amount of oxygen dissolved in water that aquatic organisms depend on.
- Conductivity (Cond): This determines to what degree water can carry an electric current, primarily due to dissolved ions.
- Temperature (T°C): Answers how hot or cold the water is, which affects several physical, chemical, and biological processes.
- Total Dissolved Solids (TDS): The sum of dissolved solutes in water, which include minerals, salts, and organic material.
  Total Alkalinity ($CaCO_3$): Measures the water's ability to resist changes in pH, primarily due to bicarbonate, carbonate, and hydroxide ions.
- Potassium (K): Represents the level of potassium ions in the water.
- Magnesium (Mg): Indicates the level of magnesium ions in the water.
- Sodium (Na): Is an important water quality parameter and gives an indication of the sodium ions found in water.
  Chlorides (Cl): Identifies levels of chloride ions, the source of which may be related to many factors, such as seawater infiltration or industrial effluents.
- Calcium (Ca): Shows the amount of calcium ions present in the water and influences hardness and alkalinity.
- Sulfate (SO_4): Refers to dissolved sulfate ions, usually given off by industrial sources or natural mineral deposits.
  Nitrate (NO_3): Indicates the presence of nitrate ions in the water, which may be due to agricultural runoff or the discharge of wastewater.
- Phosphorus (P): Indicates the concentration of phosphorus compounds in the water, which may cause an overload of algal bloom and eutrophication.

Ammonia (NH_3): Refers to ammonium nitrogen, one of the pollutants in the water environment, mainly from the decomposition of organic matter or the discharge of wastewater.

Together, these parameters summarize an integral complex of surface water physicochemical properties that are used for water quality assessment, ecological process understanding, and contaminant or pollutant source identification.

2.3 Self-organizing Maps: Methodology and Implementation

To try to create a comprehensive way of visualizing huge multidimensional datasets, Kohonen invented the topological map. Using machine learning-based methods [8], Kohonen provides a way to divide data into groups that are similar to each other and visualize them within the boundaries of a discrete low-dimensional space, which is called a so-called "topological map" [9]. Topological mapping methods provide a way to embed high-dimensional data into a lower-dimensional feature space to expose hidden structures in the data [16]. The self-organizing map (SOM) approach is a kind of unsupervised learning artificial neural network 12, which is a type of neural network architecture [14]. In self-organizing maps (SOM), input vectors or samples are given into a grid of neurons (called nodes or units) [13]. The parameter d is the dimension of the map and is fixed a priori. An input vector is connected to all neurons on the grid through d synapses and d weight vectors w. The vector' w, also called the prototype or the referent vector of the map neuron, represents the map neuron's location in data space [14]. The best matching unit (BMU) provides the most similar referent to the given data. The topological error (Te) and quantization error (Qe)15 are the standard parameters to evaluate the quantization and topological preservation capability of a self-organizing map.

The Quantization Error (Qe), which is also called the resolution measure, adequately calculates the average distance between the data points and the respective best matching units (BMU) from the data points to be clustered. It is a measure of the amount of mapping that the SOM has done on the dataset or, more precisely, the depth of the quantization 1. Hence, smaller Qe values refer to a better quality of the SOM algorithm. Mathematically, it can be written as:

$$Q_e = \frac{1}{N}\sum_{k=1}^{N}\left\|x^{(k)} - w\left(x^{(k)}\right)\right\|^2 \quad (1)$$

where N is the number of data, $x^{(k)}$ is the k-th individual, and $w(x^{(k)})$ is the BMU of the individual $x^{(k)}$.

Quantization error denotes the average distance between input vectors and best matching units (BMUs). One could also think of this as a performance measure of the SOM, where a smaller value is better since it indicates that the input data is represented well. On the contrary, a high quantization error suggests that the self-organizing map did not understand the structure of the input data, and the input vectors of the best matching unit will be separated further.

The self-organizing map (SOM) organizes input data in such a way that similar input vectors are assigned to neighboring regions on the grid while minimizing the quantization error during training. Hence, SOM could explicitly map them according to the underlying assumptions and aims so that they can provide an accurate representation in clustering visualization and classification.

Topographic error Te [16] (It measures the preservation of the topology in the SOM) [17]. The output quantities the number of data pairs for which the two closest referent neurons on the SOM grid are not neighboring map units [18]. While quantization error is concerned with how well the data is represented, Te takes into account the structure of the SOM grid [19].

Te is calculated by measuring the instances where the first winning neuron (ci) and the second winning neuron (*si*) for observation are not neighboring units on the map [20]. The second winning neuron of an observation refers to the neuron with the closest referent vector after the first winning neuron. Mathematically, the topographic error is determined as follows:

$$\text{Te} = \frac{\sum_{i=1}^{n} E}{n}$$

$$E = \begin{cases} 1\ si\, \boldsymbol{r}_{c_i} - \boldsymbol{r}_{s_i}^2 \neq 1 \\ 0\ si\, \boldsymbol{r}_{c_i} - \boldsymbol{r}_{s_i}^2 = 1 \end{cases} \quad (2)$$

where $\boldsymbol{r}_c$ and $\boldsymbol{r}_s$ are, respectively, the locations of neuron c and neuron s on the map.

The topology is perfectly preserved when this criterion is 0.

The value of the topographic error is between 0 and 1, with 0 indicating perfect topological preservation (meaning that all input vector neighbors are also neighbors on the SOM grid) and 1 indicating total topological distortion (all input vector neighbors are not SOM grid neighbors).

It is also important to minimize the topographic error during the training process, as it can indicate whether the SOM preserves the topological relationship in data [9, 10]. This means that the overall topography error is low and that the SOM preserves the structural and topological characteristics of the data domain and hence suitable for clustering, visualization, and classification functions.

The topological maps depict a paradigm shift in the use of these maps [3] as opposed to the classical linear and classification methods [21]. Figure 2: Result of C.A.(left), H.C.(right) [23]. These traditional techniques show their weaknesses in some aspects that indicate that although they have been performed in a common context regarding the grouping of the sample, these techniques are based on classical assumptions especially the nonlinear association between the variables [24].

Conversely, it has the additional advantage of being able to represent complex nonlinear relationships that can exist in the data, which gives a unique advantage in the topology mapping method. Because it uses its self-organizing map characteristics in this manner, it provides a near real-time snapshot of the relative relationships of these data points as it efficiently groups and visualizes them. Such a property of topological maps turns them into one of the most simple yet powerful exploratory and visualization tools in the researcher toolkit; it allows researchers to spot complex relationships between patterns of data that otherwise would slip past under the radar of typical linear or even some nonlinear techniques immediately.

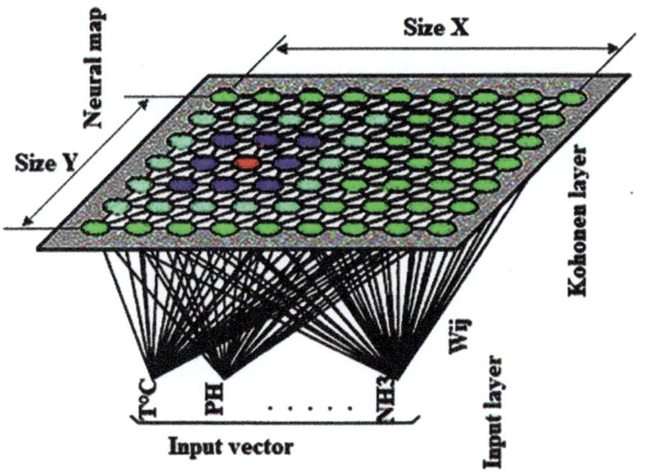

Fig. 2 Structural representation of a topological map

Topological maps reverse this trend by showcasing a solution that is both counter-intuitive and also utilizes a real-world dataset, and this accessibility can be critical for researchers searching for methods to parse through the intricacies of real-world datasets by presenting the essentials in the simplest terms.

2.4 Hierarchical Clustering Using Self-organizing Maps (SOM-CHA)

Similar to other data analysis methods, the SOM-CHA classification intends to create a brief schematic depiction [25]. The distance-based paradigm opens with a calculation of a matrix representing the mutual distances among the places that need to be classified, namely, the map nodes. As a result, this matrix clusters the closest points. Such a methodology allows deriving a hierarchical tree [26], which consists of many potential partitions, with each point being assigned to one of the groups in a given partition. After doing the hierarchical classification, the best partition is determined [27].

2.5 Algorithmic Approach: Kohonen Maps (SOM) [16]

Let $\left(w_1^t, \ldots, w_N^t\right) \in (\mathbb{R}^n)^N$ be neurons of the vector space $\mathbb{R}^n$. We designate by $V\left(w_j\right)$ the set of neighboring neurons of w_j for this Kohonen card. By definition, we have $w_j \in V\left(w_j\right)$. Let $(X_1, \ldots, X_K) \in (\mathbb{R}^n)^K$ a cloud of points. We use a sequence of positive real numbers (α_t) checking $\sum_{t\geq 0} \alpha_t^2 < \infty$ and $\sum_{t\geq 0} \alpha_t < \infty$.

Certainly! The following is a simplified description of the algorithm for the Kohonen maps:

- Initialize the Map: The first step is the initialization of the Kohonen map, usually arranged as a grid of neurons/nodes. Every neuron in the model has a weight vector corresponding to it with the same dimensions as the given input data.
- Randomize Weights: This refers to providing the random values for the weight vectors of the neurons in the map. The prototype reference vectors for the neurons are represented by these weight vectors.
- Choose Input Data: Randomly or sequentially, choose an input vector from the dataset. This input vector represents a single observation from the dataset (e.g., a row of the input matrix).
- Neuron Activation Computation: Compute the activation of each neuron in the map according to its distance from the input vector. This is mostly done by using distance metrics like Euclidean distance.
- Locate the Best Matching Unit (BMU): Identify the neuron that has the weight vector most similar to the input vector. This neuron is called the best matching unit (BMU).
- Weight Update of Neuron: Update the weight of BMU and its neighbors to come closer to the input vector. This is achieved by pulling the weight vectors in proximity to the input vector in the feature space. Updates are smaller for neighboring neurons, with the update size diminishing as a function of distance from the BMU.
- Repeat: For a given number of epochs or till convergence, repeat steps 3 to 6. Convergence can be described in terms of neuron weights or other standards.
- Plot the Kohonen Map and Other Visualization and Analysis Code: The Kohonen map can also be visualized to see clusters and patterns in the data after it is trained. Interpret the relations among the regions of the trained map and the input data.
- Optional Fine-Tuning: Optionally perform additional steps such as neighborhood shrinking or learning rate decay to fine-tune the map that had been trained in the previous iterations so that it can yield better performance.
- Take Application: Based on the purpose of the analysis, the trained Kohonen map can be used for data visualization, clustering, or classification tasks.

Through this algorithm, we can see a high level view of the steps in the training of a Kohonen map. Implementation specifics will vary based on the programming language and the specificities of an application.

Initialization

The neurons $\left(w_1^0, \ldots, w_N^0\right)$ are distributed in the space $\mathbb{R}^n$ in a regular way according to the shape of their neighborhood $t \leftarrow 0$.

Closest Neuron

We choose a point of the cloud X_i randomly; then, we define the neuron w_{k*}^t, so that:

$$w_{k*}^t - X_i = \min_{1\leq j\leq N} w_j^t - X_i.$$

Update

For each w_j^t in $V\left(w_{k*}^t\right)$

$$\begin{aligned} w_j^{t+1} &\leftarrow w_j^t + \alpha_t\left(X_i - w_j^{t+1}\right) \\ t &\leftarrow t+1 \end{aligned} \tag{3}$$

As long as the algorithm has not converged, return to the nearest neuron step.

The update step can be modified to improve the convergence speed [28]:

$$w^{t+1} \leftarrow w^t + \alpha \leftarrow (w^t, w^t *) w(X - w^{t+1}) \quad (4)$$

where h is a function with a value in the interval [0,1], which is 1 when $wt = wt$. And that decreases when the distance between these two neurons increases. A typical function is:

$$h_{ck}(\sigma(t)) = \exp\left(-\frac{d_2^2(\boldsymbol{r}_c, \boldsymbol{r}_k)}{2\sigma^2(t)}\right) = \exp\left(-\frac{\boldsymbol{r}_c - \boldsymbol{r}_k^2}{2\sigma^2(t)}\right) \quad (5)$$

where $\boldsymbol{r}_c$ and $\boldsymbol{r}_k$ are, respectively, the locations of neuron c and neuron k on the map, and (t) is the radius of the neighborhood at iteration t of the learning process.

Also known as the Kohonen map, this very effective data analysis technique projects the n-dimensional data into two-dimensional space using a nonlinear transformation based on a rectangular neighborhood. Such conversion allows easy visualization of complex data structures and patterns. However, the Kohonen maps do more than simply visualize the data—they perform unsupervised classification, and this is achieved by organizing the neurons into clusters and finding regions with a high concentration of data points.

The special aspect of Kohonen maps is their capability to capture the topological relationships existing in the original data space. In the map, neurons are connected to each other such that neighboring neurons are close to each other in the input space, which means that (this network) preserves the topology of the data. Maintaining topology is important for the overall structure and separation of the data, which is essential for good interpretation and analytical processing of data.

In addition to visualizing data and identifying clusters, Kohonen maps enable the delineation of boundaries between different classes or clusters. The edges connecting neurons, or vertices, in the map are modulated to reflect the degree of proximity or separation between classes. Narrow connections signify close relationships between neighboring neurons, indicating similar data points or clusters, while expanded connections indicate distinctions or separations between classes.

In conclusion, Kohonen maps are a robust tool to analyze and gain insights into complex datasets by showing intuitive visualizations of the high-dimensional data, performing unsupervised classification of the data and defining the boundaries of the classes. Because of this property and their capability on preserving the topological structure of the data, they are fundamental for several applications such as pattern recognition, data mining, and exploratory data analysis.

Code for MATLAB

```
% Define SOM parameters
gridSize = [10, 10]; % Size of the SOM grid
inputDim = 16; % Dimensionality of the input data
numEpochs = 100; % Number of training epochs
learningRate = 0.1; % Initial learning rate
neighborhoodSize = 3; % Initial neighborhood
size
% Generate random input data (replace this with
your actual data)
numSamples = 1000;
inputData = rand(numSamples, inputDim); %
Example: Random data
% Initialize SOM weights randomly
somWeights = rand(gridSize(1), gridSize(2),
inputDim);
% Training loop
for epoch = 1:numEpochs
% Update learning rate
currentLearningRate = learningRate *
exp(-epoch / numEpochs);
% Shuffle input data
shuffledData = inputData(randperm(size(inputData,
1)), :);
% Iterate through input data samples for i = 1:
numSamples
% Compute Euclidean distances between SOM
weights and input data
distances = sqrt(sum((somWeights -
repmat(shuffledData(i, :), [gridSize(1),
gridSize(2), 1])).^2, 3));
% Find the BMU (Best Matching Unit) [minDist,
bmuIndex] = min(distances(:));
[bmuRow, bmuCol] = ind2sub(size(distances),
bmuIndex);
% Update weights of BMU and its neighbors for row
= max(1, bmuRow -
neighborhoodSize):min(gridSize(1), bmuRow +
neighborhoodSize)
for col = max(1, bmuCol - neighborhood-
Size):min(gridSize(2), bmuCol + neighbor-
hoodSize)
% Update weight vector using neighborhood func-
tion neighborhoodDist = sqrt((row - bmuRow)^2 +
(col -
bmuCol)^2);
neighborhoodFactor = exp(-
(neighborhoodDist^2) / (2
* neighborhoodSize^2));
somWeights(row, col, :) = somWeights(row, col,
:) + currentLearningRate * neighborhoodFactor *
(shuffledData(i, :)
```

```
- somWeights(row, col, :));
end
end
end
end
% Visualize SOM grid figure;
hold on;
for i = 1:gridSize(1) for j = 1:gridSize(2)
plot(somWeights(i, j, 1), somWeights(i, j, 2),
'bo', 'MarkerSize', 10);
end
end
title('Self-Organizing Map (SOM) Grid');
xlabel('Feature 1');
hold off;
```

3 Results and Discussion

3.1 Surface Water Sample Classification Using SOM

Surface Water Classification-Based Physicochemical Characteristics: We used the self-organizing map algorithm (SOM) applied in this study to classify surface water samples according to different characteristics as determined by various tools. SOM (or self-organizing map) is a neural network method that helps us organize and visualize high-dimensional data in a lower-dimensional space. This approach facilitates the discovery of underlying patterns and relationships within a dataset, and it is especially well-suited for exploring complex environmental datasets, such as water quality parameters.

How SOMs Work: The self-organizing map technique works by iteratively adjusting a grid of neurons to create a topologically preserving representation of the input. We set a certain number of neurons in a grid as weight vectors assigned randomly or in accordance with the initialization scheme. The algorithm then passes through the input data multiple times; each time, it moves the weight vectors of neurons in the direction of where the input samples reside. As a consequence, this process induces neighboring neurons to become specialized in representing similar input patterns, leading to a topological ordering of the map.

For example, in our study, the input layer of the SOM consisted of vectors for each surface water sample, and each vector contained different measurements for 16 physicochemical parameters. Parameters monitored within these categories include pH, bicarbonate, dissolved oxygen, conductivity, temperature, total dissolved solids, total alkalinity, potassium, magnesium, sodium, chlorides, calcium, sulfates, nitrate, phosphorus, and ammoniac. By treating these parameters simultaneously, we intended to encapsulate the total physicochemical profile of the surface water samples sampling campaign in the area of study.

The surface water samples were clustered or classified based on their physicochemical similarities using the SOM algorithm. Visualization of the resulting grid for SOM allowed us to observe some spatial correlations and patterns of the surface water samples. This ease of clustering enabled the identification of uniquely identifying populations or clusters of samples with related physicochemical characteristics, informing patterns of water quality variability and distribution alongside other spatial patterns in the study region.

Conclusions: Surface water samples classified within SOM provided a powerful and fast method for unsupervised exploratory analysis of complex environmental datasets and the recognition of useful patterns in water quality features. Utilizing this methodology allowed the ability to process and analyze big data, leading to informed decision-making and targeted interventions for sustainable water resources management and environmental protection.

SOM algorithm works under nonlinear classification of complex data to find patterns closer together. Calculating Machine Learning Training X input layer vector Sample Input: In this study, the input layer consists of vectors with 16 components, each defined for each sample corresponding to their 16 physicochemical parameters of the surface waters being studied. The output layer is a grid of 10 rows by 10 columns containing 100 neurons. We chose this particular output map size since it results in the lowest values for the following two error measures: $Qe = 0.268$ (quantization error) and $Te = 0.03$ (topographic error).

The SOM component planes provide a helpful description of the dataset, and they present color patterns where dark red cells indicate high values and blue low values (Fig. 3) [16]. Using this visualization, we get to see some relationships between the physics-chemical parameters. Of particular note, any two or more variables that share a color (intensity) are, at the very least, positively correlated.

Positive correlations are present for HCO_3, Cl, Mg, Na, SO_4, Cond, NO_3, NH_3, Ca, total alkalinity (in $CaCO_3$), K, and TDS. The presence of such co-variation suggests these parameters are influenced by common and related chemical processes happening within the water samples.

Dissolved oxygen (D.O. Diss) and pH total alkalinity ($CaCO_3$) and phosphorus (P) are negatively correlated, which means variations in one parameter when changes in other parameters vary inversely. This suggests that they may play antagonistic roles or that one may regulate the other.

Similarly, PLC could not show a positive or negative correlation for some variables such as temperature (T°C), phosphorus (P), and nitrate (NO_3), which means the variations on these parameters are not correlated with each other, and

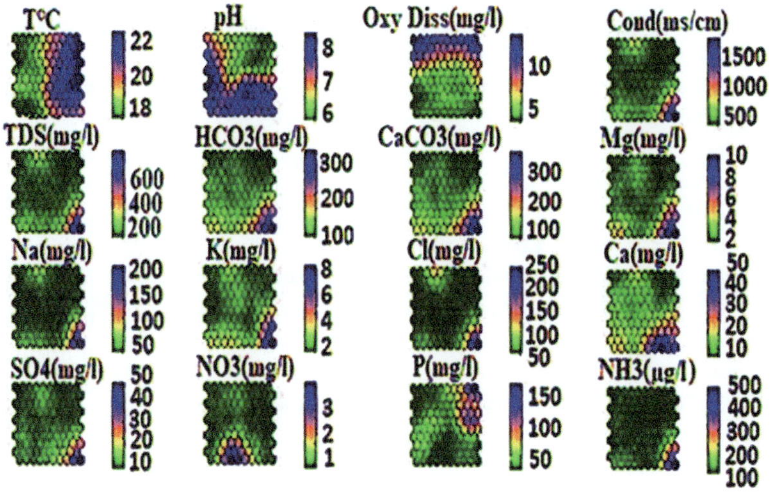

Fig. 3 Gradient of physicochemical parameter values on the Kohonen map

they appear to be independent of other parameters (Febria et al. 2015). Their independence highlights local dynamics and impacts for those fixed dimensions in the datasets whose temporal connection could be of more interest.

3.2 Exploratory Data Analysis Through Principal Component Analysis (PCA)

The PCA results reveal a score plot characterized by the first two principal components, PC1 and PC2, which are identified as the most important contributors to the total variance. Altogether, PC1 and PC2 account for 62.3% of the total variance in our investigation (PC1 = 51.1% + PC2 = 11.2%). This reinforces the actual and fundamental importance of these two elements to seize the hidden tendencies and variances in the data.

In addition, the correlation circle (Fig. 4) helps to identify the relationships between variables in PC 1 versus PC 2 subspace. A correlation circle represents the correlations among the original variables and the principal components. The representation of the variable is in the form of a vector originating from the origin angle, the length of which describes the correlation strength and direction with the principal components.

It compares the physiochemical parameters that were studied by you, and you can determine the correlation and dependency of the various physicochemical parameters studied, by looking at the factor plane defined by PC1 and PC2. In addition, the mode of statistical analysis we are conducting allows us to better formalize the relationships present in the data and the underlying patterns behind the variability we observe.

The correlation circle in the factorial diagram (PC1 x PC2) reveals interesting relations between the main components

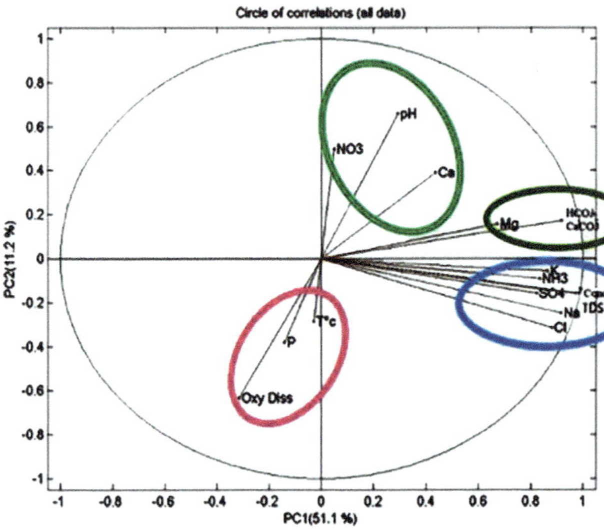

Fig. 4 Circles of correlation in principal component analysis (PCA) plot (PC1 × PC2)

and the initial variables. Particularly, the variables of Mg, HCO_3, $CaCO_3$, K, SO_3, Na, Cl, NH_3, and TDS are positively correlated with the PC1 axis. The coefficients for these variables are greater than 0.6, suggesting a strong, positive association with PC1.

On the other hand, Oxy Diss shows a negative correlation with the PC2 axis (with coefficients lower than − 0.6). Because of the inverse relationship here, this negative correlation indicates that Oxy Diss changes are inversely related to changes along the PC2 axis.

Visualization of the data structure and correlation circle allows for checking the relationships between variables and principal components. The factorial diagram reveals which of the correlations and their associated features account for most of the variability observed between the different principal components, providing an explication of the forcing mechanisms present in the system.

Therefore, Mg, HCO_3, $CaCO_3$, K, SO_4, Na, Cl, NH_3, Cond, and TDS show correlation among themselves. As for the other variables, T°, P, NO_3, and pH not show significant positive or negative correlations, they were poorly represented in the circle, indicating their autonomous variations.

3.3 Hierarchical Classification with SOM-CHA Approach

Following the Kohonen map acquisition, we perform a hierarchical classification based on the ward technique and Euclidean distance. In this way, the hierarchical approach of the SOM allows the grouping cells within the SOM map as specific clusters indicating the physicochemical characteristics of the Inaouen watershed. The dendrogram available from the SOM-CHA (Fig. 3) indicates that the 100 neurons should be clustered into five distinct clusters (see to Fig. 5).

The first class, comprising only 13% of the complete dataset, exhibits a different chemical profile characterized by slightly higher concentrations of several chemical components.

The most important among these are bicarbonates (HCO_3) (105.39 mg/l), total dissolved solids (TDS) (100.54 mg/l), calcium carbonate ($CaCO_3$) (86.38 mg/l), magnesium (Mg) (4.41 mg/l), calcium (Ca) (24.69 mg/l), and electrical conduction (201.08 µs/cm). By contrast, this class demonstrates lower values of ammonia (NH_3) at 31.23 µg/l and sodium (Na) at 11.60 mg/l, illustrating a different chemical signature in the dataset.

The second class, which is the largest class in our dataset, actually includes 48% of the whole database. It is primarily defined as a chemical type defined by low levels of certain critical chemical components. High sodium (Na) levels are low (9.00 mg/l) and potassium (K) levels are also low at 1.02 mg/l, while sulfates (SO_4) are also low at 4.61 mg/l and total dissolved solids (TDS) levels are low at 58.33 mg/l, this class also exhibits a unique feature where dissolved oxygen is high at 6.24 mg/l, which facilitates the formation of this class (i.e., it is the lowest sea level), which provides important boundary conditions for classification (Table 1).

Class 3 is the smallest, albeit most diverse component of the entire dataset, making up 13% of the entire database, one with a unique chemical profile. Of particular note, this class has significantly lower abundances of certain chemical elements, reflecting its distinct chemical composition. In specific cases, low sulfates (SO_4) (4.87 mg/l), low chloride (Cl) (3.19 mg/l), low sodium (Na) (4.91 mg/l), low magnesium (Mg) (1.81 mg/l), low potassium (K) (0.84 mg/l), and low calcium (Ca) (9.95 mg/l) were observed as well, distinguishing this class chemically. Ammonia (NH_3) is 17.69 mg/l, and total dissolved solids (TDS) 43.38 mg/l, are particularly low; the other particularly high feature of this class is phosphorus (P) at 225.39 mg/l, and dissolved oxygen is at the particularly high level of 6.30 mg/l, which provides a useful delineation of this class in chemical space and indicates that it is a prevalent feature of the population sampled.

Class 4, 23% of the database which is contained in a large class, has an identifiable chemical profile based on moderately high concentrations of numerous chemical components.

This class is particularly characterized by the presence of significant amounts of several chemical elements. And the values are; bicarbonates (HCO_3) with a high level of 117.76 mg/l, chloride (Cl) with a high level of 47.98 mg/l, magnesium (Mg) with a medium level of 4.85 mg/l, calcium carbonate ($CaCO_3$) with a significant level of 96.52 mg/l, calcium (Ca) with a high level of 29.83 mg/l, sodium (Na) with a high level of 38.90 mg/l and ammonia (NH_3) concentration with a very high level of 74.17 mg/l, as well as electrical conductivity (EC) with a high level of 375.87 µs/cm. In addition, the TDS value is particularly high at 188.74 mg/l, underscoring the distinctive chemical composition of this class within the dataset.

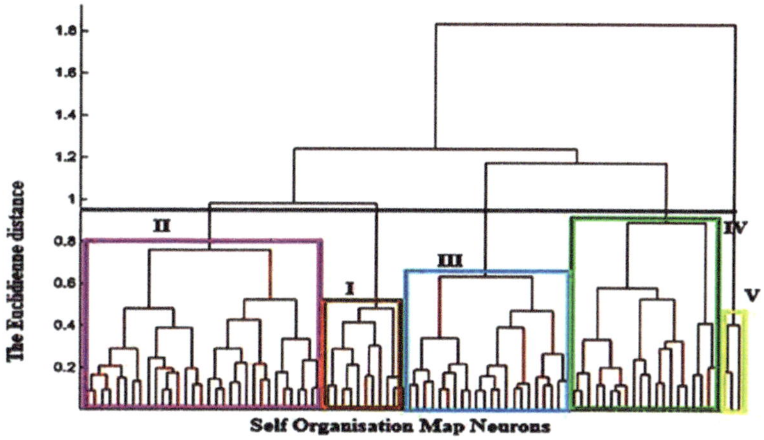

Fig. 5 Dendrogram representation of the physicochemical parameters of surface waters in the Inaouen watershed using the topological maps method (SOM)

Table 1 Summary statistics of physicochemical parameters (minimum, mean, maximum) for the entire database and each of the five classes

Stat of all data				Stat of data from cluster 1			Stat of data from cluster 2		
Name	Min	Mean	Max	Min	Mean	Max	Min	Mean	Max
T°c	16.3	19.9	23	17.1	18.9846	21	16.3	19.7042	23
pH	5.21	7.1923	9.61	6.36	7.56462	9.42	5.21	7.04458	9.61
Oxy Diss	0.09	5.0408	14.01	0.08	2.85	7.76	0.08	6.24229	13.26
Cond	20	233.69	2959	134	201.077	315	20	116.729	629
TDS	45	117.15	1480	67	100.538	157	10	58.3333	314
HCO_3	3.66	80.629	634.4	68.32	105.389	170.8	6.1	45.1654	170.8
$CaCO_3$	3	66.09	520	56	86.3846	140	5	37.0208	140
Mg	0.2	3.1735	25	1	4.40769	11	0.2	1.82542	6.8
Na	1	24.513	540	2.4	11.6	31	1	9.00208	76
K	0.1	1.7211	13	0.34	1.63615	4	0,1	1.01563	6.4
Cl	0.4	25.8394	550	0.6	5.98462	22	0.4	8.62792	99
Ca	1	17.794	110	3	24.6923	54	1	11.4333	45
SO_4	0.275	7.83175	150	1.65	7.08462	13	0.275	4.60781	15
NO_3	0.02	0.4805	4.8	0.1	1.15846	4,8	0.02	0.464167	4.5
P	10	58.5	900	20	30.7692	80	10	33.3333	80
NH_3	11	62.93	1000	11	31.2308	130	11	34.1875	130

Stat of data from cluster 3				Stat of data from cluster 4			Stat of data from cluster 5		
Name	Min	Mean	Max	Min	Mean	Max	Min	Mean	Max
T°c	18	20.8231	23	17.5	20.3174	23	19	19.8	21
pH	5.41	6.70692	8.86	5.63	7.44826	9	6.62	8.08333	9.04
Oxy Diss	1.25	6.3	14.01	1.32	3.59913	10.24	0.7	0.906667	2.2
Cond	0.09	84.8462	198	100	375.87	1118	980	1801.33	2959
TDS	23	43.3846	99	50	188.739	559	490	901	1480
HCO_3	7.32	39.3215	82.96	3.66	117.757	329.4	231.8	435.133	634.4
$CaCO_3$	6	32.2308	68	3	96.5217	270	190	356.667	520
Mg	0.53	1.80923	4.5	0.61	4.84826	23	3.1	12.4667	25
Na	1.5	4.90769	13	1.4	38.8957	180	140	303.333	540
K	0.16	0.84	2.4	0.43	2.6987	7.6	6.8	9.7	13
Cl	0.7	3.19231	11	0.4	47.9783	260	77	315.667	550
Ca	1.9	9.95385	30	14	29.8261	110	2.4	31.4	86
SO_4	1.65	4.86538	18	1.65	9.89783	37	11	59.6667	150
NO_3	0.03	0.133077	0.4	0.02	0.364348	2.4	0.1	0.2	0.4
P	20	225.385	900	20	34.7826	80	20	40	70
NH_3	11	17.6923	50	11	74.1739	250	530	770	1000

The fifth and last class, albeit small (3% of the whole database), is a defined subset. This class has the highest concentrations of different chemical elements among all classes in the dataset, making it the most extreme class in the dataset. Importantly, this class displays a highly unique chemical profile, containing very high concentrations of many important chemicals. In fact, ammonia (NH_3) values skyrocketed to an alarming 770 mg/l, demonstrating extreme pollution or contamination. In addition, conductivity exhibits an incredibly high reading of 1801.33 uS/cm, indicating a plethora of dissolved ions in the water. Likewise, total dissolved solids (TDS) have a remarkably high level of 901 mg/l, which indicates a significant amount of soluble material in the water sample.

In addition, bicarbonates (HCO_3) show a high level of 435.13 mg/l, sodium (Na) content is also high at 303.33 mg/l, and the level of chloride (Cl) is at 47.98 at a highly elevated level, which may mean salinity risks. Similarly, concentrations of magnesium (Mg) are also very high at 4.85 mg/l and calcium carbonate ($CaCO_3$), pervasive in calcareous

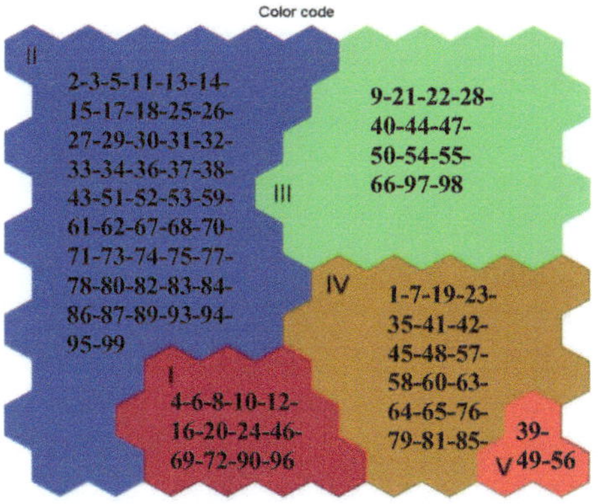

Fig. 6 Visualization of the five clusters identified through SOM hierarchical clustering, showing sample distribution

sediments—has a strong concentration (356.67 mg/l): the calcium (Ca) at this station is also notable (31.40 mg/l), as are potassium (K) concentrations (9.70 mg/l), and sulfates (SO4), at 59.67 mg/l.

The difference in chemical composition between the classes is probably due to the geological composition of the soil, differences in altitude, and local domestic discharges from adjacent localities.

One thing that matters is the geology of the surrounding landscape because it can influence the chemistry of surface waters.

Different kinds of soil, mineral deposits, and geological formations may contain other chemical constituents that can enter the water, thus changing the composition of the water. Some rock formations, for example, can have high concentrations of particular ions or minerals, which can seep into nearby water bodies and further affect differences in their chemistry.

Second, the variation of altitude within the catchment area may also influence the chemical composition of surface waters. In the case of higher elevations, we argue that hydrological processes at higher elevations, more weathering of rock formations, and greater precipitation create a partial bias in terms of the transport of chemical compounds. This leads to differences in the chemical signature of surface waters from different altitudinal zones.

Finally, the discharge of domestic wastewater from local communities may add other pollutants and contaminants to the surface waters and affect their chemical content differences as well. The physicochemical properties of receiving waters are changed upon discharging effluents from urban and rural settlements, which contain different chemical pollutants (nutrients, organic matter, and heavy metals) [8].

After all, the joint effect of geological influences, differences in elevation levels, and anthropogenic impact of neighboring communities, results in the variation of the chemistry of surface waters in terms of the classes found in the basin. Grass actually manages guarding water resources in the community is based on these root causes, and is one of the keys to success (Figs. 6, 7 and 8).

3.4 U Matrix Analysis for Classification

The U Matrix Classification—U Matrix classification, a key part of the Kohonen self-organizing map (SOM) analysis, helps to define spatial patterns and relations of the dataset. This is a crucial classification method as it helps to visualize the structure of the SOM and to find clusters or groups of similar data points.

Fig. 7 Projection of samples onto the PC1XPC2 factorial plane, showing the distribution of the five clusters

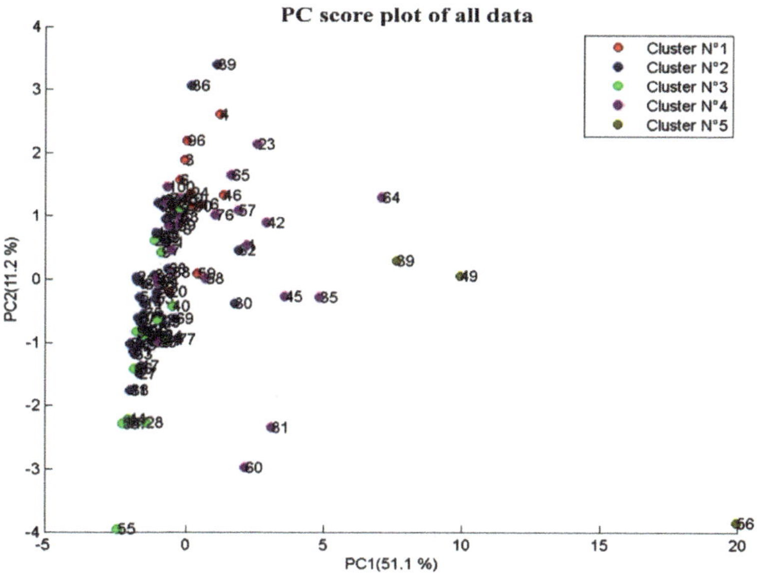

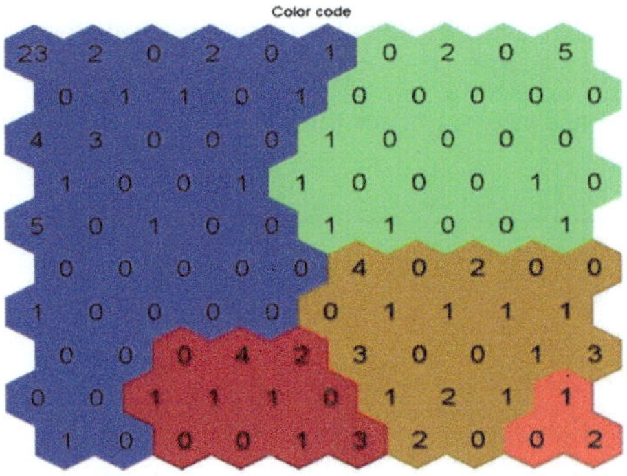

Fig. 8 Visualization of the five clusters resulting from the SOM hierarchical clustering, showing the distribution of patterns assigned to each neuron

U Matrix classification works by assigning colors to each neuron in the SOM grid to represent similarity; if two neighboring neurons are similar, they will be assigned the same or neighboring colors. Neurons with adjacent weight vectors (closely related to the training instances) are given similar colors, thus the smooth appearance across the map. The colors of the gradient place the map regions in a topological relation with similar colors for adjacent neurons, showing that they are both close and similar in the input data space.

Through the examination of the U Matrix, researchers can find clusters or areas of analogous data points, as well as delineate boundaries or transitions between distinct clusters. This information is essential for comprehending the dataset's underlying structure and seeing potential patterns or trends that may not be immediately evident from the raw data.

Additionally, the U Matrix classification shows how well the self-organizing map has mapped the input data, getting the latent characteristics of the data. An even, regular U Matrix reflects successful SOM organization of input data, while a clumpy, ragged U Matrix indicates that SOM could not map the data structure.

The U Matrix classification has emerged as an informative tool for visualizing and interpreting the results of SOM analysis to identify patterns that provide researchers with valuable knowledge about the geographical distribution and linkages of the input data. Through this classification approach, scientists can extract hidden structures, relations, and trends from the complex data, allowing us to understand better how these underlying processes drive environmental phenomena.

A hexagonal structure was used for the U Matrix classification map to improve resolution and accelerate processing. Despite the rectangular topology necessitating fewer neurons to attain minimal quantization and topographical defects, the hexagonal topology was preferred due to its enhanced performance. The resultant U matrix (refer to Fig. 9) was produced utilizing the specified SOM parameters.

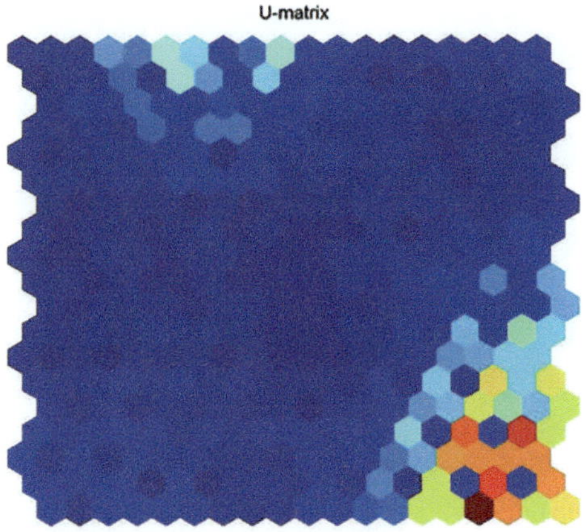

Fig. 9 U matrix depiction of the self-organizing map (SOM) illustrating the 16 physicochemical parameters from 100 sources in the Inaouen watershed

The classification precision achieved through the best matching units (BMUs) in the U matrix is deemed nearly precise, yielding a high-quality and seamlessly smooth mapping output.

4 Conclusion

In conclusion, the application of the Kohonen self-organizing map (SOM) method to the detailed dataset of physicochemical parameters from surface water samples through the entire groundwater of the Inaouen watershed collected from February 2014 to December 2015 has given us a better and deeper understanding of spatial variability and distribution behaviors of these waters. Through the application of SOM analysis, we have detected and classified both positive and negative intercorrelations between the range of parameters studied and highlighted the complexities of interactions between different markers of water quality.

Besides, the hierarchical classification of the SOM map (SOM-HC) has given more specific insight into spatial variability based on source variations in the watershed. This classification ultimately allowed for discernible patterns to surface across the chemical properties of the water samples, which indicated potential influences from the geological formation of source water, altitude, and other localized anthropogenic activities.

The noticeable difference in water quality across locations in the watershed highlights the need to appreciate and manage the complexity of the drivers of water quality change. This means that we need actions that are more targeted and can better manage systems to prevent specific sources of contamination or degradation and, at the same time, to use and protect water.

More broadly, these findings also allow us to understand better the complex interplay of influences on water quality within the Inaouen watershed. Abstract: Sustainability and sustainability have been offered as an explanation for what we need to do to promote sustainable management of our water resources. In the future, research efforts and monitoring initiatives will need to continue to clarify the impact and risk of water quality in the Inaouen watershed to continue to protect the aquatic ecosystems and communities that depend on them over the long term.

References

1. Kohonen T (2001) Self-organizing maps, 3rd edn, vol 30. Springer, Berlin, Heidelberg
2. Qu S, Shi Z, Liang X, Wang G, Han J (2021) Multiple factors control groundwater chemistry and quality of multi-layer groundwater system in Northwest China coalfield—using self-organizing maps (SOM). J Geochemical Explor 227:106795. https://doi.org/10.1016/j.gexplo.2021.106795
3. Bigdeli A, Maghsoudi A, Ghezelbash R (2022) Application of self- organizing map (SOM) and K-means clustering algorithms for portraying geochemical anomaly patterns in Moalleman district, NE Iran. J Geochemical Explor 233:106923. https://doi.org/10.1016/j.gexplo.2021.106923
4. Santos MR, Roisenberg A, Iwashita F, Roisenberg M (2020) Hydrogeochemical spatialization and controls of the Serra Geral Aquifer System in southern Brazil: a regional approach by self- organizing maps and k-means clustering. J Hydrol 591. https://doi.org/10.1016/j.jhydrol.2020.125602
5. Amiri V, Nakagawa K (2021) Using a linear discriminant analysis (LDA)-based nomenclature system and self-organizing maps (SOM) for spatiotemporal assessment of groundwater quality in a coastal aquifer. J Hydrol 603:127082. https://doi.org/10.1016/j.jhydrol.2021.127082
6. Benzougagh B, Dridri A, Boudad L, Sdkaoui D, Baamar B (2019) Contribution of GIS and remote sensing for the evaluation of the physical characteristics of Inaouene watershed (northeast Morocco) and their uses in the field of natural hazard management. Am J Innov Res Appl Sci 120–130, [Online]. Available: www.american-jiras.com
7. El Chaal R, Aboutafail MO (2021) Development of stochastic mathematical models for the prediction of heavy metal content in surface waters using artificial neural network and multiple linear regression. E3S Web Conf 314:02001. https://doi.org/10.1051/e3sconf/202131402001
8. Olubi O, Oniya E, Owolabi T (2021) Development of predictive model for Radon-222 estimation in the atmosphere using stepwise regression and grid search based-random forest regression. J Niger Soc Phys Sci 3(2):132–139. https://doi.org/10.46481/jnsps.2021.177
9. Kohonen T (1998) The self-organizing map. Neurocomputing 21(1–3):1–6. https://doi.org/10.1016/S0925-2312(98)00030-7
10. Opeoluwa Oyewola D, Dada EG, Ndunagu JN, Abubakar Umar T, A SA (2021) COVID-19 risk factors, economic factors, and epidemiological factors nexus on economic impact: machine learning and structural equation modelling approaches. J Niger Soc Phys Sci 3(4):395–405. https://doi.org/10.46481/jnsps.2021.173
11. Umarani V, Julian A, Deepa J (2021) Sentiment analysis using various machine learning and deep learning techniques. J Niger Soc Phys Sci 3(4):385–394. https://doi.org/10.46481/jnsps.2021.308
12. Rougier NP, Detorakis GI (2021) Randomized self-organizing map. Neural Comput 33(8):2241–2273. https://doi.org/10.1162/neco_a_01406
13. Olszewski D (2021) A data-scattering-preserving adaptive self-organizing map. Eng Appl Artif Intell 105:104420. https://doi.org/10.1016/j.engappai.2021.104420
14. Vassilas N (2011) Self-organization of the batch Kohonen network under quantization effects. Int J Comput Math 88(17):3586–3612. https://doi.org/10.1080/00207160.2011.620094
15. Oyelade J et al (2019) Data clustering: algorithms and its applications. In: 2019 19th international conference on computational science and its applications (ICCSA), pp 71–81. https://doi.org/10.1109/ICCSA.2019.000-1
16. Ponmalai R, Kamath C (2019) Self-organizing maps and their applications to data analysis. U.S.
17. Cabanes G, Bennani Y (2010) Learning topological constraints in self-organizing map. In: Neural information processing: models and applications, PT II, vol 6444. 17th International conference on neural information processing, pp 367–374
18. Ritter H (1999) Self-organizing maps on non-Euclidean spaces. In: Kohonen Maps. Elsevier, pp 97–109
19. Uriarte EA, Martin FD (2006) Topology preservation in SOM. In: Proceedings of world academy of science, engineering and technology, vol 15. Conference of the world-academy-of-science-engineering-and- technology. Univ Deusto, Fac Engn, Bilbao, Spain NR—20 PU—WORLD ACAD SCI, ENG and TECH-WASET PI—CANAKKALE PA—PO BOX 125, CANAKKALE, 17100, TURKEY. WE-Conference Proceedings Citation Inde, pp 187–190
20. Kiviluoto K (1996) Topology preservation in self-organizing maps. In: ICNN—1996 IEEE international conference on neural networks. IEEE International conference on neural networks (ICNN 96), vol 1–4, pp 294–299
21. Yusuf AB, Dima RM, Aina SK (2021) Optimized breast cancer classification using feature selection and outliers detection. J Niger Soc Phys Sci 3(4):298–307. https://doi.org/10.46481/jnsps.2021.331
22. Umar D, Omonona OV, Okogbue C (2021) Groundwater quality assessment using multivariate analysis and water quality index in some saline fields of central Nigeria. J Niger Soc Phys Sci 3(4):267–277. https://doi.org/10.46481/jnsps.2021.183
23. Nakano FK, Cerri R, Vens C (2020) Active learning for hierarchical multi-label classification. Data Min Knowl Discov 34(5):1496–1530. https://doi.org/10.1007/s10618-020-00704-w
24. Giraudel JL, Lek S (2001) A comparison of self-organizing map algorithm and some conventional statistical methods for ecological community ordination. Ecol Modell 146(1–3):329–339. https://doi.org/10.1016/S0304-3800(01)00324-6
25. Serrano-Pérez J, Sucar LE (2021) Artificial datasets for hierarchical classification. Expert Syst Appl 182:115218. https://doi.org/10.1016/j.eswa.2021.115218
26. Luo C, Li T, Chen H, Fujita H, Yi Z (2018) Incremental rough set approach for hierarchical multicriteria classification. Inf Sci (Ny) 429:72–87. https://doi.org/10.1016/j.ins.2017.11.004
27. Zhao H, Guo S, Lin Y (2021) Hierarchical classification of data with long-tailed distributions via global and local granulation. Inf Sci (Ny) 581:536–552. https://doi.org/10.1016/j.ins.2021.09.059
28. Lo Z-P, Bavarian B (1991) On the rate of convergence in topology preserving neural networks. Biol Cybern 65(1):55–63. https://doi.org/10.1007/BF00197290
29. Roux M (2018) A comparative study of divisive and agglomerative hierarchical clustering algorithms. J Classif 35(2):345–366. https://doi.org/10.1007/s00357-018-9259-9
30. Randriamihamison N, Vialaneix N, Neuvial P (2021) Applicability and interpretability of ward's hierarchical agglomerative clustering with or without contiguity constraints. J Classif 38(2):363–389. https://doi.org/10.1007/s00357-020-09377-y

Sustainable Building Materials: Enhancing Clay Blocks with Natural Waste

Said Bajji, Youssef Naimi, and Ahmed Saba

Abstract

The incorporation of natural waste into clay blocks enhances their properties. This article presents the findings of research on clay blocks enriched with natural waste from traditional Moroccan industries, specifically broken pottery waste and wood ash. The study emphasizes the thermal properties of these blocks, particularly comparing their thermal conductivities and diffusivities. Mechanical properties are assessed through compressive strength measurements, while physical properties are determined by density. The findings show a significant improvement in the insulating capacities of the analyzed blocks. A mixture of 50% clay, 30% wood ash, and 20% broken pottery waste provided optimal thermal resistance, allowing for the construction of environmental barriers with enhanced heat resistance akin to conventional houses.

Keywords

Earthen bricks · Organic residues · Thermal properties · Moroccan pottery heritage

1 Introduction

The adoption of resource-efficient and environmentally responsible building materials is a key factor in advancing sustainable development. Among these, ecological bricks made from clay stand out due to their abundant availability, minimal environmental impact, and low carbon footprint. However, ensuring their durability, performance, and suitability for various climatic conditions requires a comprehensive understanding of their thermomechanical behavior. This study aims to bridge the knowledge gap regarding the performance of ecological bricks, particularly in response to fluctuations in humidity and temperature [1, 2].

As climate change challenges conventional construction techniques, there is an urgent need to reassess building materials with reduced environmental impact. Historically linked to low-cost housing, earthen materials are gaining renewed interest in sustainable construction due to their affordability, low energy requirements, and ability to enhance indoor thermal comfort. Despite often being associated with traditional architecture, nearly one-third of the global population still resides in structures made from earth-based materials (Fig. 1) [2, 3].

Although modern materials such as cement and lime are widely used for their strength, their high greenhouse gas emissions necessitate a transition toward greener alternatives. This shift has sparked renewed interest in earth-based construction, aligning with efforts to reduce environmental footprints and improve building energy efficiency [4, 5]. In pursuit of sustainable construction, researchers have explored ways to enhance the energy performance of earthen buildings by incorporating hydraulic binders or organic fibers, often sourced from local environments [6].

Various studies have examined modifications to traditional clay bricks through the addition of organic and inorganic stabilizers. Findings suggest that integrating agricultural by-products like rice husk ash and sawdust can enhance the thermal insulation of bricks, contributing to better indoor temperature regulation [7, 8]. Similarly, the inclusion of industrial by-products such as fly ash and silica fume has been shown to strengthen the mechanical properties of clay bricks while simultaneously reducing reliance on natural clay resources [9, 10].

S. Bajji (✉) · A. Saba
Information Processing Metrology Laboratory (LMTI), FSA Agadir, Agadir, Morocco
e-mail: bsaidfssm@gmail.com

Y. Naimi
Physical Chemistry of Materials Laboratory (LCPM), FSBM Casablanca, Casablanca, Morocco

A. Moumen et al. (eds.), *Artificial Intelligence and Data Analytics for Innovative Applications in Engineering, Sustainability and Technology*, Sustainable Artificial Intelligence-Powered Applications, https://doi.org/10.1007/978-3-031-90318-2_12

Fig. 1 Building on the locally available clay in the Tiznit-Agadir region of Morocco

This study builds upon previous findings by proposing a novel composite material integrating clay, wood ash, and crushed pottery waste. The objective is to enhance the thermomechanical properties of clay-based materials, thereby improving their thermal performance. The research focuses on two key aspects: (1) optimizing the composite's thermal and mechanical behavior by adjusting the proportions of stabilizing components, and (2) assessing its durability and performance across different Moroccan climatic conditions through extensive thermo-physical analyses, including measurements of thermal conductivity, heat capacity, and thermal diffusivity. Ultimately, this work aims to develop a sustainable, energy-efficient, and long-lasting construction material adapted to specific environmental challenges [11].

Fig. 2 Picture of wood ash and pottery waste

2 Materials and Methods

2.1 Used Materials

By-products from traditional pottery production, combined with wood ash, have been effectively utilized to improve the stability of clay bricks. This innovative approach, which integrates these materials into clay-based blocks tailored for local applications, promotes sustainability by facilitating recycling and reducing environmental impact.

The study involves fabricating various clay brick samples, including reference specimens, and conducting thermal characterization to assess their performance. The main objective is to evaluate the potential of crushed pottery waste and wood ash as stabilizing agents while analyzing the thermo-physical properties of the modified bricks.

The research methodology encompasses the production of different brick formulations, followed by mechanical and thermal testing. By comparing the enhanced bricks with conventional samples, the study aims to determine the influence of these additives on thermal behavior and overall material performance. Ultimately, this investigation seeks to provide valuable insights into the role of these by-products in improving the properties of clay-based construction materials (Fig. 2).

2.2 Manufacturing Test Samples

After selecting the appropriate components for each mixture, the test samples were prepared using the following procedure:

1. **Water Incorporation**—Adding the optimal amount of water to achieve the desired consistency.
2. **Homogenization**—Manually mixing the ingredients to ensure even distribution of water throughout the mixture.
3. **Molding and Compaction**—Filling the molds and compacting the material to achieve uniform density.
4. **Demolding and Curing**—Carefully removing the samples from the molds and allowing them to cure under controlled conditions.

2.3 Experimental Protocol

2.3.1 Sample Preparation

The materials were first dry-mixed until a uniform consistency was achieved, ensuring even distribution of components. Water was then gradually added, followed by thorough kneading to achieve complete homogenization. Hand mixing was essential in minimizing the risk of cracks during the drying phase. The prepared mixture was immediately placed into molds and compacted.

To ensure consistency across samples, the molds were carefully filled in layers, with each layer compacted using a metal pestle. Additionally, the molds were cleaned before use and between different samples to prevent cross-contamination.

Before conducting measurements, the composite material underwent an initial preparation phase. Two identical samples were created from each mixture using separate molds. After a 24-h setting period, the samples were demolded. To optimize contact for thermo-physical characterization, the surfaces were smoothed as needed to eliminate air gaps between the samples and the measurement probe.

Test samples were prepared in two forms: brick blocks and cylindrical specimens measuring 5 cm in diameter and 10 cm in height. These samples were stored under stable laboratory conditions at **20 ± 2 °C**. A drying period of 14 days was maintained before performing compressive strength tests (Table 1).

The experimental trials were conducted using six different concentrations of shredded pottery waste mixed with clay, as well as six different concentrations of wood ash mixed with clay. In addition, seven different concentrations were tested using a combination of shredded pottery waste and wood ash mixed with clay (Fig. 3).

Table 1 Compositions of manufactured clay bloc

Mixtures
Mixture M1: Clay alone
• 100A: 100% Clay, 0% Wood Ash, 0% Crushed Pottery Waste
Mixture M2: Clay + Crushed pottery waste
• 95A5P: 95% Clay, 0% Wood Ash, 5% Crushed pottery waste • 90A10P: 90% Clay, 0% Wood ash, 10% Crushed pottery waste • 80A20P: 80% Clay, 0% Wood ash, 20% Crushed pottery waste • 70A30P: 70% Clay, 0% Wood ash, 30% Crushed pottery waste • 60A40P: 60% Clay, 0% Wood ash, 40% Crushed pottery waste • 50A50P: 50% Clay, 0% Wood ash, 50% Crushed pottery waste
Mixture M3: Clay + Wood ashes
• 95A5C: 95% Clay, 5% Wood ash, 0% Crushed pottery waste • 90A10C: 90% Clay, 10% Wood ash, 0% Crushed pottery waste • 80A20C: 80% Clay, 20% Wood ash, 0% Crushed pottery waste • 70A30C: 70% Clay, 30% Wood ash, 0% Crushed pottery waste • 60A40C: 60% Clay, 40% Wood ash, 0% Crushed pottery waste • 50A50C: 50% Clay, 50% Wood ash, 0% Crushed pottery waste
Mixture M4: Clay + Crushed pottery waste + Wood ashes
• 40A30C30P: 40% Clay, 30% Wood ash, 30% Crushed pottery waste • 50A30C20P: 50% Clay, 30% Wood ash, 20% Crushed pottery waste • 50A20C30P: 50% Clay, 20% Wood ash, 30% Crushed pottery waste • 60A20C20P: 60% Clay, 20% Wood ash, 20% Crushed pottery waste • 70A20C10P: 70% Clay, 20% Wood ash, 10% Crushed pottery waste • 70A10C20P: 70% Clay, 10% Wood ash, 20% Crushed pottery waste • 70A15C15P: 70% Clay, 15% Wood ash, 15% Crushed pottery waste

2.4 Thermo-Physical and Mechanical Tests

2.4.1 Thermo-Physical Characterizations

The thermo-physical analysis of samples is conducted using the "two-box method" or EI700 device. This apparatus enables the simultaneous measurement of thermal conductivity and thermal diffusivity in both solid and liquid materials. It is designed to handle large samples and perform measurements under real-world conditions, making it ideal for evaluating the thermo-physical properties of construction materials. The method is user-friendly and delivers precise measurements, offering performance comparable to traditional techniques like the hot wire and hot disk methods [7].

When comparing the thermal conductivity results from both methods, the difference is less than 9%, with the method's accuracy ranging between 3 and 5%.

The image below illustrates the testing apparatus for thermal conductivity, including the hot plate, insulation materials, and sample positioning. This configuration ensures uniform temperature gradients across the sample, enabling accurate heat transfer measurement through the material.

Crucial factors influencing the size of the discouragement include the presence of fire and the absence of activity in the first box (B1), as shown in Fig. 5. The heat flow occurs between the device's isothermal cold reservoir and a heat source with a constant heat flux $Q = U^2R^{-1}$, which is controlled by a rheostat. The objective is to maintain the temperature in box B1, represented as TB1, slightly lower than the ambient temperature, with a temperature difference of less than 0.1 °C, i.e., TB1-Tamb < 0.1 °C) [11].

Figure 5 illustrates the mechanical press used to measure compressive strength, showing how samples are arranged and the direction of the applied force. Important components, including the load cell and data acquisition system, are highlighted to demonstrate how the results are captured.

Figures 4 and 5 provided essential visual context for the methodologies employed in this study. The thermal conductivity results demonstrated that composites with higher wood

Fig. 3 Images of specimens prepared for tests

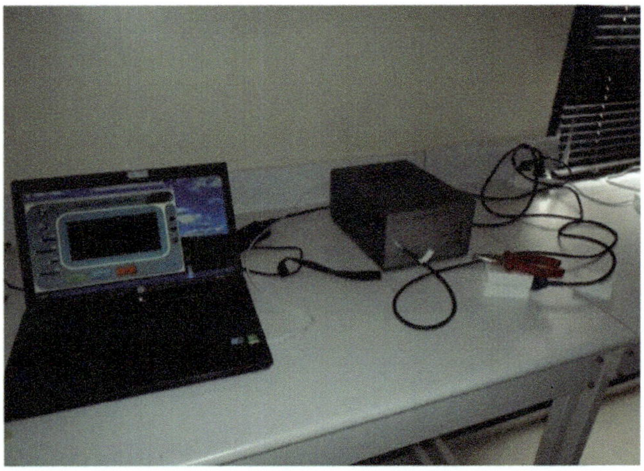

Fig. 4 Apparatus used to measure conductivity

ash content had lower conductivity values, as tested using the equipment shown in Fig. 4. Meanwhile, the compressive strength tests indicated that samples with increased pottery waste content exhibited higher resistance to deformation, consistent with the setup illustrated in Fig. 5.

The measurement principle involves applying a small thermal heat input to the material, just above the ambient temperature, at a localized point, and monitoring the resulting temperature increase over a set period, typically lasting a few minutes. Through mathematical analysis within the corresponding software programs, the required value is then determined. Once a steady state is reached, indicated by stable temperature readings over a prolonged period, the temperature gradient across both sides of the sample's mid surfaces,

Fig. 5 Composite material and the hot wire probe

the ambient temperature, and the heating voltage are recorded. The thermal conductivity value is then calculated using the following formula:

$$\lambda = \frac{e}{S(T_H - T_C)}(Q - \beta(T_{B1} - T_{\text{amb}})) \qquad (1)$$

The measurement procedure starts by applying a small temperature increase to the material, typically just a few degrees above the ambient temperature, at a specified location. This temperature rise is closely monitored over a defined period, usually lasting a few minutes. Subsequently, mathematical algorithms are used to analyze the collected data, and the results are processed through specialized software platforms to calculate the thermal conductivity of the material. This approach ensures precise determination of the thermal conductivity properties of the material under examination.

2.5 Mechanical Tests

2.5.1 Dry Compressive Strength

Brick blocks and cylindrical specimens with a diameter of 5 cm and a height of 10 cm were fabricated. The compressive strength of these cylindrical specimens was evaluated through compression tests, where a progressively increasing load was applied to a section until failure occurred (Fig. 6).

In accordance with the standard procedure for clay blocks [8], this test involves positioning two halves of the sample, joined together with mortar, in a simple compression test, as illustrated in Fig. 6. The compressive strength is then determined using the following method:

$$R_c = 10 \times \frac{F}{S} \qquad (2)$$

Fig. 6 Compressive strength test

R_c Compressive strength of the blocks, measured in megapascals (MPa).

F Maximum load supported by the two half-blocks, expressed in kilonewtons (kN).

S Average surface area of the test faces, measured in square centimeters (cm^2).

3 Results and Discussion

3.1 Thermo-Physical Characterizations

The test results for the clay blocks are presented in Figs. 7, 8, and 9. These results were compared with those of the cement blocks, and measurements were repeated three times to evaluate the margin of error.

The property in focus refers to the material's ability to conduct heat. A lower thermal conductivity signifies better insulation properties.

As shown in Fig. 8, the thermal conductivity decreases from 3.7512 kJ/hmK for 100 A to 1.5462 kJ/hmK for the sample (50A30C20P), representing a reduction of approximately 40.2%. This decrease in thermal conductivity with varying loads can be attributed to the lower thermal conductivity of the material compared to the clay matrix.

Thermal resistance, which measures a material's insulation efficiency, indicates that walls with higher thermal resistance provide better insulation.

The thermal conductivity (λ) values for different composite mixtures of clay, wood ash, and pottery waste reveal that increasing the proportion of wood ash results in lower thermal conductivity, due to the formation of air

Fig. 7 Blocks volumic mass studied and cement brick

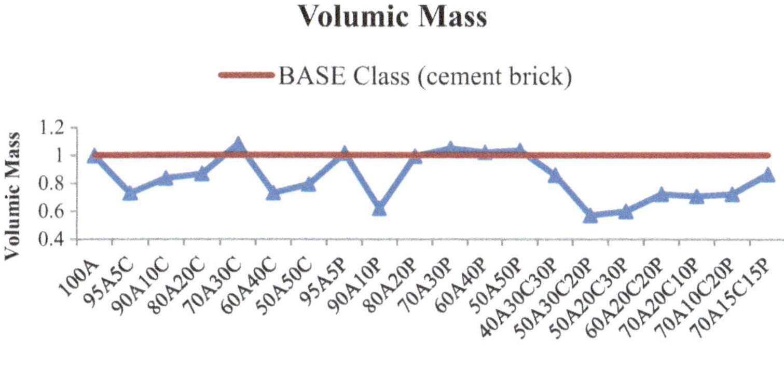

Fig. 8 Clay blocks thermal conductivity of the studied and cement brick

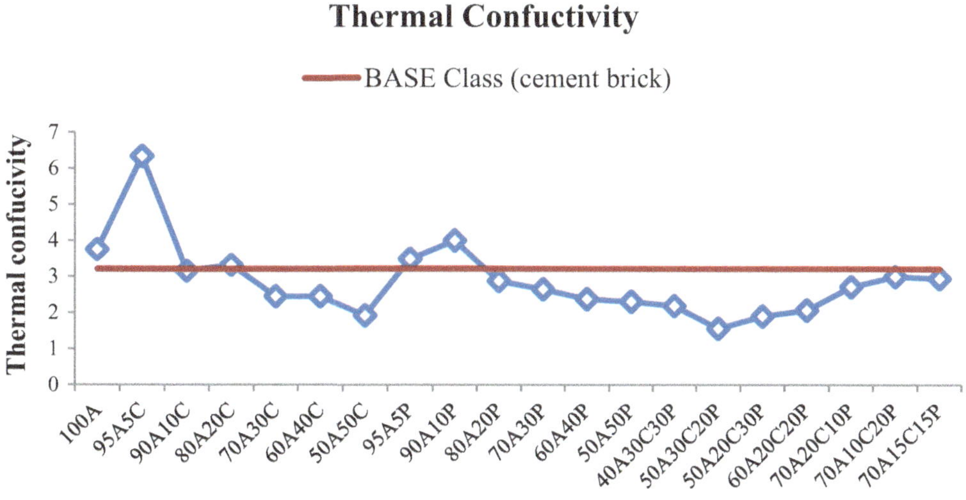

Fig. 9 Clay blocks heat capacity of studied and cement brick

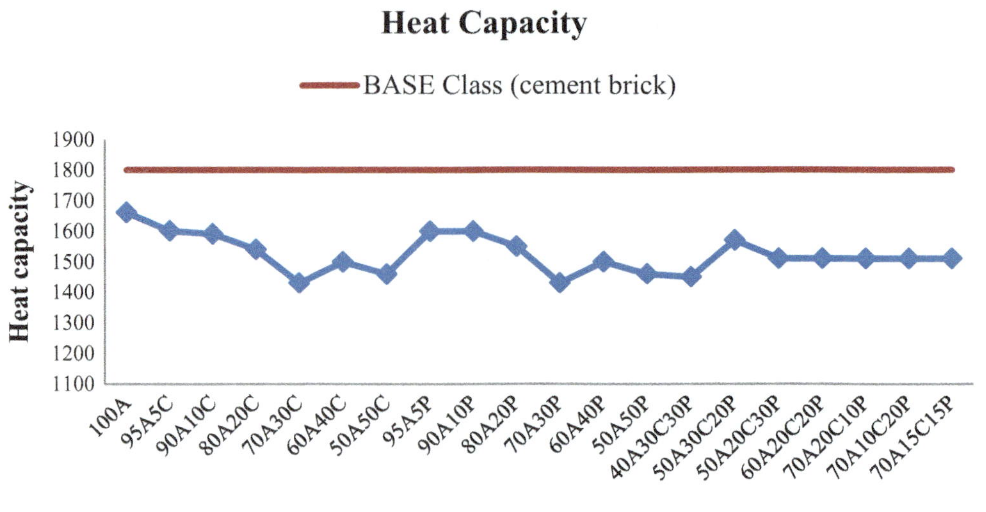

pockets and higher porosity. Mixtures with a greater content of pottery waste, while still offering adequate thermal insulation, exhibit slightly higher conductivity compared to those with more wood ash, indicating a denser, less porous structure. These findings highlight the need for a balance between the two additives to achieve optimal thermal insulation without sacrificing mechanical strength.

3.2 Dry Compressive Strength

The results of the dry compressive strength of each sample studied are shown in (Fig. 10).

The results presented here align with the findings of Khedari et al. [12], suggesting that an increase in fiber content leads to a decrease in the binding force within the specimens,

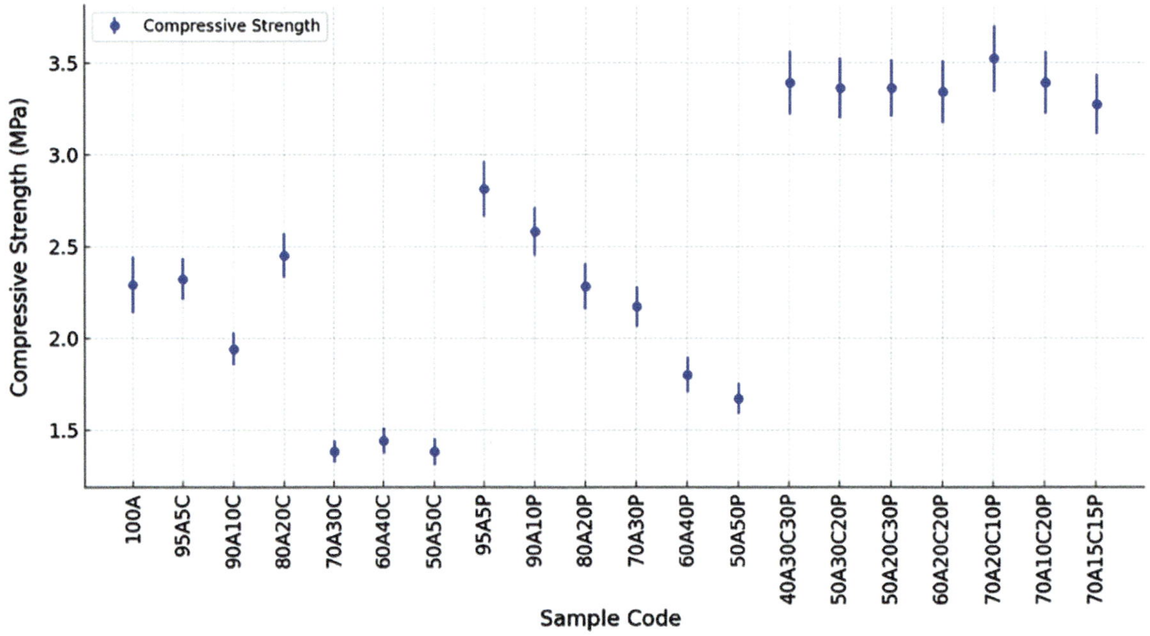

Fig. 10 Compressive strength of the different clay mixtures with various additives

which in turn reduces compressive strength. This decline in compressive strength can be attributed to the dominant effect of fiber content on the material, as the formation of hydration products is relatively limited compared to the voids created within the mixture.

4 Conclusion

In this article, we investigated the effect of waste from the manufacture of translational pottery, ground pottery waste, and wood ash on the thermal and mechanical properties of clay blocks. The results lead to the following conclusions:

- The mechanical tests conducted through compression provided satisfactory results. The mix (50A20C30P), consisting of 50% clay, 20% wood ash, and 30% crushed pottery waste, exhibited optimal compressive strength.
- The compressive strength of the bricks decreases with the inclusion of crushed pottery waste, due to the weak bond between the particles of this waste and the clay matrix.
- The best mechanical compressive strength was observed in the mix (clay + pottery waste + wood ash) compared to the other three block mixes examined.
- The addition of filler significantly enhances the thermal insulation of the blocks. The more the blocks are filled, the more their thermal conductivity decreases and their thermal resistance improves.
- Thermal conductivity is particularly low for all the mixes tested.
- A strong correlation exists between the density of the blocks and their thermal conductivities: the lower the density, the better the thermal insulation of the blocks.

The ecological bricks studied promote energy savings in line with prescribed standards, supporting sustainable construction practices by reducing energy consumption and contributing to environmental preservation.

References

1. Brown LR (2013) Eco-economy: building an economy for the earth. Routledge
2. Minke G (2013) Building with earth. Birkhäuser
3. Boumhaout M, Boukhattem L, Nouh FA, Hamdi H, Benhamou B (2013) Energy efficiency in buildings: thermophysical characterization of building materials. In: 2013 international renewable and sustainable energy conference. IEEE, pp 391–395
4. Ddani M, Meunier A, Zahraoui M, Beaufort D, El Wartiti M, Fontaine C, Boukili B, El Mahi B (2005) Clay mineralogy and chemical composition of bentonites from the Gourougou volcanic massif (northeast Morocco). Clays Clay Miner 53(3):250–267
5. Hajjaji M, Mezouari H (2011) A calcareous clay from Tamesloht (Al Haouz, Morocco): properties and thermal transformations. Appl Clay Sci 51(5):507–510
6. El Cadi H, El Bouzidi H, Selama G, Ramdan B, El Majdoub YO, Alibrando F, Arena K, Lovillo MP, Brigui J, Mondello L, Cacciola F, Salerno TMG (2021) Elucidation of antioxidant compounds in Moroccan *Chamaerops humilis* L. fruits by GC–MS and HPLC–MS techniques. Molecules 26(9). https://doi.org/10.3390/molecules26092710
7. Boumhaout M, Boukhattem L, Hamdi H, Benhamou B (2017) Thermomechanical characterization of a bio-composite building material: Mortar reinforced with date palm fibers mesh. Constr Build Mater 135:241–250. https://doi.org/10.1016/j.conbuildmat.2016.12.217
8. NF XP P13-901 (2001) Blocs de terre comprimée pour murs et cloisons: definitions—Spécifications-Méthodes d'essais—Conditions de réception. AFNOR, p 35
9. Salih TWM, Jawad LAA (2022) Evaluating the thermal insulation performance of composite panels made of natural Luffa fibres and urea-formaldehyde resin for buildings in the hot arid region. Adv Build Energy Res 16(6):696–710
10. Ghosh A, Ghosh A, Neogi S (2018) Reuse of fly ash and bottom ash in mortars with improved thermal conductivity performance for buildings. Heliyon 4(1)
11. Bajji S, Bahammou Y, Bellaziz Y, Saba A, Naimi Y (2024) Thermophysical characterizations and simulation study of an energy-efficient building material: clay stabilized by wood ashes or crushed waste from traditional. Heat Mass Transf
12. Khedari J, Watsanasathaporn P, Hirunlabh J (2005) Development of fibre-based soil–cement block with low thermal conductivity. Cem Concr Compos 27(1):111–116

The Perception of the Integration of Artificial Intelligence in the Supply Chain: The Case of an Automotive Industrial Company in Morocco

Samia Haman, Younes El Bouzekri El Idrissi, Brahim El Bhiri, and Aniss Moumen

Abstract

The complexity of the supply chain processes has increased exponentially. To cope with this complexity, integrating new technologies is much recommended. The integration of Industry 4.0 technologies, especially artificial intelligence and machine learning, in the supply chain processes of industrial companies, can help achieve the company's objectives. Through this research, we have tried to probe the contribution of Industry 4.0 technologies in the supply chain of Moroccan automotive industries, through a qualitative study. This qualitative study was conducted using NVIVO. This analysis is based on interviews conducted with a global automotive cable industry leader based in the Atlantic free zone of Kenitra in Morocco. This work demonstrates the perception of the actors of the supply chain regarding the challenges of integrating artificial intelligence in the supply chain processes, its benefits, and its disadvantages.

Keywords

Industry 4.0 · Supply chain · Automotive cables industry · Artificial intelligence · Qualitative study · Morocco

1 Introduction

For several years, the interest in the integration of Industry 4.0 technologies has been rising due to changes in the supply chain. The latter has undergone major changes that increased the amount of data produced. Nowadays, supply chain processes produce a voluminous set of data that includes details about processes, events, and alerts found throughout an industrial production continuum [1]. When meticulously collected and evaluated, this data store can reveal priceless insights into industrial processes and the inherent dynamics of the system [1] and it can also help the company increase its productivity or optimize its processes. Moreover, leveraging this data can empower companies to enhance productivity and streamline their processes.

Industry 4.0 leverages data to foster digitalization and automation within the manufacturing sector, thereby establishing a digital framework to facilitate seamless interaction across all facets of a company [2]. This paradigm shift toward Industry 4.0 not only necessitates the integration of advanced technologies but also underscores the importance of harnessing data as a strategic asset. By embracing Industry 4.0 principles, organizations can unlock new avenues for innovation, efficiency, and competitiveness in the global marketplace.

The advent of Industry 4.0 heralds a new era characterized by interconnectedness, agility, and data-driven decision-making. Through the amalgamation of cutting-edge technologies such as the Internet of Things (IoT), artificial intelligence (AI), and machine learning (ML), companies can orchestrate a synchronized ecosystem where data flows seamlessly across various touchpoints. This interconnectedness enables real-time monitoring, predictive analytics, and adaptive control mechanisms, empowering organizations to anticipate and respond to dynamic market demands with unparalleled agility.

S. Haman (✉) · Y. El Bouzekri El Idrissi · A. Moumen
Laboratory of Engineering Sciences, National School of Applied Sciences, Ibn Tofail University, Kenitra, Morocco
e-mail: samia.haman@uit.ac.ma

B. El Bhiri
Smartilab Laboratory Emsi, Rabat, Morocco

A. Moumen et al. (eds.), *Artificial Intelligence and Data Analytics for Innovative Applications in Engineering, Sustainability and Technology*, Sustainable Artificial Intelligence-Powered Applications, https://doi.org/10.1007/978-3-031-90318-2_13

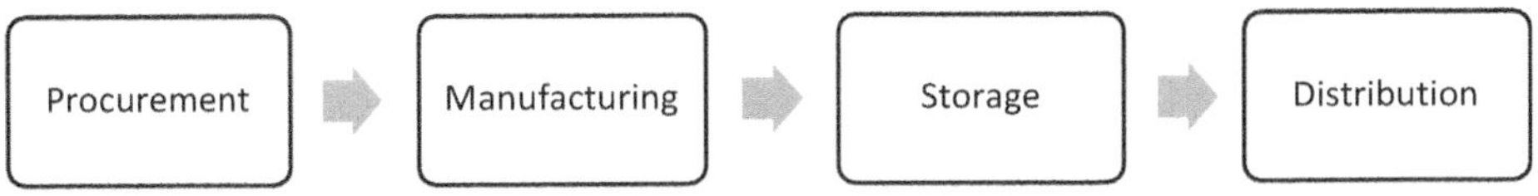

Fig. 1 Different processes of supply chain

Recent researches in this area show that integrating advanced technologies represents an opportunity for the supply chain. Recognizing the significance of implementing machine learning techniques within the supply chain, organizations are integrating its methodologies into their supply chain operations [3]. However, many companies are still afraid of these technologies. They must, however, decide whether to carry out a digital transformation while also ensuring cost and time-to-market optimization, productivity, and continual improvement of all business processes [4]. A firm's ability to profit from the ongoing transformation demands skilled adaptation from a range of angles [5]. Research on the integration of Industry 4.0 in the supply chain is minimal and is still in the initial phase. In light of this uncertainty, this paper aims to discover the different perceptions of the actors.

This study thus investigates the following research questions:

1. What are the perceived benefits, challenges, and disadvantages of integrating artificial intelligence in the supply chain by the actors of the industrial companies?
2. What is the perceived contribution of integrating machine learning in the supply chain by actors of industrial companies?

This research represents a work that was initiated in the context of an internship in an automotive company. The automotive industry is a key sector that relies on complex partnerships, influencing its upstream and downstream operations [6]. The objective is to discover the perception of the actors of the supply chain toward the integration of artificial intelligence, and to identify the advantages, disadvantages, and challenges, according to their opinion, especially the perceived benefits of machine learning.

The rest of the article is arranged as follows. In the next section, we present a literature review on the integration of Industry 4.0 technologies. In Sect. 3, we describe the methodology followed and the materials used in this study. Section 4 reports the results of our interviews and a discussion based on the literature. Then we propose in Sect. 5 our conceptual model. Finally, we conclude our paper, in Sect. 6, with a conclusion and perspectives.

2 Supply Chain and Industry 4.0

2.1 Supply Chain

In the literature, there are several definitions of supply chain, but they are quite similar. A supply chain is a global network that works together to enhance the flow of materials and information between suppliers and customers at the lowest possible cost and maximum speed [7].

As shown in Fig. 1, the supply chain includes four processes. Procurement, manufacturing, storage, and distribution. The supply chain starts with the procurement of raw materials from different suppliers, the manufacturing phase includes transforming the raw material into finished goods, storing the latter then distributing it to the customer. The primary goal of a supply chain is to maximize the coordination of these operations to ensure that items or services are delivered to clients on time, in the correct quantity, location, and period, all while reducing costs and increasing overall value.

In the supply chain, the information flows from customers to retailers, manufacturers, and logistics and raw material suppliers [7]. This information flow between different partners and departments generates a tremendous amount of data. This data concerns the quantity of raw material ordered, the price, the date, the supplier, and the material ordered from each supplier. In addition, the quantity manufactured the date of production, and the number of quality problems. The quantity delivered, the date of the delivery, and the price of the finished product delivered to each.

2.2 Industry 4.0

Industry 4.0 represents the fourth revolution in the industrial world. It integrates numerical tools and new technology across the whole manufacturing value chain [8]. Today, the industry is largely reliant on interconnectivity, automation, artificial intelligence, machine learning, and real-time data [9]. The integration of industrial technologies represents an opportunity for industrial companies. It has many benefits thanks to the different technologies of Industry 4.0.

Industry 4.0 emerged in the twenty-first century. It represents the fourth industrial revolution. The notion of Industry 4.0 was originally supposed to explain the impact of developing technologies in the realm of manufacturing [10]. It has been described in a variety of ways, such as a vision, a paradigm, a scenario, or a digital revolution in manufacturing and service firms [11].

Industry 4.0 integrates new technologies into the supply chain such as artificial intelligence, machine learning, big data, the internet of things, augmented reality, cloud computing, additive manufacturing, etc. In this paper, we focus essentially on artificial intelligence and machine learning.

Artificial intelligence is defined as a model that has the ability to think, learn, and act autonomously like human behavior [12]. It refers to the simulation of human intelligence processes by machines, especially computer systems. It allows machines to execute tasks like a human being. The goal of AI is to construct machines that can mimic human cognitive functions and perform activities independently, eventually increasing efficiency and production in numerous industries. AI technologies represent one of the most significant advancements in technological, digital, and computational fields, as they simplify and enhance everyday life [13]. The path to fully embracing Industry 4.0 in Morocco comes with a unique mix of challenges and opportunities [14].

Machine learning is a subfield of artificial intelligence that focuses on the development of algorithms and statistical models that allow computers to learn and improve their performance on a given task without being explicitly programmed [15]. The ultimate goal of machine learning is to enable computers to solve complicated problems or make judgments in a way that replicates human intelligence, albeit frequently with higher speed and scalability.

3 Methodology

3.1 Method

A qualitative study is an investigation that aims to obtain detailed information about users' perceptions in a precise field. In our paper, we conducted qualitative research to obtain detailed information about the users' perception of integrating artificial intelligence in the supply chain of a manufacturing organization in the automotive cable industry in Morocco.

We started by preparing our interview guide based on a literature review. We chose the company and then conducted interviews with the actors in the supply chain. We transcribed then the interviews and analyzed the data. The data analysis refers to calculating several variables, such as the saturation and the coverage rate. After the analysis, we presented the results. The results in our case represent the verbatim of each interviewee.

Fig. 2 Research methodology

Figure 2 presents the details of the methodology followed.

3.2 Interview Procedures

We used a semi-structured interview guide. The interview guide contains open questions. The authors prepare in advance the questions. These questions ask about the perception of the integration of artificial intelligence in the supply chain.

For our interviews, we used a guide containing three themes; we added the interview guide in the appendix.

The first theme concerns the digitalization of the supply chain processes. In this theme, we asked general questions about the supply chain processes, the maturity level of their digitalization, and the process of decision-making. The second concerns the information system and artificial intelligence where the questions were about the information system used in the organization and general questions about business intelligence. The third concerns the integration of artificial intelligence and machine learning in the supply chain. The questions concerned the integration of AI and ML, such as the perceived benefits, challenges, and disadvantages of integrating AI and the perceived benefits of integrating ML in the supply chain.

This work aims to study the application of advanced technologies in Moroccan industrial companies. In our case, we focused on only one manufacturing company.

The company studied is a global leader in the automotive cables industry based in the Atlantic free zone of Kenitra. In this company, the main departments are the human resources

Table 1 Sample of respondents interviewed

Type of the respondents	Number of interviews	Average duration
Customer service engineer	3	1 h
Transport engineer	2	1 h 30

Table 2 Details about the sample of the engineer

Engineer	Gender	Experience in the supply chain	Diploma
ENG1	M	11 years	Master's degree
ENG2	M	10 years	Master's degree
ENG3	M	6 years	Master's degree
ENG4	M	4 years	Master's degree
ENG5	M	9 years	Master's degree

department, the logistics department, the quality department, the purchasing department, the information technology department, the production department, and the maintenance department.

Table 1 present a synthesis of the interview information.

We conducted a qualitative study to explore the application of Industry 4.0 technologies in Moroccan industries by interviewing five actors in the logistics department of the same automotive company. We targeted logistics engineers for our interviews. Three of these interviewees were customer service engineers, and the other two were transport engineers. The interviews took place on the company's site. Each interview took approximately an hour and a half as shown in Table 1. The interviewees all have a master's degree, and more than 4 years of experience in the supply chain as shown in Table 2.

3.3 Data Analysis

We transcribed the interviews. The first step is coding each interview and theme, by identifying where each of the three themes appears in the responses of each interview. Then we calculated semantic saturation using Pearson's correlation coefficient in NVIVO. The average value in our corpus is 81%. The value of this coefficient means that the similitude level in our corpus is 81%.

After that, we calculated the coverage rate of each theme used in our interview using NVIVO. The coverage rate refers to the extent to which each of the three themes is represented in the interview responses of the five individuals. It presents how each theme appears in each interview.

In theme 1, the coverage rate is between 20% for the ENG4 and 15% for the ENG3, its average is 17.5%. For theme 2, its value is between 17% for ENG1 and 10% for ENG3, the average value is 13.5%. The coverage rate for theme 3 value is between 20 and 12%, the average value is 16%.

Table 3 presents the average coverage rate for each theme.

Table 3 Average coverage rate of each theme

Theme	Average coverage rate (%)
Theme 1	17.5
Theme 2	13.5
Theme 3	16

4 Findings and Discussion

In this section, we present the content analysis of the interviews conducted in this study. We report in what follows, the results concerning the most important topics in our interviews. These results represent the interviewers' perceptions. We will also compare and discuss our findings in this research with the existing literature.

4.1 Perceived Benefits of Artificial Intelligence in the Supply Chain

The emergence of Industry 4.0 encouraged the actors of the industries to explore the different technologies that could lead to the optimization of the supply chain. Artificial intelligence is one of the most known technology.

The utilization of AI-related technologies has led to significant transformations in the operations of industries. These changes encompass enhancements in maintenance and control, process monitoring, production process optimization, service and technique management, and simplification of change implementation. Artificial intelligence is used in the industry for these advantages: the assistance of monitoring systems, and continuous analysis is conducted through integrated monitoring, it also aids in identifying problems before errors occur, ultimately reducing the cost associated with managing them [16]. Artificial intelligence improved data collection and inventory processes [17]. In addition, artificial intelligence contributes to warehouses by improving the service quality of the distribution network [17]. Moreover, it helps minimize downtime, optimize production processes, and boost operational efficiency [14].

The interviewees agreed that including artificial intelligence in the supply chain has many benefits. The engineer based on past data plans the forecast in the supply chain. The engineers analyze this data and anticipate future orders from clients. This analysis of the data can be done using artificial intelligence. It represents an opportunity for the supply chain to more accurate forecasts. Another advantage

Table 4 Verbatim of interviewees concerning the advantages of using artificial intelligence in the supply chain

Verbatim	
ENG 1	Benefits of integrating artificial intelligence in the supply chain include more accurate forecasts, reduced costs, and improved customer satisfaction through smoother operations
ENG 2	The Optimization of tasks is one of the most valuable advantages of integrating artificial intelligence in the supply chain, which will facilitate tasks. Speed of access to information
ENG 3	For the integration of artificial intelligence advantages, we can talk about saving time and also money that is due to an optimization of tools
ENG 4	For the advantages, artificial intelligence allows more visibility in the supply chain: monitoring products and machines. Establish predictive maintenance for machines to avoid machines downtime
ENG 5	Many advantages such as saving time, or increasing productivity by avoiding production stoppages that are caused by machine shutdowns, and avoiding shortages of raw materials or finished products

of using artificial intelligence, it allows more visibility into the supply chain, by monitoring the products and machines. The monitoring of machines can help establish predictive maintenance to avoid machine downtime. This helps gain time and money. Artificial intelligence also serves as a complement to performing difficult, repetitive, and time-consuming tasks.

Table 4 presents the interviewees verbatim.

4.2 Perceived Barriers and Challenges of Artificial Intelligence in the Supply Chain

Integrating artificial intelligence into the supply chain is a big decision that an organization can take. It has many advantages as we talked about in the previous section, but at the same time, it represents a challenge to the organization. To integrate artificial intelligence, the organization needs to prepare in advance.

According to the interviewees' perception, the main barriers to the integration of artificial intelligence are ensuring data quality and availability, people management, the integration of artificial intelligence into the existing systems, and cost requirements.

Many challenges can slow down the integration process, and ensuring data quality and availability is a big one. Industries must establish a comprehensive data management strategy that encompasses data collection, cleansing, organization, storage, governance, and security [16]. Moreover, they should prioritize ensuring data interoperability and accessibility across various systems and platforms.

Another challenge is the people management. Control and monitoring involve a collaborative approach, necessitating human expertise to manage effectively artificial intelligence solutions. The human factor represents a big challenge. It is necessary to maintain the interaction between people with artificial intelligence tools.

In addition, the integration of AI into the existing systems needs to be seamless into the existing workflows and processes of the industries. Businesses should prioritize upgrading their IT infrastructure and equipment to integrate effective artificial intelligence solutions. By ensuring smooth integration, organizations can enhance efficiency, productivity, and overall performance. It is important to carefully plan and implement the integration process to ensure a successful outcome.

Generally, managers avoid investing in artificial intelligence systems because of the cost requirements. There is the initial cost that represents the purchase of the technology, then the cost to maintain it. The potential benefits and return on investment are uncertain at this time, and the duration for realizing any payoff remains unknown [2].

In addition, there is a lack of experts and knowledgeable employees in this area to initiate a new system or revamp the current system for optimal results [2]. Changes in the workplace, such as digital transformation, notably in Industry 4.0, necessitate unique skills that are not always taught in educational institutions or cultivated in the labor market [9].

Our research revealed that among the challenges found in the literature, the interviewees perceive the lack of training of staff in AI skills as also a challenge. Artificial intelligence technologies are new, and working with them requires new training for professionals.

Table 5 presents the interviewees verbatim.

4.3 Perceived Disadvantages of Artificial Intelligence in the Supply Chain

After citing the perceived advantages of integrating artificial intelligence according to the interviewees and literature, we present the perceived drawbacks.

The recurring disadvantage of integrating artificial intelligence according to our interviewees is the lack of security and the risk of data leakage. As things become more advanced, the risk of security issues also rises, so it is crucial to safeguard data and smart systems from cyber-attacks [18]. The increasing use of artificial intelligence technologies brings up important issues regarding trust, risk, and security. Trust in artificial intelligence pertains to the confidence users and stakeholders have in the reliability, integrity, and ethical use of

Table 5 Verbatim of interviewees about the challenges of using artificial intelligence in the supply chain

Verbatim	
ENG 1	Challenges include data quality, supply chain complexity, organizational changes, and initial costs
ENG 2	The human factor will create problems in integrating this technology. The initial costs are for sure the biggest challenge
ENG 3	The difficulties would be concerning reliability, and whether this tool is capable of satisfying the needs of the organization. Without forgetting the lack of employee training for this tool, which is considered a blocking point. As well as the integration of this technology in the existing systems
ENG 4	Concerning the difficulties, we could cite the lack of training of the staff, given that they will have to work with tools that they have never used, without forgetting the initial costs following the integration of AI into the new system established
ENG 5	I think that the lack of staff training would be the first difficulty to face. As well as the costs generated following the purchase and integration of these tools

Table 6 Verbatim of interviewees concerning the disadvantages of using AI in the supply chain

Verbatim	
ENG 1	For the drawbacks, we can talk about the need for specialized training for the employees to work with AI systems
ENG 2	Job displacement is one of the significant drawbacks of integrating artificial intelligence
ENG 3	For the disadvantages, we could talk about problems of lack of security and data leakage. For example, in the event of a hack, hackers can collect important data for the company and can even delete it
ENG 4	The disadvantages of integrating artificial intelligence: the risk of data leakage specific to the organization, and security attacks
ENG 5	The disadvantages of integrating artificial intelligence are the unreliability and insecurity of data

artificial intelligence systems [19]. Every component in cyber-physical system-enabled smart manufacturing is linked to its operating systems or applications without sufficient security protection [20].

There is also the need for specialized training for employees and job displacement. To work with artificial intelligence tools, employees need training to have special skills that are needed in this case. Also, integrating artificial intelligence into the industry would reduce the need for human participation and interactions within the system [2].

Table 6 presents the verbatim of interviewees concerning the perceived disadvantages of artificial intelligence in the supply chain.

4.4 Perceived Contribution of Artificial Intelligence in the Supply Chain of the Organization

According to the interviewees, AI has many benefits such as avoiding stock shortages by predicting demand and shortages in finished products, optimizing delivery routes, managing stock more efficiently, and preventing breakdowns with predictive maintenance. They also ensured that AI could help gain time and money by giving examples.

According to the interviewees, artificial intelligence can also contribute to their organization by avoiding stock-outs and breakdowns, increasing efficiency, saving time and money, and improve supply chain performance. The interviewees gave many examples of the perceived contribution of AI in the supply chain. They presented many problems that they deal with in their organization and that AI could solve in their opinion.

In this section, interviewees give examples of the perceived contribution of artificial intelligence in their organization.

Table 7 presents the examples proposed by the participants.

4.5 Perceived Contribution of Machine Learning in the Supply Chain of the Organization

Machine learning is a branch of artificial intelligence. It is giving computers the ability to learn without being explicitly programmed. Over the past decade, the field of machine learning has witnessed remarkable advancements that have revolutionized the accuracy and capabilities of predictive models [6]. These developments have not only enhanced the efficiency of algorithms but have also expanded the scope of applications across various industries. As a result, the accuracy and reliability of predictive models have significantly improved, enabling more precise and insightful predictions to be made.

All interviewees agreed that integration of machine learning in the supply chain would be beneficial. It has many advantages for the supply chain such as demand forecasting, optimizing routes, planning resources, machine failure prediction, and decision-making.

For demand forecasting, machine learning algorithms analyze large datasets based on the demand in the past years, to predict future demands and it could also detect seasonal orders. It helps organizations prepare the master production

Table 7 Verbatim of interviewees concerning the contribution of artificial intelligence in the supply chain of their organization

Verbatim	
ENG 1	Artificial Intelligence optimizes the supply chain by: • Predicting demand to avoid stock-outs • Optimizing delivery routes in real-time • Managing stock levels more efficiently • Preventing breakdowns with predictive maintenance
ENG 2	As already mentioned, AI would be very beneficial for the supply chain As an example: in our organization for the detection of copper coils in stock entry Optimization of space given that at the level of our organization, given that there is a problem of over-stocking when our customers' production is stopped and our production is still running.
ENG 3	AI can be used to optimize the use of stackers, i.e. it will improve the performance of stackers during order preparation, which means that we could increase the number of orders prepared during a shift, so it saves time. This can also reduce the order preparation workforce, in which case it will save money
ENG 4	The use of artificial intelligence will be very beneficial for the supply chain Whether for the prediction of orders, for planning the production plan for each week based on customer orders and backorders from the week before, or for the allocation of manufacturing orders for each production line This can lead to greater efficiency, reduced costs, and overall improvement in supply chain performance
ENG 5	Artificial intelligence can help predict shortages in finished products given that copper coils are not always delivered following FIFO (First In First Out). A second example is to have alerts for obsolete coils, given that a coil has a lifespan of 6 months so have an alert when this duration approaches. For the production planning process: automate the processing of forecasts (according to the season and the history of previous years). Another example is to have a Truck and trailer supervision system: route and location verification, identify the shortest and least expensive path

schedule. This planning is used to specify what is going to be produced, when, and the quantity produced. By having advanced knowledge of the quantities they need to produce, businesses can reduce overproduction [21]. Also, machine learning techniques have superior accuracy and robustness in automotive demand forecasting compared to traditional mathematical models [6].

For the realm of machine failure prediction, machine learning models can be used for the prediction and prevention of potential downtime by identifying anomalies that occurred in the past that may indicate a problem in the machines. This approach minimizes production shutdown. Leveraging machine learning techniques enhances maintenance planning by providing insights into the timing of potential breakdowns, thereby minimizing downtime risks. It enables more precise scheduling of maintenance activities and allows for proactive anticipation of failures. This, in turn, facilitates targeted repair interventions, resulting in faster and more efficient execution of maintenance tasks [22].

In terms of optimizing routes, machine learning can give the shortest path to follow to the truck drivers to deliver the finished goods at the least cost. Furthermore, machine learning can help make decisions that are more informed by analyzing complex datasets to assess potential risks.

Table 8 presents the verbatim of the interviewees concerning the contribution of machine learning in the supply chain.

5 Proposed Conceptual Model of Artificial Intelligence in the Supply Chain

In this section, we summarize our findings and we present our proposed conceptual model based on the interviewees' perception. This conceptual model explains the motivation for integrating artificial intelligence into the supply chain.

Our proposed conceptual model explores the multifaceted motivations behind the integration of artificial intelligence in the supply chain. At its core, the model posits that organizations are driven to integrate artificial intelligence technologies in the supply chain by a combination of factors, including the pursuit of increased efficiency, cost reduction, improved customer satisfaction, more visibility, and enhanced decision-making processes.

According to our participants, several factors motivate the decision-makers to integrate artificial intelligence tools into their organizations' supply chain. Improving efficiency

Table 8 Verbatim of the perceived contribution of machine learning in the supply chain

Verbatim	
ENG 1	Machine learning is widely used to predict demand, optimize routes, and plan resources in the supply chain
ENG 2	In my opinion, ML has a very important role in optimizing the supply chain. For example: For order preparation: Having the shortest path using Machine learning for stackers for order preparation For production planning: use of machine learning to create the Master Production Plan
ENG 3	Machine learning is the ability of the machine to make decisions based on several data. Machine learning can help us anticipate orders. It can detect seasonal orders, so when these orders approach we can schedule to produce before these dates
ENG 4	Machine learning is a tool that is used for everything related to prediction, whether it is order prediction or machine failure prediction
ENG 5	Machine learning is widely used to predict demand, optimize routes, and plan resources in the supply chain

is an essential factor. Artificial intelligence can streamline processes and optimize operations across the supply chain, resulting in greater efficiency. Machine learning algorithms can evaluate large amounts of data to detect patterns, estimate demand, optimize inventory levels, and reduce waste. "... *saving time, or increasing productivity by avoiding production stoppages that are caused by machine shutdowns, and avoiding shortages of raw materials or finished products*" ENG5.

As a second factor, we can cite improving customer satisfaction. Artificial intelligence-powered supply chain systems can enhance customer experience by assuring on-time delivery, reducing stock-outs, and offering personalized product suggestions. Organizations may improve the overall customer experience and increase customer loyalty by better knowing their preferences and behavior. "*Benefits of integrating artificial intelligence in the supply chain include more accurate forecasts, reduced costs, and improved customer satisfaction through smoother operations*" ENG1 AND "*Artificial Intelligence optimizes the supply chain by:—Predicting demand to avoid stock-outs.—Optimizing delivery routes in real-time.—Managing stock levels more efficiently.—Preventing breakdowns with predictive maintenance.*" ENG1.

Integrating artificial intelligence into the supply chain allows more visibility. Artificial intelligence helps in providing real-time visibility into inventory levels, demand fluctuations, and market trends. This enables organizations to quickly adjust their strategies and respond to changing customer needs and market conditions. "*For the advantages, artificial intelligence allows more visibility in the supply chain: monitoring products and machines*" ENG4.

In addition, artificial intelligence enhances the decision-making process. AI-powered analytics provide valuable insights that enable better decision-making at every stage of the supply chain. Organizations can make more informed decisions about inventory management, production planning, transportation routes, and supplier selection, by leveraging real-time data and predictive analytics. "... *Machine Learning can help us anticipate orders; it can detect seasonal orders, so when these orders approach we can schedule to produce before these dates*" ENG3.

In addition, artificial intelligence has a huge impact on reducing costs. By automating repetitive tasks and reducing errors, organizations lower labor costs and improve overall profitability. Artificial intelligence helps reduce costs by improving resource allocation, lowering inventory carrying costs, and identifying cost-cutting options. "*The use of artificial intelligence will be very beneficial for the supply chain. Whether for the prediction of orders, for planning the production plan for each week based on customer orders and back-orders from the week before, or for the allocation of manufacturing orders for each production line. This can lead to greater efficiency, reduced costs, and overall improvement in supply chain performance*" ENG4.

Technological capabilities, organizational culture, market dynamics, and regulatory requirements, these internal and external factors influence these motivations.

We hypothesize that the integration of artificial intelligence technologies leads to improvements in supply chain performance, including enhanced decision-making, streamlined processes, and better alignment with customer needs.

By elucidating the motivations and implications of artificial intelligence integration in the supply chain, our proposed conceptual model provides a comprehensive framework for understanding and analyzing this critical phenomenon.

In our proposed conceptual model, we proposed five factors that motivate artificial intelligence integration in the supply chain: improving efficiency, improving customer satisfaction, allowing more visibility to the supply chain, enhancing decision-making, and reducing cost.

In Fig. 2, we present the proposed conceptual model. In Fig. 3, we present the proposed conceptual model.

Fig. 3 Proposed conceptual model

6 Conclusion

In summary, this study offers valuable insights into the perspectives of actors in the supply chain regarding the integration of artificial intelligence in the supply chain. The interviewees believe that integrating artificial intelligence could offer tremendous benefits to the supply chain. The perceived benefits encompassed the ability to optimize the supply chain, reduce time and cost, improve customer satisfaction, more accurate forecasts, allow more visibility into the supply chain, and avoid machine downtime. Artificial intelligence has many benefits for the supply chain, but many organizations are still afraid. The challenges are ensuring data quality and availability, people management, the integration of AI into the existing systems, cost requirements, and the lack of training of staff in these new technologies. The study has also shed light on the interviewees' perception of the disadvantages of the integration of artificial intelligence. The disadvantages encompass job displacement, the need for special skills, the lack of security, and the risk of leakage. Concerning the contribution, the actors of the supply chain believe that artificial intelligence can contribute to their organization by avoiding stockouts and breakdowns, increasing efficiency, saving time and money, improving supply chain performance. According to the interviewees, the contribution of integrating machine learning helps in demand forecasting, optimizing routes, planning resources, and machine failure prediction. The use of machine learning could improve the decision-making process.

The findings of this study indicate that the majority of the participants held a positive perception regarding the integration of artificial intelligence tools in the supply chain. However, it is essential to pay attention to the drawbacks. In addition, many challenges may appear in the integration phase. However, the contribution of artificial intelligence and machine learning to the supply chain is huge compared to the drawbacks and barriers.

We conducted a qualitative study in the automotive sector in Morocco; the main goal was to have insights into integrating artificial intelligence into the supply chain. In this work, we collected information and based on that information we presented the proposed conceptual model and our hypotheses.

We presented our proposed conceptual model. The conceptual model serves as a framework for understanding and analyzing the motivations and implications of AI integration in the supply chain. It provides a structured approach to exploring this complex phenomenon, helping stakeholders gain insights into why organizations may choose to adopt artificial intelligence technologies and how they influence supply chain performance.

Our hypothesis proposes that the motives for AI adoption result in actual benefits for supply chain operations. Future work consists of conducting a quantitative study within the supply chain of the automotive organization to validate our hypothesis.

Appendix

Interview Guide:

Name:

Organization:

Department:

Job:

Gender:

Experience in the supply chain (number of years):

Theme 1: The Digitalization of the Supply Chain Processes

1. How do you define the supply chain?
2. Can you describe the supply chain in your organization?
3. What do you think of the processes of the supply chain in your organization? Do they have to be improved?
4. What do you think of the digitalization of the supply chain processes?
5. Can you tell us about your organization's experience with the digitalization of supply chain processes?
6. What do you think of the level of maturity of this digitalization?
7. In your opinion, what would be the role of digitalization in decision-making in the supply chain?
8. What would be, in your opinion, the recommendation to improve the decision-making process in the supply chain?

Theme 2: The Information System and Artificial Intelligence

1. Do you have only one global information system or an information system for each department?
2. What do you think of the information system in your organization?
3. What do you think of artificial intelligence in the supply chain?
4. What do you think of business intelligence for the supply chain?
5. Do you use business intelligence for the decision-making process in your organization? Can you tell us about your feedback?

Theme 3: The Integration of Artificial Intelligence and Machine Learning in the Supply Chain

1. What do you think of the use of artificial intelligence in the supply chain? Can you give us some examples? In your opinion, what would be the contribution?
2. What would be the challenges, in your opinion, in the integration of artificial intelligence in the supply chain?
3. What would be the advantages of the integration of artificial intelligence in the supply chain?
4. What can you tell us about the disadvantages of the integration of artificial intelligence in the supply chain?
5. If you have any experience or feedback, can you share it with us?
6. What do you think of the application of machine learning tools in the supply chain? Can you share with us your feedbacks?

References

1. Alshboul O, Al Mamlook RE, Shehadeh A, Munir T (2024) Empirical exploration of predictive maintenance in concrete manufacturing: harnessing machine learning for enhanced equipment reliability in construction project management. Comput Ind Eng 110046. https://doi.org/10.1016/j.cie.2024.110046
2. Abdirad M, Krishnan K (2021) Industry 4.0 in logistics and supply chain management: a systematic literature review. Eng Manag J 33(3), 187–201. https://doi.org/10.1080/10429247.2020.1783935
3. Haman S, Moumen A, Jenoui K, Elbhiri B, El Bouzekri El Idrissi Y (2024) Machine learning techniques in supply chain management: an exploratory literature review. In: El Bhiri B, Saidi R, Essaaidi M, Kaabouch N (ed) Smart mobility and industrial technologies. Springer Nature Switzerland, Cham, pp 155–161. https://doi.org/10.1007/978-3-031-46849-0_17
4. Abouzid I, Saidi R (2023) Digital twin implementation approach in supply chain processes. Sci Afr 21:e01821. https://doi.org/10.1016/j.sciaf.2023.e01821
5. Pedota M, Grilli L, Piscitello L (2023) Technology adoption and upskilling in the wake of Industry 4.0. Technol Forecast Soc Change 187:122085. https://doi.org/10.1016/j.techfore.2022.122085
6. Kim S (2023) Innovating knowledge and information for a firm-level automobile demand forecast system: a machine learning perspective. J Innov Knowl 8(2):100355. https://doi.org/10.1016/j.jik.2023.100355
7. Govil M, Proth JM (2002) 2—definition of a supply chain. In: Govil M, Proth JM (ed) Supply chain design and management. Academic Press, San Diego, pp 7–16. https://doi.org/10.1016/B978-012294151-1/50002-3
8. Haman S, Tajmout Y, El Bouzekri El Idrissi Y, El Bhiri B, Moumen A (2023) Adoption of advanced technologies in industrial companies: a bibliometric analysis. In: présenté à ACM international conference proceeding series. https://doi.org/10.1145/3607720.3607794

9. Rikala P, Braun G, Järvinen M, Stahre J, Hämäläinen R (2024) Understanding and measuring skill gaps in Industry 4.0—a review. Technol Forecast Soc Change 201:123206. https://doi.org/10.1016/j.techfore.2024.123206
10. Culot G, Nassimbeni G, Orzes G, Sartor M (2020) Behind the definition of Industry 4.0: analysis and open questions. Int J Prod Econ 226:107617. https://doi.org/10.1016/j.ijpe.2020.107617
11. Rupp M, Schneckenburger M, Merkel M, Börret R, Harrison DK (2021) Industry 4.0: a technological-oriented definition based on bibliometric analysis and literature review. J Open Innov Technol Mark Complex 7(1):68. https://doi.org/10.3390/joitmc7010068
12. Ofosu-Ampong K (2024) Artificial intelligence research: a review on dominant themes, methods, frameworks and future research directions. Telemat Inform Rep 14:100127. https://doi.org/10.1016/j.teler.2024.100127
13. Ouidani RE (2023) Factors influencing the appropriation of artificial intelligence technologies by the banking sector: proposal of a theoretical model. Moroc J Quant Qual Res 5(2), Art. no 2. https://doi.org/10.48379/IMIST.PRSM/mjqr-v5i2.35100
14. Benabdelouahab A, Chaibi A, Haman S, Bentassil Z, Elouadi A, Moumen A (2024) Artificial Intelligence and Industry 4.0 trends: a comprehensive review and case study analysis in the Moroccan context. In: 2024 international conference on intelligent systems and computer vision (ISCV), pp 1-8. https://doi.org/10.1109/ISCV60512.2024.10620117
15. Goodfellow I, Bengio Y, Courville A (2016) Deep learning. MIT Press
16. Alenizi FA, Abbasi S, Hussein Mohammed A, Masoud Rahmani A (2023) The artificial intelligence technologies in Industry 4.0: a taxonomy, approaches, and future directions. Comput Ind Eng 185:109662. https://doi.org/10.1016/j.cie.2023.109662
17. Aravindaraj K, Rajan Chinna P (2022) A systematic literature review of integration of industry 4.0 and warehouse management to achieve Sustainable Development Goals (SDGs). Clean Logist Supply Chain 5:100072. https://doi.org/10.1016/j.clscn.2022.100072
18. El khediri S et al (2024) Integration of artificial intelligence (AI) with sensor networks: trends, challenges, and future directions. J King Saud Univ Comput Inf Sci 36(1):101892. https://doi.org/10.1016/j.jksuci.2023.101892
19. Habbal A, Ali MK, Abuzaraida MA (2024) Artificial Intelligence trust, risk and security management (AI TRiSM): frameworks, applications, challenges and future research directions. Expert Syst Appl 240:122442. https://doi.org/10.1016/j.eswa.2023.122442
20. Cao, Xi, Smith (2003) A reinforcement learning approach to production planning in the fabrication/fulfillment manufacturing process. In: Proceedings of the 2003 winter simulation conference, vol 2, pp 1417–1423. https://doi.org/10.1109/WSC.2003.1261584
21. Rodrigues M, Miguéis V, Freitas S, Machado T (2024) Machine learning models for short-term demand forecasting in food catering services: a solution to reduce food waste. J Clean Prod 435:140265. https://doi.org/10.1016/j.jclepro.2023.140265
22. Steurtewagen B, Van den Poel D (2021) Adding interpretability to predictive maintenance by machine learning on sensor data. Comput Chem Eng 152:107381. https://doi.org/10.1016/j.compchemeng.2021.107381

Contribution of a Web-Based GIS to Groundwater Resource Management: The Sminja–Oued Rmel Aquifer System in the Zaghouan Region, Tunisia

Hela Smiri, Sonia Gannouni, and Noamen Rebai

Abstract

This work consists of the development of a geoportal and a dashboard bringing together quantitative and qualitative indicators on water resources and also hydrological and hydrogeological indicators characterizing the Aquifer System of Sminja–Oued Rmel in the sector of study of Zaghouan. The information aggregated to build these dashboards comes from official sources. The objective of this work is to design a simple system based on free and open-source tools and web mapping which represents a decision support tool to meet the needs of users (experts, researchers, and large public) in monitoring the quality and quantity of water resources in the governorate of Zaghouan. The design of a web GIS must face multiple obstacles, mainly the complexity of the configurations and the high cost of the cartographic servers. After having collected and modulated our database through DBMS (postgreSQL, Postgis). We also used GeoServer, as cartographic server, the HTML, CSS, JavaScript/php programming language for the dynamic display of web pages and maps, and Apache as web server. It should be noted that all these tools are free and downloadable on the internet. This prototype is mainly used to illustrate both the issue of accessibility and reuse of public data in Zaghouan but also to demonstrate the contribution of GIS tools, Open Source.

Keywords

Web mapping · Open source · Aquifer system · Web GIS · OGC · Zaghouan

1 Introduction

In the current global context, due to increasing rapid urbanization and climate change, water resources management has become a primary concern. Tunisia, like many other countries in North Africa, is faced with these problems. The country faces specific geoclimatic challenges, particularly erratic rainfall in arid to semi-arid regions [1]. These conditions induce significant water stress, recalling the crucial need for integrated and informed management of water resources. The continued exploitation of water resources increases with demographic growth and rapid urbanization (domestic and industrial) [2]. In addition, irrigated agriculture also consumes a significant part of water, this has an impact on its quality and quantity [3]. In addition, pollution of groundwater and surface water further aggravates the situation. Water availability is also affected by changes in precipitation, making the situation critical [4]. The amount of fresh water per capita in Tunisia fell from 3650 m3 per year in 2002 to just 2670 m3 per year in 2017.

Tunisia is thus experiencing a state of imbalance between the need for water and production and to remedy this state, thus water resources are both rare and unevenly distributed in time and space.

It adopted a strategy for the development of water resources including one of its important axes is the preservation of

H. Smiri (✉)
Faculty of Sciences from Bizerte, Carthage University, Carthage, Tunisia
e-mail: smirihala112@gmail.com

S. Gannouni
The Georesources Laboratory, Water Research and Technologies Center (CERTE), Technopark of Borj Cedria, Soliman, Tunisia
e-mail: gannounisonia2017@gmail.com

N. Rebai
University of Tunis El Manar, National School of Engineering of Tunis, Geotechnical Engineering and Georisk Research Laboratory (LR14ES03), Tunis, Tunisia
e-mail: noamen.rebai@enit.utm.tn

A. Moumen et al. (eds.), *Artificial Intelligence and Data Analytics for Innovative Applications in Engineering, Sustainability and Technology*, Sustainable Artificial Intelligence-Powered Applications, https://doi.org/10.1007/978-3-031-90318-2_14

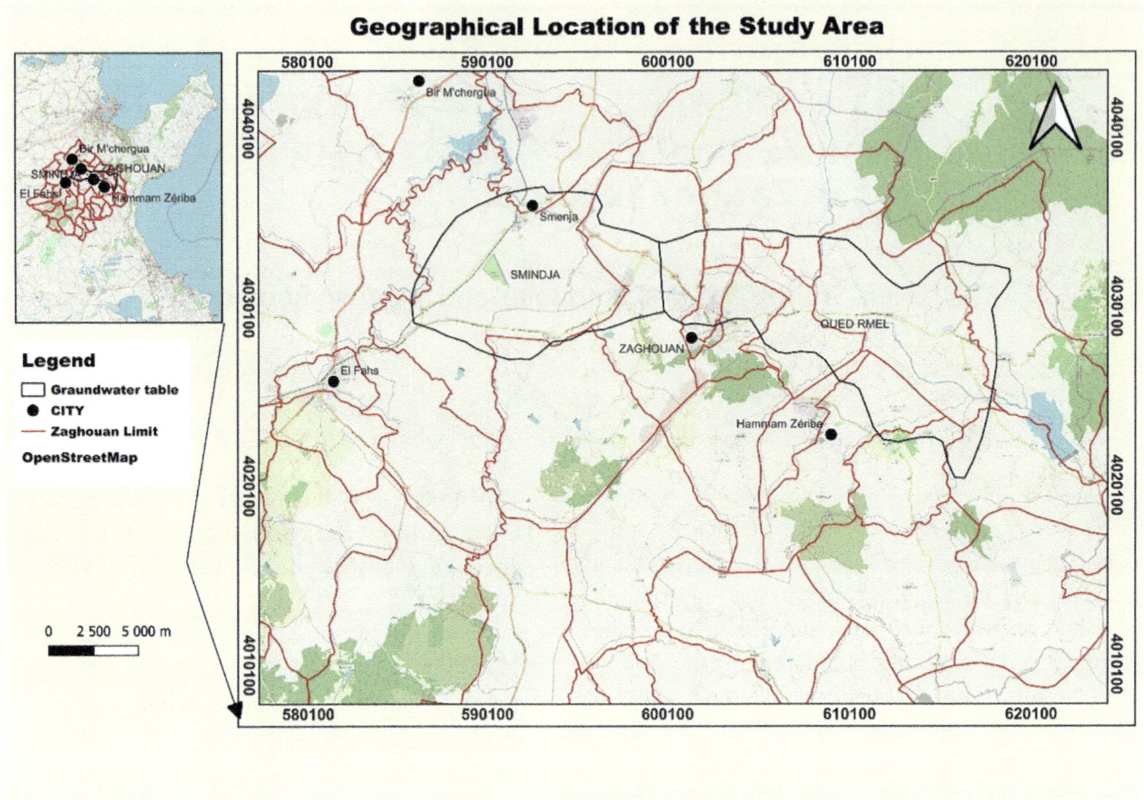

Fig. 1 Zaghouan study area. *Source* open streetMap

groundwater through rational exploitation. However, the success of such a strategy requires a good understanding of deep structuring and its effect on the characteristics of aquifers aimed at ensure sustainable and efficient management of water resources [5].

In this context, this work concerns the Sminja–Oued Rmel Aquifer systems, which is located in the Zaghouan Region (Fig. 1); climate is arid to semi-arid [4]. The water resources of the Aquifer system of the Sminja aquifer and the Oued Rmel aquifer are under the influence of overexploitation [6] and the effect of intense agricultural activity, which can be the cause of the deterioration of water quality [7].

Currently, the use of new technologies such as the geographic information system and web mapping which provide effective and more advanced tools to facilitate the management of water resources [8–10], to better evaluate and understanding the hydrological and hydrogeological behavior of the Aquifer system which is becoming a primordial necessity [11], and improving access to water data, which is essential for an effective management of water resources [12] in the Zaghouan region.

GIS-Web platforms have been used by many countries for water resources management. For example, in the United States, the Groundwater Information System (GWSI) is an online GIS platform that allows easy access to groundwater data [13]. The National Water Portal (NWP) in Australia provides real-time information on water quality, availability, and drought forecasts [14]. In India, the Water Resources Information System (WRIS) is a GIS-Web platform that offers data on water availability across the country [7]. These examples show the importance of platforms GIS-Web in global water resources management. Additionally, these technologies are used in other fields, such as forest management [15], coastal zone management [16], and disaster management natural [17]. They make it possible to collect data on the spatial distribution of natural resources, analyze trends and changes, and facilitate decisions for more efficient and sustainable management of resources [18]. In this context, emerging technologies play a vital role in improving the ability to anticipate outcomes. A recent study [19] demonstrates how integrating GIS with machine learning algorithms, such as support vector machines (SVM) and neural networks, can significantly improve early flood detection in Tunisia. This method achieves an accuracy rate of approximately 93.1 percent by analyzing aerial images of risk areas. In conclusion, the management of water resources in Tunisia constitutes a significant challenge which requires an approach to the integration of technological tools such as GIS and web mapping. To ensure sustainable management

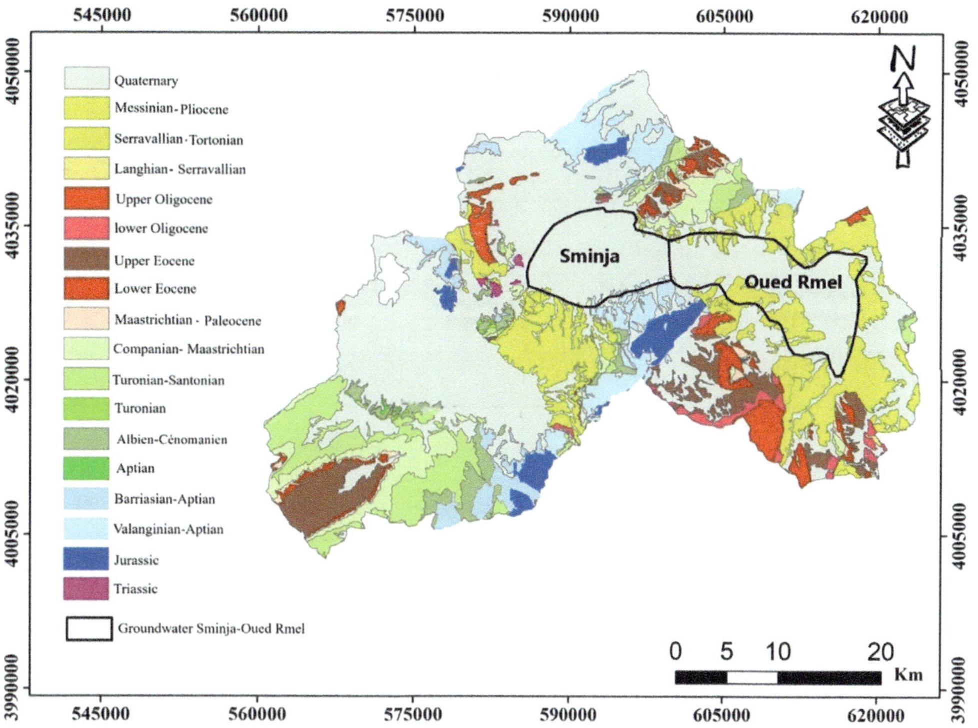

Fig. 2 Geological map of the study area (extract from the geological map of Tunisia, scale 1/200000, National Office of Mines Tunisia, Personal Development)

of water resources and preserve this essential resource for future generations, the Tunisian authorities must continue their efforts.

2 Materials and Methods

2.1 Study Area

Zaghouan region is located in the northeastern part of the country and covers an area of 2820 km^2, or 1.7% of the country's surface area. This region is located between Altitude 36°24'North and Longitude 10° 09′ East. It opens onto the governorates of Ben Arous and Manouba in the north, the governorates of Béja and Siliana to the west, the governorates of Nabeul and Sousse to the east, and the governorate of Kairouan to the south, characterized by Mediterranean continental climate. The Mograne rainfall station made it possible to monitor annual precipitation for a period from 2000 to 2018. They show great variability. Indeed, the rainfall was only 220 mm in 2008, while it reached 823.3 mm in 2011 [17].

Based on data acquired at the Mograne station for the period extending from 1991 to 2020, the region is experiencing significant seasonal variations in average temperature are of the order of 25 °C in summer (from June to September) and between 10 °C and 16 °C for winter (December, January, and February) [17, 18].

In the study area, the aquifer systems extend from the Pliocene to the Quaternary (Fig. 2), including the aquifers of Oued Rmel and Sminja, characterized by significant hydraulic exchanges and varied geological formations [19]. The Oued Rmel aquifer is located within a graben structure and is distinguished by extensive outcrops of Quaternary, Pliocene, and Miocene formations. In contrast, the Sminja aquifer system is situated in a tectonic basin adjacent to the anticlinal dome of Djebel Ouest. The underlying bedrock of this aquifer system is thought to consist of upper Eocene clays; however, drilling operations have yet to penetrate beyond the Plio–Miocene formations (Fig. 3) [19]. The aquifers in the study area are multi-layered systems, characterized by significant hydraulic exchanges between the different layers [19].

3 Methodology

The main objective of this work is to establish a client–server Geographic Information System (GIS) that incorporates dynamic mapping to facilitate the management of water resources in the study area. This approach aims to address specific objectives and provide insights into the challenges posed by enhancing access to and visualization of geographical data. Moreover, it enables more efficient management of water resources within the study area. The methodology adopted for this purpose is outlined below.

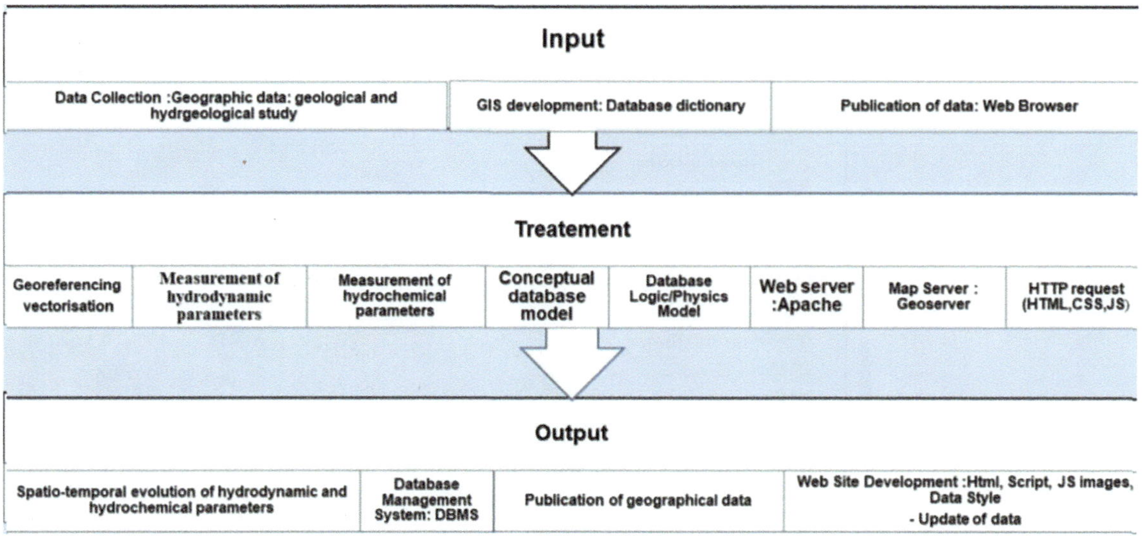

Fig. 3 Methodological work flowchart developed

3.1 Data Collection

The materials and data utilized in the web GIS for water resource management include spatial data, renderings, agricultural, and geological maps, as well as alphanumeric information related to surface and groundwater, topography, and land use.

We also used GeoServer, as a cartographic server, the HTML programming language, CSS, JavaScript/php for the development and dynamic display of web pages and maps, and Apache as web server.

3.2 Design of Spatial Databases

By definition, the implementation of a web GIS is a "computer system of hardware, software, and processes designed to enable collection, management, manipulation, analysis, modeling, and the display of spatially referenced data in order to solve complex management problems [20]."

It therefore relies on a certain number of geographic databases, which it allows to integrate, manage, process, and represent in the form of maps for a strong correlation between the accuracy of geographic data and the success of programs water resources management [21]. In order to analyze the existing selection and the actors and respond to the need, datasets on these main Aquifer systems were tested by the IWRM tool to know the capabilities of the software. Our goal is to develop a database to store all the information regarding our study. To begin, it is essential to go through a database modeling step. Its objective is to correctly describe the architecture of the database.

3.3 Database Design

- Identification of Various Stakeholders

The main users are Internet users who are among the final beneficiaries of this project. This is for the general public, its function is limited to consulting data and locating hydrological and hydrogeological data for the understanding of water resources in this region.

- Identification of Use Cases

Use cases make it possible to express the needs of users of a system, they are therefore a user-oriented vision of this need unlike a computer vision [22], associated by scenarios which explain how users interact with the system to accomplish particular tasks, such as tracking the qualitative and quantitative state of water, in the context of a web GIS for water resources management [23] (Tables 1 and 2).

Use Case Descriptions: Scenarios

4 Results

4.1 Website Concept and Interface

The platform developed is mainly focused on the management of water resources using the Dynamics Platform. The homepage of the IWRM platform is designed to highlight user security and ease of use, emphasizing a simple registration process. Users provide essential details such as a

Table 1 Various use cases for different system stakeholders

Actors	Use cases
User	• Manage aquifer systems: – View geological and stratigraphic overview – View hydrographic network – View aquifer – View characteristics of water points and their inventories – View climate overview of the region – View maps and dynamic map – Monitor water quality and environmental impact – Develop policies and regulations related to water use – Research and analyze water resource data • Note: The user can view and manage various information about aquifer systems, water quality, and more
Administrator	• Manage access rights to information • Manage user accounts • User training and awareness on system usage • Technical maintenance of the system • Technical support for users • Analysis of water resources data needs • Report and analysis development based on data

Table 2 Description of use cases in the system

Use case	Purpose	Actor		Precondition	Scenario
View maps	View maps	User		The actor is authenticated	1. The system displays the map 2. The user can freely navigate on the map
Search	Perform search	User		The actor is authenticated	1. The user clicks on the search button 2. The system displays the search window 3. The user fills in the data and launches the search 4. Multiple search results are displayed
Consult	Add, modify, delete user	User trator)	(Adminis	The actor is authenticated	1. User adds, modifies, or deletes a user
View water resources history	View water resources history for Zaghouan region	User		The actor is authenticated	1. The user selects the Zaghouan region 2. The system displays the water resources history for the region
View water points characteristics and inventories	View water points characteristics and inventories	User		The actor is authenticated	1. The system displays the characteristics and inventories of water points
View dynamic map	View dynamic map	User		The actor is authenticated	1. The system displays the dynamic map
Water quality monitoring	Monitor water quality and environmental impacts	User		The actor is authenticated	1. The user selects the water quality monitoring feature 2. The system displays water quality and environmental data
Update data	Update data in the system	User trator)	(Adminis	The actor is authenticated	1. The user selects the update data feature 2. The system displays options for updating 3. User updates data in the system

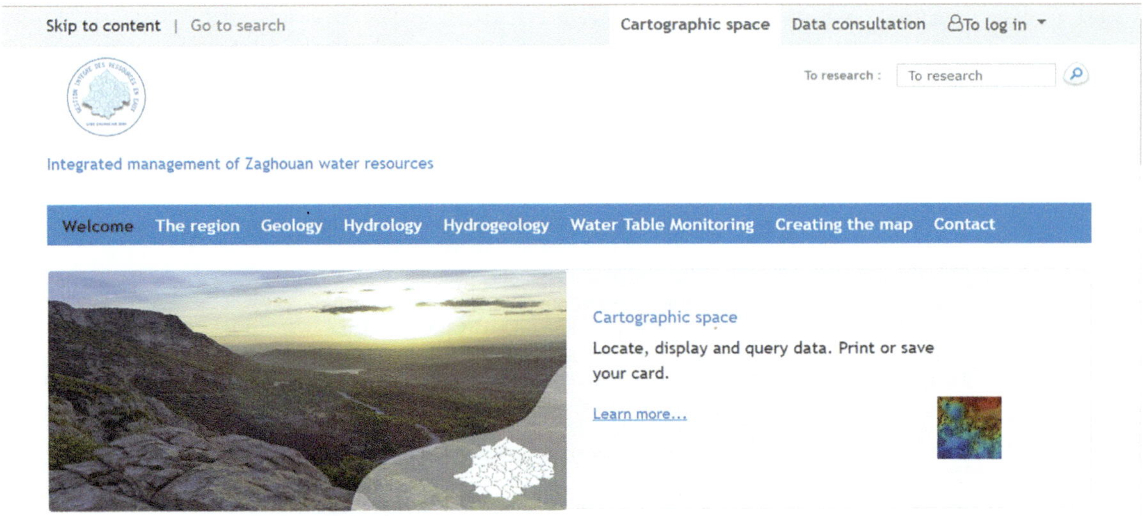

Fig. 4 Home page

Fig. 5 Extract from the source code of the main interface of the "IWRM" website. *Source* Screenshot derived from my own work

```
<divid="sub-nav">
          <ulclass="tabbed">
            <liclass="current-
tab"><ahref="index.html">Accueil</a></li>
            <li><ahref="La-region.html">La
région</a></li>
            <li><ahref="Geologie.html">Géologie</a></li>

<li><ahref="Hydrologie.html">Hydrologie</a></li>

<li><ahref="Hydrogéologie.html">Hydrogéologie</a></li>
            <li><ahref="Surveillance-des-
nappes.html">Surveillance des Nappes </a></li>
            <li><ahref="Carte.html">Création de la
carte</a></li>
            <li><ahref="Contact.html">Contact</a></li>
```

username, email, and a secure password. To personalize their experience, additional information such as the user's role is collected. After successful registration, a robust login system, including encryption and possibly multi-factor authentication, ensures secure access. User roles, like that of an expert, define specific features, thereby enhancing the overall user experience (Fig. 6).

The platform is structured around seven major components, starting with the homepage serving as the entry point. It provides an overview of the platform's diverse functionalities, preparing users to explore key sections such as "The Region," "Geology," "Hydrology," "Hydrogeology," "Monitoring of the Water Table," and finally, a page dedicated to map creation (Figs. 4 and 5).

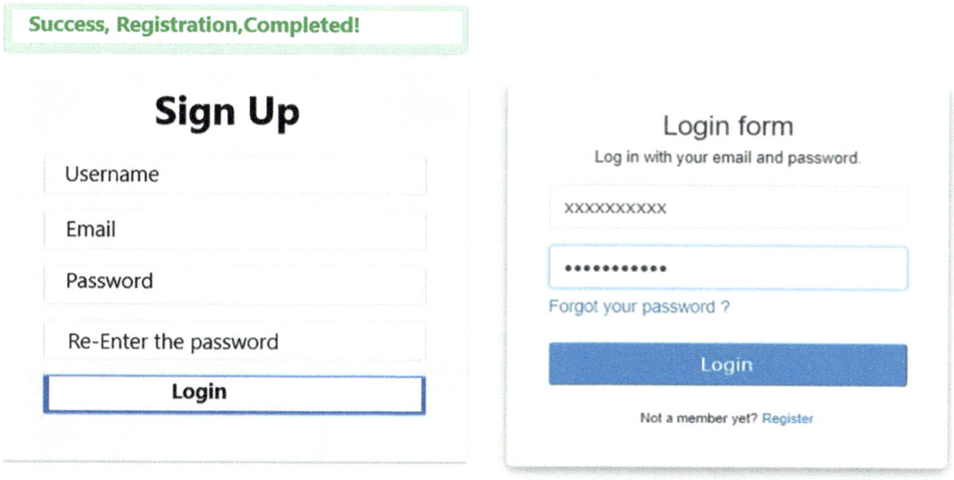

Fig. 6 Registration form. *Source* The Integrated Water Resources Management (IWRM) website derived from my own work

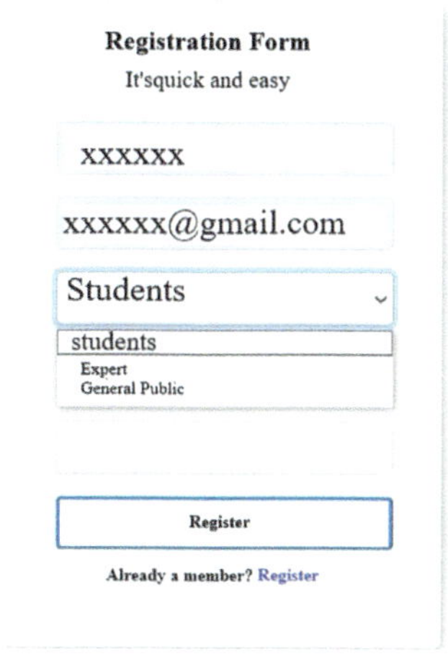

Fig. 7 Registration form by selection. *Source* The Integrated Water Resources Management (IWRM) website derived from my own work

Firstly, other users (the general public, researchers, and experts) need to register. Afterward, they can log in as members. The figures below illustrate the steps to follow (Fig. 6).

- Selection-Based Registration Form

The selection-based registration form displays information according to the user's choice. This functionality in our system enables a visitor to search for information based on their specific preferences or criteria (Fig. 7).

4.2 Development of the Mapping Interface

Interface is aimed at providing access to all information related to the published regional data. This interface allows the display of georeferenced objects for selecting entities based on combined geographical criteria. With this tool, users can navigate within the specified area and query the objects that make it up. In accordance with the specifications, it is presented as follows (Fig. 8).

The principle of the dynamic map relies on the ability to display geographic data interactively and in real-time, allowing users to explore, analyze, and understand changing spatial phenomena. The central idea is to provide a continually evolving visual representation, offering updated information on various subjects such as weather, traffic, or other real-time data.

In the context of the Integrated Water Resources Management (IWRM) website, the integration of an interactive map can be crucial for visualizing and analyzing hydrological data in real-time (Fig. 9). This approach would enable users to explore the dynamics of water resources, track climate changes, and obtain updated information on resource availability, thereby enhancing decision-making within the framework of integrated water resources management (Figs. 10 and 11).

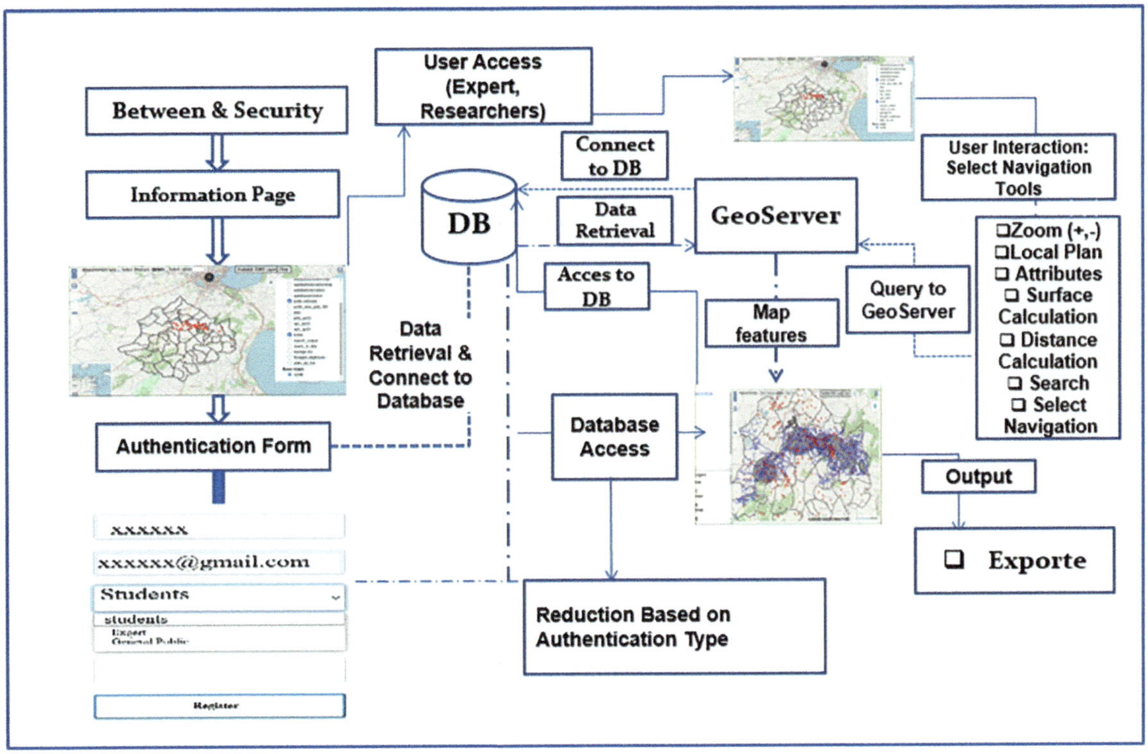

Fig. 8 Development organizational chart for the map interface

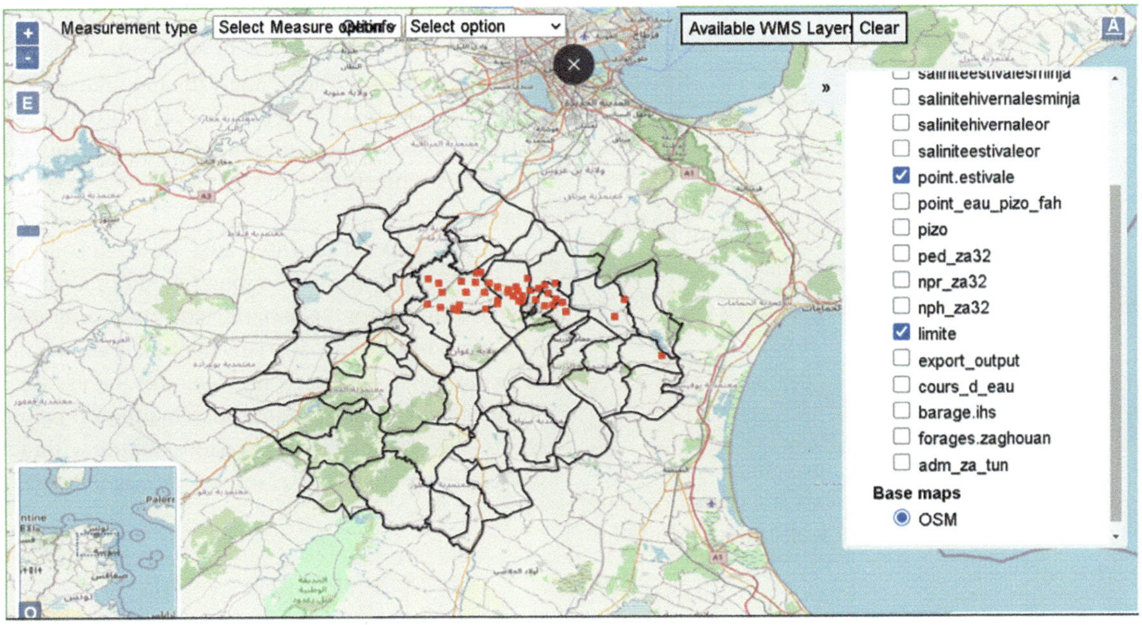

Fig. 9 Dynamic map page. *Source* The Integrated Water Resources Management (IWRM) website derived from my own work

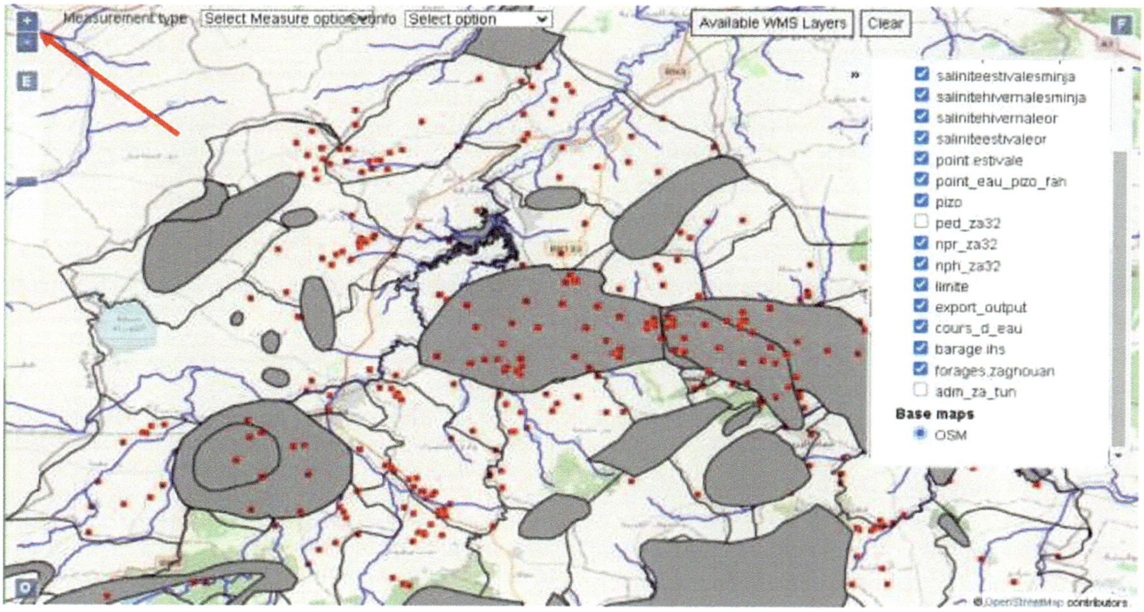

Fig. 10 The use of layers. *Source* The Integrated Water Resources Management (IWRM) website

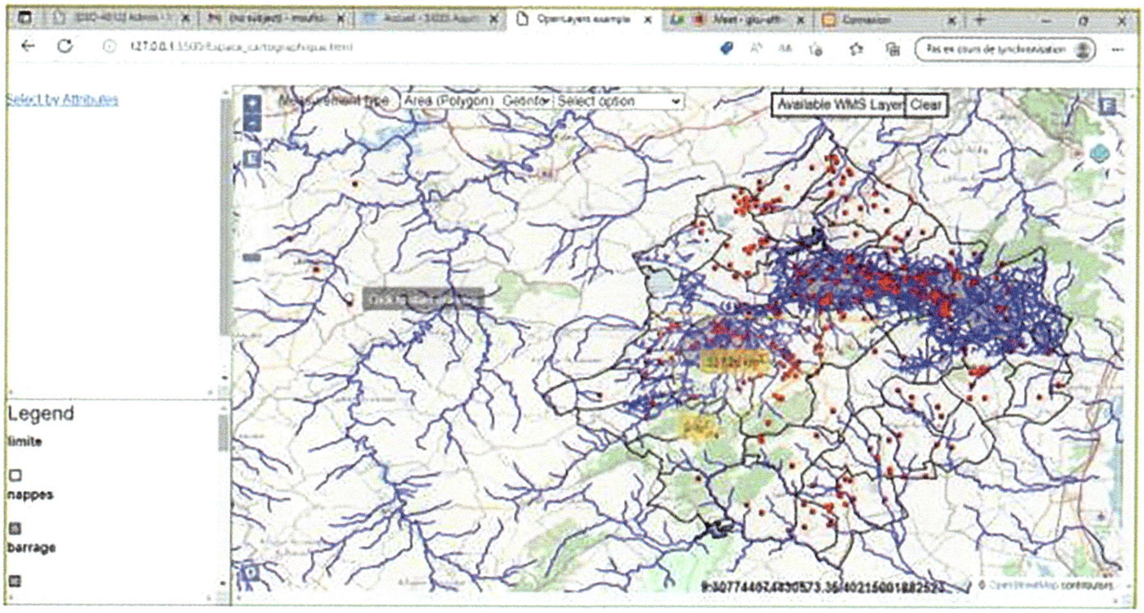

Fig. 11 Attributes of geovisualization. *Source* The Integrated Water Resources Management (IWRM) website derived from my own work

5 Conclusion

The climatic and geological characteristics of the Zaghouan region negatively impact the water balance. This deficiency is primarily due to three reasons: the decrease in water supply, overexploitation of aquifers, and high water demand, especially in the agricultural sector. Consequently, we have developed a methodology to establish a dynamic mapping dashboard accessible via the internet, addressing the water resource needs of Zaghouan. The objective of this work is to design a simple system based on free and open-source tools. Additionally, the web mapping will serve as a decision support tool to meet the needs of users (experts, researchers, and the general public) and monitor the quality and quantity of Zaghouan's water resources.

The application we developed offers several significant advantages to users, especially those involved in water resource management in Zaghouan. Firstly, it provides an interactive and real-time platform for experts, researchers, and the general public to visualize and analyze data related to the quality and quantity of water resources in the region. This facilitates a better understanding of current trends, variations, and water needs.

Regarding the optimization of the infrastructures, several technical considerations were taken into account. The compatibility of tools was carefully assessed to ensure smooth integration between different components of the system. We also used GeoServer, as a cartographic server, the HTML programming language, CSS, JavaScript/php for the development and dynamic display of web pages and maps, and Apache as web server. It should be noted that all these tools are free and downloadable from the internet.

The results of this work are an online information system with an ergonomic interface, allowing the management of information developed by several water resource managers. This will consist of analyzing the needs of users, providing a solution by offering suitable tools. More concretely, we have built a database which contains information concerning the quality and quantity of groundwater and their exploitation.

References

1. FAO (2021) Tunisia: country water resources assistance strategy
2. Abida H, Aloui F, Boudabous A (2020) Assessment of water resources in Tunisia: a review. Heliyon 6(1):e03134
3. Amdouni Z, Agoubi B, Oueslati W (2019) Assessment of groundwater quality in agricultural areas of Tunisia using multivariate statistical analysis. Environ Monit Assess 191(6):356
4. CRDA (2014) Étude de caractérisation de l'aquifère de Sminja-Oued Rmel. Rapport final.Commissariat régional au développement agricole de Zaghouan
5. Moumen A, Mansouri R (2012) Application des systèmes d'information géographiquepour la gestion de l'eau en Tunisie. Revue de Géographie et d'Aménagement 5:13–24
6. DGRE. Bilans des eaux souterraines des nappes aquifères en Tunisie. Directiongénérale des ressources en eau, Ministère de l'Agriculture, des Ressources Hydrauliques et de la Pêche (1985–2020)
7. CGWB (2021) Water resources information system. Retrieved from http://wrisonline.cgwb.gov.in/
8. Open Source Geospatial Foundation (2022) GRASS GIS. Retrieved from https://grass.osgeo.org/
9. Peña-Haro S, Martínez-Carreras N, Pulido-Velazquez M, Andreu J (2022) Advances in hydrological data management: the role of geographic information systems and web mapping. Water 14(6):1506
10. Peña-Haro S, García-Bartual R, García-García JA, Estrela T (2022) Web GIS application for supporting integrated water resources management in semiarid regions. Environ Model Softw 142:105100
11. MAPR (2019) Stratégie nationale de l'eau. Retrieved from https://www.eaupotable.tn/images/strategieeau.pdf
12. Faunt CC, Sneed M, Traum J, Brandt JT, Chartier K (2016) Ground water availability of the United States: U.S. Geological Survey Professional Paper 1824. U.S. Geological Survey
13. CSIRO (2016) Australia's National Water Portal. Retrieved from https://www.csiro.au/en/research/natural-resources/water/Water-portal
14. Zipf A, Hahmann S (2014) Towards intrinsic accessibility—analysing the relation between urban form and existential accessibility. Procedia Soc Behav Sci 74:197–206
15. Sui D, Goodchild M (2011) The convergence of GIS and social media: challenges for GIScience. Int J Geogr Inf Sci 25(11):17371748
16. Elwood S, Goodchild MF, Sui D (2013) Researching volunteered geographic information: spatial data, geographic research, and new social practice. Ann Assoc Am Geogr 102(3):571–590
17. Chrigui M (2019) Groundwater quality degradation in Tunisia: a review of the current situation and future challenges. Environ Sci Pollut Res 26(28):28653–28666
18. Satauri I, Ohamouddou M, Hajji S, Zouiten M, Chaouan J (2023) New GIS approach using machine learning algorithm for early floods detection. Moroc J Quant Qual Res 5(1)
19. Bajanik A, Zouari K, Ayadi A (1977) Étude hydrogéologique du système aquifèreplio-quaternaire de la région de Zaghouan. Office National des Mines, Tunisie
20. Kheirallah AM, Omar MF, Khedr MA (2021) The role of GIS in water resources management: a review of the Egyptian experience. Arab J Geosci 14(6):1–17
21. Thapa R, Murthy MSR, Babel MS (2019) Challenges and opportunities of using geospatial technology for water resources management in developing countries: a review. Environ Sci Pollut Res 26(25):25473–25487
22. Jin S, Yang X, Chen B (2017) Design and implementation of web-based GIS for water resource management. J Hydrol 552:226–237. https://doi.org/10.1016/j.jhydrol.2017.06.05
23. CGWB (2021) Water resources information system. Retrieved from http://www.cgwb.gov.in/Wris.aspx
24. Gharbi A, Hajji F, Kharroubi A, Chkir N (2020) Water scarcity and its management in Tunisia. In: Handbook of climate change resilience, pp 255–275
25. ANPE (2021) National agency for the protection of the environment: water resources. Retrieved from https://www.anpe.nat.tn/eau/
26. LNCS Homepage. http://www.springer.com/lncs. Accessed 25 Oct 2023

Addressing the Learning Gap with Adaptive Learning

Soukaina Hakkal and Ayoub Ait Lahcen

Abstract

As a consequence of ChatGPT's emergence, the education field faces a major challenge. The need for a flexible and adaptive educational system is increasingly pressing. While education remains to be a place of innovation and progress, artificial intelligence and big data have emerged as two pivotal technological advancements. Through the application of artificial intelligence and big data analytics, which model patterns of our intelligence as humans to infer, adjudicate, or predict, Intelligent Tutoring Systems can analyze student performance data. This scrutiny provides an adaptive learning path for each learner based on their individual skills and abilities. This not only enhances learner performance and engagement but also contributes to bridging the learning gap for students. The aim is to guarantee equitable access to all the personalized resources for mastering their learning journey. Real-time adaptation is facilitated through student modeling where inferences are drowned from the learner's actions during a learning process. This paper discusses the concept of adaptive learning and its transformative potential in the field of education. It also underscores the significance of AI and big data in facilitating adaptive learning and addressing the achievement gap.

Keywords

Adaptive learning · AI · Big data · Learning gap

S. Hakkal (✉) · A. A. Lahcen
Engineering Sciences Laboratory, National School of Applied Sciences, Ibn Tofail University, Kenitra, Morocco
e-mail: soukaina.hakkal@uit.ac.ma

A. A. Lahcen
e-mail: ayoub.aitlahcen@uit.ac.ma

1 Introduction

Adaptive learning, tailored to each individual learner's journey, has revolutionized education by delivering large-scale learning personalization. This approach enables learners to advance at their rate and concentrate on areas where they needed the most support. By creating more engaging learning experiences through increased interactivity and personalized content, learning achievement can be improved. By dynamically adjusting the content, the learning method, and the course pacing to suit the unique require of each learner, adaptive learning ensures satisfying learning experience. Leveraging AI and big data technologies, adaptive learning enhances the learning process.

Moreover, adaptive learning offers remarkable flexibility and accessibility for learners. It allows them to access learning resources anytime, anywhere, tailored to their schedules and learning priorities. Furthermore, adaptive learning can assist teachers in optimizing their time by providing crucial insights into learner's needs. This empowers teachers to focus on their interventions when they are most needed.

Additionally, adaptive learning can contribute to a reduction in learning costs by efficiently using resources and decreasing reliance on traditional teaching materials such as textbooks and paper-based course materials.

This enables the provision of a learning experience that is not only adapted to the learner's level but also enriching and equitable.

Each learner possesses a unique set of characteristics, including their natural pace and learning abilities. This diversity underscores the limitations of standardized, uniform, and uniform learning approaches, which may fail to cater to the varied learner's needs. As a result, the quest for a more equitable and effective learning experience has become paramount.

A. Moumen et al. (eds.), *Artificial Intelligence and Data Analytics for Innovative Applications in Engineering, Sustainability and Technology*, Sustainable Artificial Intelligence-Powered Applications, https://doi.org/10.1007/978-3-031-90318-2_15

The education field stands as a vast source of data, known as the educational data explosion [1]. Academic institutions, including universities, colleges, and schools, amass large amounts of data on both learners and educators, through assessments and performance records. This data presents an opportunity for exploration and analysis, facilitated by AI and big data technologies.

The incorporation of big data into the e-learning models [2] holds promise for refining pedagogical methods and informing strategic educational decisions. This ensures learner satisfaction by aligning learning content with their skills needs and addressing their challenges [3]. Online assessment, an essential element of e-learning, plays a crucial role in driving personalized learning paths (PLP) within e-learning systems. These PLPs are customized to each learner, accommodating their unique learning trajectories and requirements [4].

The implementation of artificial intelligence and big data in education encompasses several key areas, including learning analytics (LA), educational data mining (EDM), multimodal learning analytics (MMLA), and artificial intelligence in education (AIED).

These domains [5] have been used to explore and construct models covering several issues within e-learning systems. One such area of focus is user modeling and profiling, which both suggest real-time adaptations [6].

To avoid all the challenges facing the education sector, such as student disengagement and dropout rates, the incorporation of big data and artificial intelligence is indispensable. These technologies have demonstrated significant results, especially in terms of adapting learning content to cater to individual learner characteristics and independent learning [7].

Adaptive learning, fueled by the fusion of AI and big data, has brought about profound transformations in education. This review contributes significantly to the current research landscape by synthesizing various studies on adaptive learning systems, integrating insights from educational data mining, learning analytics, and artificial intelligence. It addresses existing gaps by providing a comprehensive analysis of how big data and AI can be combined to enhance personalized learning experiences, a perspective often overlooked in prior research. By highlighting practical strategies for implementation and discussing the challenges ahead, this review offers valuable insights that can inform both educators and researchers, ultimately advancing the understanding and effectiveness of adaptive learning in educational contexts. As far as we know, this study is the first comprehensive endeavor to cover the various perspectives, methodologies, and applications of adaptive learning.

This paper focuses on adaptive learning systems, beginning with an overview that highlights their significance and evolution in education. It examines the architecture of these systems and how big data and AI enhance their effectiveness. A review of existing adaptive learning platforms is provided, followed by the key features analysis. Finally, the paper discusses the challenges faced in implementing adaptive learning and outlines future directions for its development.

2 Overview of Adaptive Learning Systems

2.1 Adaptive Learning in Education

In the literature, the term "adaptive" typically tends to be used to refer to the ability to change when necessary to address different situations [8]. Brusilovsky and Maybury [9] initially conceptualized the adaptive web as the ability to deliver personalized applications and services to individuals in the knowledge society [10]. Within the learning context, an adaptive learning system (ALS) [11], gathers data on students' interaction with the system, builds a learner model [12], and subsequently utilizes it to tailor the presentation of course material [10].

Adaptive learning, therefore, defined as the process of creating a model based on learner's goals, knowledge, and preferences and continuously employing it throughout the learner's experience to deliver personalized feedback or adjust content and the interface to suit the learner's educational needs.

There are two types of adaptive learning technology, designed adaptivity and algorithmic adaptivity [13]. Designed adaptivity entails computer-based learning that provides students with the ability to choose when and how they learn content, as well as which learning style is the most comfortable for them. On the other hand, algorithmic adaptivity involves the automatic recognition of a user's particular needs and behaviors and adapting to them seamlessly.

Traditional learning [14] involves teachers delivering uniform, fixed, and standardized courses, and all learners are expected to follow the same trajectory and fixed course content. This one-size-fits-all approach is used in traditional learning systems. This approach necessitates a considerable amount of teacher time to develop and adjust course content to suit each student's learning level. Furthermore, it fails to account the human forgetfulness, which some students may suffer during homework or assignments, resulting in falling behind and potential disengagement.

In contrast, adaptive learning [15] has emerged as a critical educational paradigm in the twenty-first century, employing the power of AI and big data to introduce flexibility into education through innovative pedagogical methods. It also demonstrates the ability to expedite learning compared to traditional methods (see Fig. 1).

The data collected on each student can offer valuable insights into their performance, aiding teachers in determining

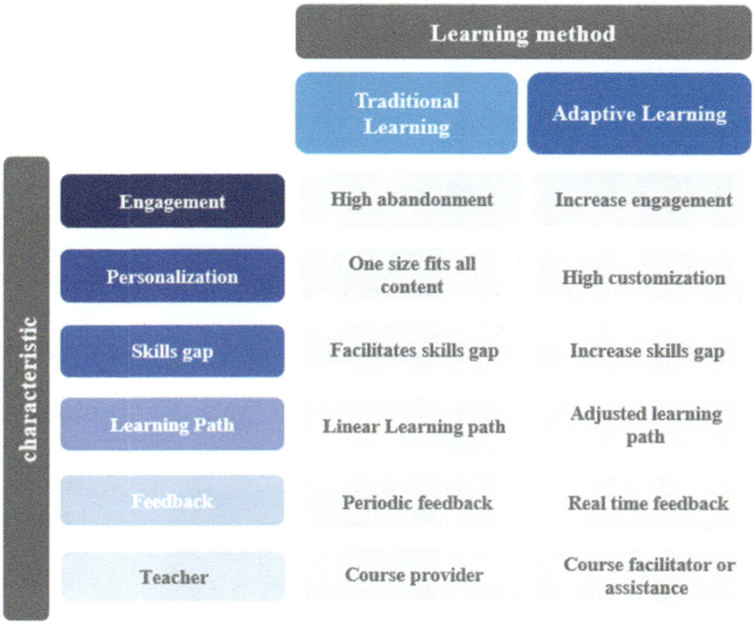

Fig. 1 Main differences between traditional and adaptive learnings

how and what to teach. Such methodologies promote student productivity and motivation, leading to increased engagement. Consequently, adaptive learning, supplemented with personalized teacher assistance, represents an effective way of improving the learning experiences for all learners.

2.2 Types and Development of Adaptive Learning Systems

Adaptive learning systems exist in several types, each attracting considerable the attention from researchers. These types, as identified in the literature, include macro-adaptive system, Adaptive Educational Hypermedia System, micro-adaptive system, Intelligent Tutoring System, aptitude-treatment interaction system, and adaptive learning platform [16] (see Fig. 2).

The macro-adaptive system is based on a macro-adaptive approach dating back to the 1900s, accommodating adaptation based on general objectives, general abilities, and student levels [17]. Learners are grouped according to their aptitude test scores to receive personalized learning treatment tailored to their learning pace [18].

The aptitude-treatment interaction system, as defined by Cronbach, focuses on individual characteristics (aptitude) that influence a learner's likelihood of success with a particular treatment. The "treatment" here refers to variation in the pacing or style of instruction [19]. Interaction systems for aptitude processing are designed to identify learners' aptitudes and adjust the learning process accordingly [18].

The micro-adaptive system is based on a micro-adaptive instructional approach [20], this system choose the most suitable learning material following the learner's dynamic quantitative abilities, prior knowledge, and motivation. It relies on measuring performance during the task [18].

Intelligent Tutoring System represents a hybrid system combining micro-adaptive and aptitude-treatment interactions developed by artificial intelligence [21]. Intelligent Tutoring System monitors learners' psychological (learner modeling) to respond adaptively respond to these states in a flexible manner [20]. Intelligent Tutoring Systems are usually used to provide personalized instruction in specific domains such as mathematics or engineering, leveraging data mining or cognitive modeling techniques.

Adaptive Educational Hypermedia System (AEHS) results from the fusion of adaptive pedagogical systems and hypermedia systems [22], using artificial intelligence and integrated user models to cater to learner's state of knowledge [18, 21]. AEHS systems offer customized educational experiences by characterizing each learner's needs, thereby enhancing user satisfaction and minimizing time and costs [23].

Adaptive learning platform (ALP) utilizes an artificial intelligent algorithm to offer customized learning experiences by tracking learning achievement and adapting content to individual learner needs and levels. Suitable for various educational backgrounds, from elementary to higher education, ALP platforms offer tailored learning experiences across diverse subjects and domains.

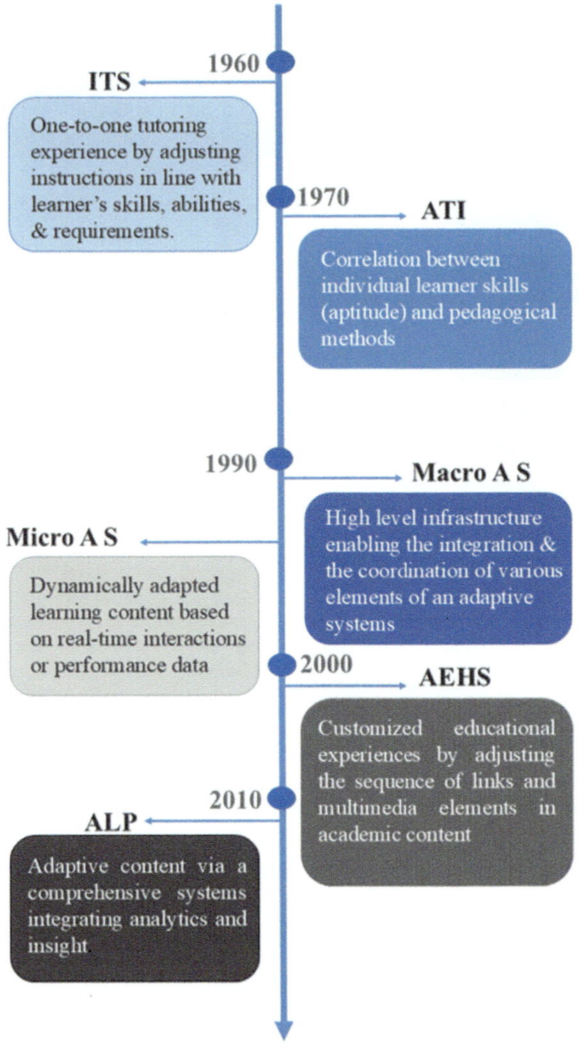

Fig. 2 Adaptive learning systems evolution

2.3 Architecture of Adaptive Learning System

The adaptive learning system architecture comprises six interconnected components, which are fundamental to its functionality. These components include the domain model, pedagogical model, adaptive engine, analytic engine, learner model, and learner interface (see Fig. 3). The domain model designs the skills and competencies that the system aims to impart to the learner [24]. It includes concepts, relevant properties, requirements, and objectives within a particular domain, serving as a cornerstone for storing and structuring learning content. This model guides the selection and provision of appropriate content to the learner [25]. The learner model reflects of the progressive state of the learner's knowledge and abilities. It stores pertinent information concerning each learner, such as their learning style, preferences, and disabilities facilitating personalized learning experiences [26].

The pedagogical model embodies a comprehensive framework of educational principles and methodologies. Within an adaptive learning system, the pedagogical model includes information about the learner's performance and progress to provide it to the adaptive engine for personalization [27]. The adaptive engine combines data from the pedagogical model, learner model, and domain model from which personalized learning recommendations are generated, including tailored activities and resources [28]. The analytic engine is the component responsible for collecting and analyzing pertaining data to learner performance, engagement behavior, and other relevant parameters. This data is used to provide continuous feedback, thereby enhancing the system's ability to personalize learning for each individual learner. Finally, learner interface serves as the primary point of interaction between the system and learners. It can involve features such as personalized dashboards, tutorials, quizzes, exercises, progress tracking, and feedback mechanisms [28].

2.4 Implementation of Big Data in Adaptive Learning

According Russom definition, "big data analytics is the application of the advanced techniques used on big data" [29]. On the other hand, Brynjolfsson and McAfee declared that "technology such as the analysis of big data, high-speed communications and rapid prototyping have augmented the value of abstract and data-driven reasoning, thereby increasing the value of individuals with the right skills in creativity, engineering, or design" [30].

In the realm of educational, big data analytics is used to personalize learning experiences for individual students. This is achieved by understanding the educational journey of learners and synthesizing learning patterns to develop customized learning programs based on their knowledge needs and existing skills [31]. Therefor big data in education serves as a catalyst for effective understanding and enhanced student performance.

Learning analytics (LA) defined as the analysis and visualization of learner data, employing big data techniques to improve learning. It is an approach that enables teachers to gain insights into education by leveraging increased amounts of learner data and employing management approaches focused on quantitative measures [32].

The use of big data analytics in adaptive learning is exemplified by the work of [33], where they proposed an adaptive learning service recommendation algorithm which based on big data, boasting higher coverage, accuracy, strong reliability, and recall rate. Meanwhile [34] focused on the social context, particularly the interaction between learning objects and learners to enhance personalized and effective e-learning.

Fig. 3 Simplified adaptive learning system architecture

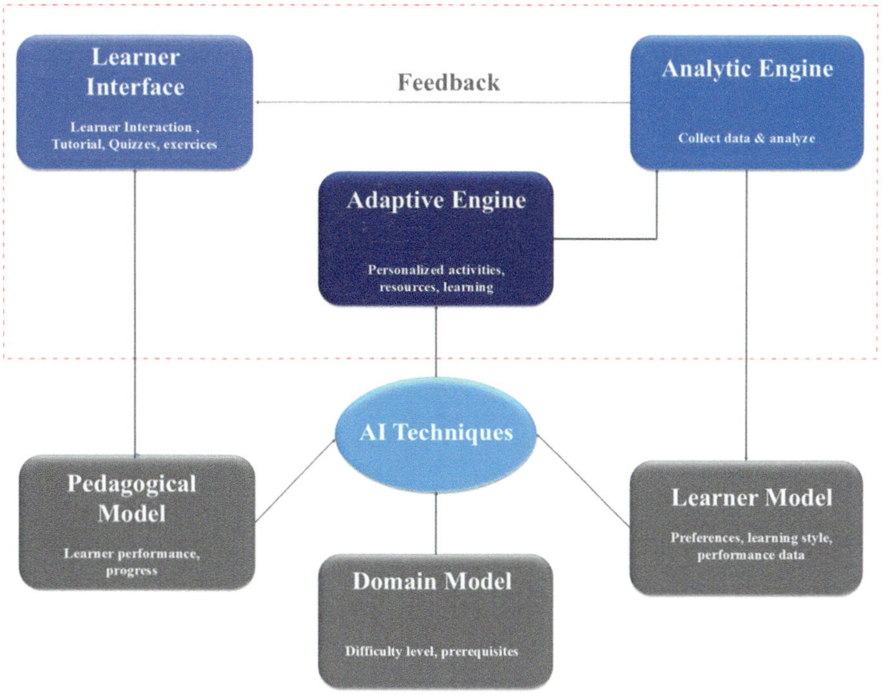

Research findings indicate that the system can achieve admissible classification accuracy and recommend a customized learning path for learners.

Additionally, [35] introduced lightweight domain modeling for adaptive web-based educational systems, demonstrating its effectiveness in recommendation and its represents pivotal role in fostering interaction and collaboration, thus contributing to the development of adaptive systems.

2.5 Implementation of AI in Adaptive Learning

Educational Data Mining (EDM)

According to [36], the primary goal of AI is to create an intelligent machine, while a secondary goal is to explore the nature of intelligence. AI is defined as the science that focused on enabling programs to enhance themselves autonomously based on their experiences.

Educational data mining (EDM) is an approach that integrates machine learning, statistics, psychometrics, and data mining techniques. It serves as a field of research aimed at gaining deeper insights into the learning process by deconstructing it into smaller components for analysis, thereby facilitating adaptation to individual learner needs [37].

Baker identified several key categories within EDM, including prediction, relationship exploration, model-based discovery, and data distillation for human judgment (Table 1). These types are instrumental in predicting students' future learning behavior, improving domain models, investigating pedagogical support methods, and advancing scientific research on learning and learners [37].

AI in Education (AIED)

The growing of interest in AI and its increasing integration into education has led the development of the "AI in Education" (AIED) research domain in recent years [38]. AIED focuses on implementing AI in the education field, characterized by three main paradigms, as outlined by [39] (see Fig. 4).

The first paradigm involves utilizing AIED to construct knowledge models and facilitate cognitive learning, with learners benefiting from tailored educational services. In addition, the second approach, AIED is utilized to build knowledge models students during their interaction with AI tools.

Finally, the third paradigm aims to enhance learning experiences as students actively participate in the learning process. In short, the three paradigms are designed to empower learners to become recipients, collaborators, and leaders in their educational journey.

Overall, the trajectory of AIED development is geared toward learner empowerment and personalization of the learning process. This enables learners to reflect on their educational journey and provide feedback to AI systems to

Fig. 4 Hierarchical tree of AI technique used in education

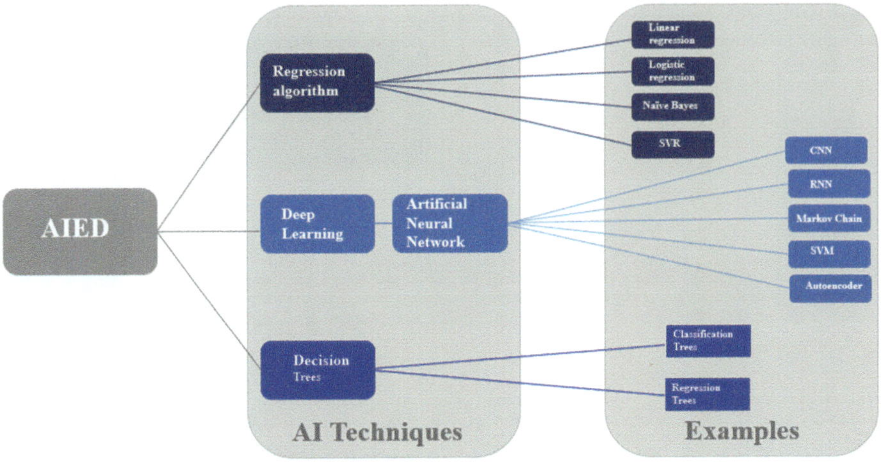

adaptive adjustments. The results are an iterative process that fosters personalized, learner-centered, data-driven learning [40].

In various research fields, several AI family techniques are employed, typically grouped into three categories, predictive, descriptive, and prescriptive. Within the context of education, predictive analytic emerges as the most leveraging statistical modeling and forward-looking insights to enhance learner performance and tackle achievement disparities. This approach leads the way for adaptive learning, exemplified in models like Bayesian networks, naive Bayes, fuzzy logic, Bayesian knowledge tracing (BKT), and neural networks [41].

Multimodal Learning Analytics (MMLA)

As an emergent field within learning analytics, multimodal learning analytics (MMLA) encompasses a variety of techniques designed to gather, synchronize, and analyze diverse sources of high-frequency data within ecologically valid, realistic, social, multimedia learning environments [42] (see Fig. 5). By combing the different data sources, MMLA seeks to offer a comprehensive understanding of learners' behaviors, interaction in addition to learning mechanisms at a micro level. MMLA intersects with various disciplines including behaviorism, cognitive science, multimodal interaction, computer vision, natural language processing.

A range of analytical techniques are used in multimodal learning analysis, including machine learning, natural language processing [43], computer vision [44], sentiment analysis [45], and data mining [46]. Such techniques enable the processing, interpretation, and extraction of valuable insights from cross-modal data related to learner engagement, cognitive and emotional states, collaboration styles, learning strategies, and performance metrics.

These insights serve to reinforce adaptive learning approaches. Through the multimodal data analysis, the adaptive systems can refine the learning experience and improve the learner engagement by better understanding individual needs and preferences.

Data Used in EDM, LA, AIED, MMLA

Data plays an increasingly pivotal role in education, emerging as a cornerstone for providing pedagogical support to students. The data collected from student learning interactions is dynamic, precise, extensive, and personalized [7]. This wealth of information sheds light on each student's learning process, enabling a nuanced understanding of individual progress and facilitating prompt intervention where necessary. Essentially, it allows for the assessment of student performance, identification of difficulties, and provision of targeted assistance to address any challenges they may encounter. Within an adaptive learning framework, diverse types of data, both dynamic and static, serve crucial roles in domain modeling, user profiling, and adaptation and personalization (Table 2).

Fig. 5 Conceptual framework to address multimodal learning analysis

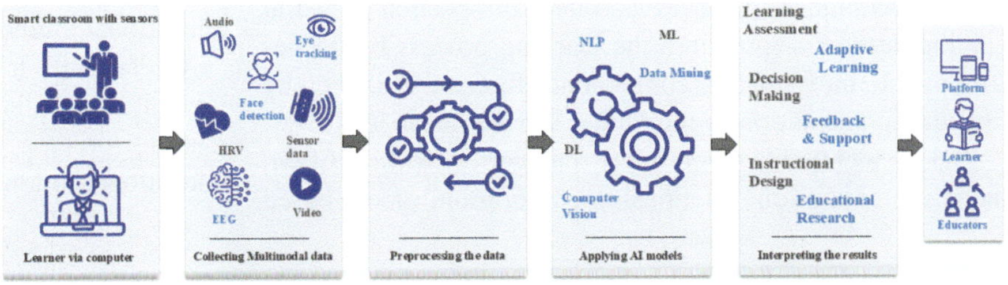

Table 1 EDM methods [46]

Technical method	Description	Applications
Prediction	Design a model able to predict one aspect of the data (predicted variable) in the absence of other features of the data (the input variables)	Predict student achievement and identify student behavior
Clustering	Identify similarities between data points based on their features	Group students according to their learning disabilities and interaction patterns
Relationship mining	Search for links between variables in a dataset and encode these links in the form of rules	Discover pedagogical strategies and the relationship between student results and course structures
Discovery with models	Analyze a phenomenon using an already validated model	Incorporate psychometric design frameworks into machine learning models to identify correlations between student attitudes and characteristics
Distillation for human judgment	Process data according to a human ability to quickly determine or categorize data patterns	Visualize to analyze the various actions taken by students and their use of information

Table 2 Data type by application area

Application examples	Question	Required data type
Domain modeling	What is the right level for dividing subjects into modules? How these can be sequenced?	Student answers (correct, incorrect), time stamps, hints, repetition of wrong answers, errors Relationships within problems and between skills and problems
User profiling	Which groups do learners belong to?	Student answers (correct, incorrect), time stamps, hints, repetition of wrong answers, errors
Adaptation and personalization	What future actions are suggested to the learner? How can we enhance the learner experience in real-time?	Student's academic achievements record

This data contains a spectrum of student responses (both correct and incorrect), timestamps, hints, repetitions, errors, and relationships between skills and items, constituting a repository of historical learner's performance records [6].

Static data consists of outcomes from standardized tests focusing on particular academic content. Unlike dynamic data, no learning takes place during the test, leaving the student's state of knowledge remains unchanged.

This data form, often referred to as cross-sectional data, is typically represented as a binary matrix, commonly known to as a response matrix [47]. Well-known examples of static datasets include TIMSS [48], comprising more than 23 mathematical problems from the 2003 TIMSS assessment (trends in international mathematics and science study), and Fraction [49], a collegiate-level test containing more than 10,000 responses.

On the other hand, dynamic data is collected in real-time as students engage with the learning system. This type of data involves the notion of sequence and is referred also as longitudinal data. This data type, often characterized by sequences, is instrumental in monitoring knowledge acquisition as part of formative assessment [47]. Among the best-known datasets in this category include the KDD dataset [50], introduced for an 'educational data mining' competition, which aims to forecast the accuracy of a student's response in an Intelligent Tutoring System (ITS) based on the tutor cognitive framework [51]. Additionally, the ASSISTments dataset [52], released by Niel Heffernan from the ASSISTments online learning platform focusing on mathematics education, serves as a valuable resource in this domain [53].

When it comes to multimodal learning analytics (MMLA), data collection varies depending on whether the learner is engaged in digital or physical activities. In the digital realm, data is gathered through clickstream, log data [54], mouse [55] and keyboard strokes [56], or via qualitative data, and then it is collected from text [57], handwriting [58], and digital footnotes [59]. In contrast, when a learners are engaged in physical activities, data collection extends to include eye movements [60], eye contact [61], audio cues [62], facial expressions [63], head movements [64], hand gestures [65], arm movements [66], and body posture [67] (Table 3).

3 Review of Existing Adaptive Learning Systems

Table 4 is a comprehensive analysis of 12 existing adaptive learning systems, focusing on their key information and characteristics. These systems vary in their categorizations, courseware features, and levels of adaptivity.

Most of the systems are adaptive learning platform (ALP), while others are Adaptive Educational Hypermedia System (AEHS) and Intelligent Tutoring System (ITS). The level

Table 3 Comparative table of MMLA data type and their applications

Application examples	Description	Data type
Using hand motion to understand embodied mathematical learning [68]	Analyzing text handwritten or typewritten by apprentices	Text data
Think-aloud protocols used in cognitive and metacognitive activities [69]	Audio recordings of apprentices	Audio data
Engagement detection using video [70]	Visual recordings detecting facial interactions or gestures	Video data
Emotion recognition using heart rate variability [71]	Quantifying physiological behavior via sensor data	Sensor data
Combined visual attention [60]	Recording of eye variations across learning activities	Eye-tracking data
Engagement detection using clickstream [54]	Analyzing learner interaction with learning environment	Interaction data
Predicting learning performance using step count [72]	Analyzing body movement	Gesture data
Analysis of co-located collaborative learning groups [61]	Analyzing information concerning inter-learner connections	Social network data
Detecting cognitive load using EEG [73]	Analyzing cognitive state using electroencephalogram (EEG) recordings of brain activity	Brainwave data
Emotion recognition using blood volume pulse (BVP) [74]	Detection of stress or excitement levels during the learning process	Physiological response data

of adaptivity can vary from system to other, depending on whether it has a high level of adaptivity or a low one. While most of the system offer English course content, several others do so in other languages. It is also noticed that the majority are hosted in the USA, with some others hosted in Canada or Australia such as Smart Sparrow or Classcraft. The table is designed to be a useful overview of some existing adaptive learning systems. This overview could be used by learners and teachers to determine which system would be most beneficial to them.

A notable concentration of these platforms is found in the USA, with some originating from Australia, Canada, and Ireland, indicating the US's leadership in educational technology. While most systems primarily offer content in English, the inclusion of multilingual support reflects a growing recognition of the need for accessibility in diverse educational contexts. Most platforms operate on a freemium model, enhancing accessibility for various users, though further expansion of free offerings could benefit wider adoption. The courseware features target various educational levels, from K-12 to higher education, with some systems specializing in niche areas like language learning and gamified classroom management.

The varying levels of adaptivity highlight differences in how personalized the learning experience can be, with many systems providing high adaptivity to tailor learning pathways effectively. Furthermore, the integration of Learning Management Systems (LMS) across most platforms indicates a trend toward creating cohesive educational ecosystems. Overall, this analysis underscores the potential for continued growth and innovation in adaptive learning technologies, emphasizing their capacity to meet diverse learner needs in both informal and formal educational environments.

4 Key Features of Adaptive Learning Systems

Adaptive learning systems have emerged as a pivotal component of modern educational technology, facilitating personalized learning experiences that address the unique needs of each student. By utilizing algorithms and data analytics to tailor educational content, pacing, and assessment, enhancing engagement and improving outcomes. Our analysis of real world examples of adaptive learning systems reveals several key features (Table 4).

The primary characteristic of adaptive learning systems is their ability to adjust the experience of learning in response to real-time assessments of student preferences and performance. For example, Knewton provides personalized learning pathways that adapt in response to the learner's interactions and progress, making the material more relevant and engaging. Many adaptive systems offer a range of courseware features, from basic assessments to interactive modules. ALEKS employs a mastery-based approach, enabling students to gain a comprehensive understanding of

Table 4 Review of existing adaptive learning systems

System	Website	Created by	Year of cre	Country	System type			Charges		Courseware features	Ada. level			LMS		Type
					Cloud	Mobile	Local	Free	Pay		High	Med	Low	Yes	No	
ALEKS	aleks.com	McGraw Hill Education	1994	USA	+	–	–	–	+	K-12, Higher Ed	+	–	–	+	–	ALP
McGraw Hill Connect	mheducation.com/highered	McGraw Hill Education	2009	USA	+	+	–	–	+	K-12, Higher Ed	+	–	–	+	–	ALP
Knewton	knewton.com	Jose Ferreira	2008	USA	+	+	–	–	+	Test Prep, K-12, Higher	+	–	–	+	–	ALP
DreamBox	dreambox.com	Lou Gray and Ben Slivka	2006	USA	+	+	–	–	+	K-8 Math	–	+	–	+	–	ALP
Smart Sparrow	smartsparrow.com	University of New South Wales	2011	Australia	+	+	–	–	+	Customizable	+	–	–	+	–	ALP
Fishtree	fishtree.com	Fishtree	2012	Ireland	+	+	–	+	+	K-12	+	–	–	+	–	ALP
ScootPad	scootpad.com	Bharat Kumar and Maya Gadde	2011	USA	+	–	–	–	+	K-8 Math, ELA	+	–	–	+	–	ALP
Classcraft	classcraft.com	Shawn and Devin Young	2013	Canada	+	–	–	+	+	Classroom Management	–	–	+	+	–	Gamified LMS
Duolingo	duolingo.com	Luis von Ahn and Severin Hacker	2011	USA	–	+	–	+	–	Language Learning	–	–	+	–	+	ITS
MindEdge	mindedge.com	Harvard and MIT educators	1998	USA	+	+	–	–	+	Higher Edu	–	–	+	–	+	AEHS
GIFT	gifttutoring.org	U.S. Army Research Laboratory	2009	USA	+	+	+	+	–	Course builder	–	+	–	+	–	ITS
Author	muzzylane.com	Muzzy Lane Software	2002	USA	+	+	–	+	+	Course builder	–	+	–	+	–	ITS

topics before advancing. Similarly, DreamBox focuses on K-8 math, utilizing interactive visuals and game-like elements to make learning more engaging.

Adaptive learning systems often integrate with existing Learning Management Systems to streamline administrative tasks and provide a cohesive learning environment. Platforms like McGraw Hill Connect and Smart Sparrow illustrate this trend by providing tools for educators to manage and track student progress within a familiar framework. Most systems adopt a freemium model, providing users with access to basic features at no cost while offering premium features for a fee. This approach increases accessibility, particularly in K-12 education. Duolingo, for example, offers free language learning resources with optional premium features, expanding its reach to a global audience.

Some adaptive systems, such as Classcraft, incorporate gamification elements to enhance student engagement and motivation. This platform uses game mechanics to manage classroom behavior and learning, providing an interactive and rewarding experience for students. Notable examples of these systems include ALEKS, created by McGraw Hill Education in the USA, which primarily targets K-12 and higher education by identifying students' knowledge gaps and offering personalized instruction in mathematics. Duolingo, founded in 2011, has revolutionized language learning through its user-friendly app that tailors lessons based on user performance. Smart Sparrow, developed by the University of New South Wales, allows educators to create customizable learning experiences that adapt to student needs, emphasizing formative assessment and feedback. DreamBox Learning focuses on K-8 mathematics, utilizing adaptive technology to provide personalized instruction that adjusts the complexity of tasks based on student performance. Classcraft merges educational content with gamification, enabling teachers to foster a collaborative and engaging learning environment while allowing for personalized learning experiences that motivate students.

5 Challenges and Future Direction in Implementing Adaptive Learning

In recent years, significant number of publications have related to adaptive learning. However, despite this growth, there still needs for improvement in this field, with numerous challenges still need to be addressed.

Various research studies have highlighted the obstacles associated in adopting adaptive learning systems in higher education [38]. Notably, with reference to [75], the primary obstacles encompass technology, management, and pedagogical dimensions. While these studies provide a broad overview of the challenges, they often fail to offer nuanced insights into how these obstacles affect different institutional contexts. For instance, ALEKS faces challenges related to seamless integration with existing Learning Management Systems and ensuring effective user engagement.

Implementing and utilizing adaptive learning systems [76] require meticulous planning, continuous support, and adequate training for both teachers and learners. However, the literature often lacks empirical data demonstrating the effectiveness of such training programs, creating a gap in understanding their actual impact on educators' readiness to adopt these systems. This gap is evident in systems like Smart Sparrow, which provides customizable learning experiences but may struggle if educators are not adequately prepared to leverage its features.

Johnson [77] has identified a challenge related to pedagogy, particularly disengagement of educators due to their unfamiliarity with such platforms. This points to a significant limitation in the existing research: while it acknowledges the issue, it does not investigate the root causes of this disengagement. Specific factors, such as age, prior technology experience, or institutional support, may contribute to educators' reluctance to engage with adaptive learning tools. For example, Knewton, if educators lack confidence in using such technologies, their potential remains underutilized. Additionally, Zliobaite et al. [78] outlined six common problems encountered in designing and implementing adaptive learning system, spanning issues of scalability, realistic data, usability, expert knowledge, diversity of needs, and application areas, as well as switching from adaptive algorithms to adaptive tools. Their analysis would benefit from a deeper examination of how these issues interact. For instance, scalability may exacerbate usability challenges if systems cannot adapt to diverse user needs.

Another challenge is related to the integration of several technologies, as seen with McGraw Hill Connect, from Learning Management Systems to analytics tools and content management systems [75]. Maintaining seamless interoperability among these technologies can prove to be complex and time-consuming. While the intention for adaptive learning systems is to offer interactive and enjoyable learning experiences, the persistent challenge of learner dropouts is always present due to the lack of personalization [79]. The literature often discusses dropout rates without fully addressing the demographic factors at play, such as whether certain groups of students are more likely to disengage and the underlying reasons for this behavior. The scarcity of publicly available datasets in the sector of education exacerbates the difficulty of this task, therefore, underscoring the need to intensify in collecting learner-centered data for future researchers [79].

A prevalent issue affecting not only adaptive learning systems but also the majority of recommendation systems is the cold start problem. In adaptive learning, this issue arises when encountering a new item or user, rendering the system incapable of recommending customized content for the user [80].

While several solutions, such as content-based filtering and collaborative filtering, are suggested to mitigate this challenge, the literature rarely explores the effectiveness of these methods in real-world settings, particularly in specific contexts where these techniques may fail to deliver satisfactory results.

Ethical considerations [81] must not be overlooked, encompassing concerns regarding privacy and data security, equity, access, and pedagogical integrity. The discussions in the literature often lack depth, failing to adequately address how educational institutions can navigate these ethical dilemmas while implementing adaptive learning technologies.

Many students consider ChatGPT as an effective tool for adaptive learning, thanks to its capability to learn from user interactions and understand student questions through NLP use. However, despite all these capabilities, ChatGPT still grapples with limitations that pose significant challenge in the field of education [82]. One major limitation is its limited knowledge base, leading to the generation of erroneous and biased information. As the authors suggested in [83, 84], one potential solution is to employ ChatGPT as a tool for educators to craft learning materials and foster dialogue to aid students, as demonstrated in an English learning context. While it can assist educators in creating learning materials, there are concerns regarding the relevance and accuracy of AI-generated content, as seen in platforms like MindEdge, which targets higher education but must address issues of bias in its resources.

6 Conclusion

Leveraging big data and AI in the field of education has led to the emergence of several new areas of research, including AIED, EDM, MMLA, and big data analytics. Each of these areas has significantly contributed to understanding and adapting learning processes effectively. Adaptive learning, as delineated in the literature, provides a personalized educational experience tailored to the individual needs and abilities, thereby enabling progress at their own pace within an enriched and effective learning environment. By offering highly customized learning experiences, adaptive learning has the potential to bridge the learning gap, reduce costs, and enhance engagement, despite traditional learning approaches.

Since its inception at the early-twentieth century, adaptive learning, along with AI techniques, has undergone several innovations. These range from early systems like the aptitude-treatment interaction system proposed by Cronbach to the more recent, developments such as Adaptive Educational Hypermedia Systems, micro-adaptive systems, and Intelligent Tutoring Systems. These innovations have culminated in the creation of learning platforms in the last decade. The strategies used by these systems typically revolve around three main components: domain modeling, user profiling, and user adaptation/personalization, all aimed at delivering tailored learning experiences.

In conclusion, this paper has reviewed the up-to-date landscape of adaptive learning, emphasizing the pivotal role played by AI and big data in addressing learning gaps. Furthermore, it has identified several challenges that need resolution. These challenges mainly refer to the implementation and application of a new adaptive learning system, pedagogy, teacher disengagement, learner dropout, and the limited availability of education-related public datasets. Addressing these challenges is crucial for researchers to build new models related to student performance to enhance adaptation.

This paper has examined the changing landscape of adaptive learning, highlighting the role of big data and AI in creating personalized educational experiences tailored to individual learners' needs. The analysis of existing adaptive learning systems revealed that while many platforms excel in customization, challenges such as teacher disengagement, learner dropouts, and the cold start problem hinder their effectiveness. Furthermore, the scarcity of public datasets limits research opportunities and the potential for innovation in this field.

The importance of this research is based on its identification of critical obstacles that educators and developers face when implementing adaptive learning systems. By addressing these challenges, future studies can concentrate on developing more robust models to improve student engagement and performance. Practical implications include the need for comprehensive training for educators, improved data collection methodologies, and the establishment of ethical guidelines to safeguard learner privacy. Overcoming these challenges will enhance the efficiency of adaptive learning systems and play a key role in bridging learning gaps and fostering equitable access to education in diverse contexts.

Acknowledgements This work was funded by the Al-Khwarizmi Program supported by Morocco's Ministry of Education, Ministry of Industry, and the Digital Development Agency (ADD) under Project No. 451/2020 (Smart Learning).

References

1. Birjali M, Beni-Hssane A, Erritali M (2018) Learning with big data technology: the future of education. In: Proceedings of the third international Afro-European conference for industrial advancement—AECIA 2016. Springer International Publishing, pp 209–217
2. Liu J-H, Ruan L-X, Zhou Y-Y (2019) Application of big data on self-adaptive learning system for foreign language writing. In: Xiong N, Xiao Z, Tong Z, Du J, Wang L, Li M (eds) Advances in computational science and computing. Advances in intelligent systems and

computing. Springer International Publishing, pp 86–93. https://doi.org/10.1007/978-3-030-02116-0_11
3. Coccoli M, Maresca P, Stanganelli L (2017) The role of big data and cognitive computing in the learning process. J Vis Lang Comput 38:97–103. https://doi.org/10.1016/j.jvlc.2016.03.002
4. Krynicki K, Jaen J, Navarro E (2016) An ACO-based personalized learning technique in support of people with acquired brain injury. Appl Soft Comput 47:316–331. https://doi.org/10.1016/j.asoc.2016.04.039
5. Baker RS, Martin T, Rossi LM (2016) Educational data mining and learning analytics. In: The Wiley handbook of cognition and assessment: frameworks, methodologies, and applications, pp 379–396. https://doi.org/10.1002/9781118956588.ch16
6. Bienkowski M, Feng M, Means B (2012) Enhancing teaching and learning through educational data mining and learning analytics: an issue brief. Office of Educational Technology, US Department of Education, pp 77
7. Khan MA, Khojah M, Vivek U (2022) Artificial intelligence and big data: the advent of new pedagogy in the adaptive e-learning system in the higher educational institutions of Saudi Arabia. Education Research International 2022:e1263555. https://doi.org/10.1155/2022/1263555
8. Froschl, C. (2005) User modeling and user profiling in adaptive e-learning systems
9. Brunsilovsky P, Maybury MT (2002) From adaptive hypermedia to the adaptive web. Commun ACM 45(5):30–33. https://doi.org/10.1145/506218.506239
10. Brunsilovsky P, Karagiannidis C, Sampson D (2004) Layered evaluation of adaptive learning systems. Int J Contin Eng Educ Life-Long Learn 14(4/5):402. https://doi.org/10.1504/IJCEELL.2004.005729
11. Peylo C (2000) W2—Adaptive and intelligent web-based education systems. In: Gauthier G, Frasson C, VanLehn K (eds) Intelligent tutoring systems. ITS 2000. Lecture notes in computer science, vol 1839. Springer, Berlin, Heidelberg, pp 663. https://doi.org/10.1007/3-540-45108-0_86
12. Brusilovsky P (1999) Adaptive and intelligent technologies for web-based education. KI 13:19–25
13. Inspired eLearning (2023) What is adaptive learning?. Retrieved from https://inspiredelearning.com/blog/what-is-adaptive-learning/
14. Hallahan DP, Keller CE, McKinney JD, Lloyd JW, Bryan T (1988) Examining the research base of the regular education initiative: Efficacy studies and the adaptive learning environments model. J Learn Disabil 21(1):29–35. https://doi.org/10.1177/002221948802100106
15. Messner W, Horowitz R, Kao W-W, Boals M (1990) A new adaptive learning rule. In: Proceedings of the IEEE international conference on robotics and automation. IEEE, pp 1522–1527. https://doi.org/10.1109/ROBOT.1990.126223
16. Hakkal S, Lahcen AA (2021) An overview of adaptive learning fee-based platforms. https://doi.org/10.5220/0010731400003101
17. Rasco RW, Tennyson RD, Boutwell RC (1975) Imagery instructions and drawings in learning prose. J Educ Psychol 67(2):188–192. https://doi.org/10.1037/h0077014
18. Mödritscher F, Garcia-Barrios VM, Gütl C (2004) The past, the present, and the future of adaptive e-learning
19. Cronbach LJ, Snow RE (1977) Aptitudes and instructional methods: a handbook for research on interactions. Irvington, Oxford, England
20. Graesser AC, Conley MW, Olney A (2012) Intelligent tutoring systems. In: Harris KR, Graham S, Urdan T, Bus AG, Major S, Swanson HL (eds) APA educational psychology handbook, vol 3: application to learning and teaching. American Psychological Association, pp 451–473. https://doi.org/10.1037/13275-018
21. Osadcha K, Osadchyi V, Chemerys H (2020) The review of adaptive learning systems for the formation of individual educational trajectory. In: Research and industrial applications. integration, harmonization, and knowledge transfer 2020: proceedings of the 16th international conference on ICT in education, research and industrial applications, vol 2732, pp 547–558
22. Brusilovsky P (1998) Methods and techniques of adaptive hypermedia. In: Brusilovsky P, Kobsa A, Vassileva J (eds) Adaptive hypertext and hypermedia. Dordrecht: Springer Netherlands, pp 1–43. https://doi.org/10.1007/978-94-017-0617-9_1
23. Mulwa C, Lawless S, Sharp M, Arnedillo-Sánchez I, Wade V (2010). Adaptive educational hypermedia systems in technology enhanced learning: a literature review, p. 84. https://doi.org/10.1145/1867651.1867672
24. Utterberg M, Tallvid M, Lundin J, Lindström B (2021) Intelligent tutoring systems: why teachers abandoned a technology aimed at automating teaching processes. https://doi.org/10.24251/HICSS.2021.186
25. Thapliyal M, Ahuja NJ, Shankar A, Cheng X, Kumar M (2022) A differentiated learning environment in domain model for learning disabled learners. J Comput High Educ 34(1):60–82. https://doi.org/10.1007/s12528-021-09278-y
26. MacLellan CJ, Koedinger KR (2022) Domain-general tutor authoring with apprentice learner models. Int J Artif Intell Educ 32(1):76–117
27. Hertz B et al (2022) A pedagogical model for effective online teacher professional development—findings from the Teacher Academy initiative of the European Commission. Eur J Educ 57(1):142–159. https://doi.org/10.1111/ejed.12486
28. Gumbheer CP, Khedo KK, Bungaleea A (2022) Personalized and adaptive context-aware mobile learning: review, challenges, and future directions. Education Tech Research Dev 27(6):7491–7517. https://doi.org/10.1007/s10639-022-10942-8
29. Russom P (2011) Big data analytics. TDWI Best Pract Rep 19(4):1–34
30. Brynjolfsson E, McAfee A (2014) The second machine age: work, progress, and prosperity in a time of brilliant technologies. W. W. Norton & Company
31. Lei G, Luo X, Yang S, Xiao K (2021) Adaptive online learning model based on big data. In: Application of intelligent systems in multi-modal information analytics: proceedings of the 2020 international conference on multi-model information analytics (MMIA2020), vol 1. Springer International Publishing, pp 643–649
32. Clow D (2013) An overview of learning analytics. Teach High Educ 18(6):683–695
33. Yang Y, Zhong Y, Woźniak M (2021) Improvement of adaptive learning service recommendation algorithm based on big data. Mob Netw Appl 26(5):2176–2187. https://doi.org/10.1007/s11036-021-01772-y
34. Intayoad W, Becker T, Temdee P (2017) Social context-aware recommendation for personalized online learning. Wireless Pers Commun 97(1):163–179. https://doi.org/10.1007/s11277-017-4499-2
35. Simko M, Bielikova M (2019) Lightweight domain modeling for adaptive web-based educational system. J Intell Inf Syst 52(1):165–190. https://doi.org/10.1007/s10844-018-0518-3
36. Schank RC (1987) What is AI, anyway? AI Magazine 8(4), Article no 4. https://doi.org/10.1609/aimag.v8i4.623
37. Baker RSJD, Yacef K (2009) The state of educational data mining in 2009: a review and future visions. J Educ Data Min 1(1), Article no 1. https://doi.org/10.5281/zenodo.3554657
38. Martin F, Chen Y, Moore RL, Westine CD (2020) Systematic review of adaptive learning research designs, context, strategies, and technologies from 2009 to 2018. Education Tech Research Dev 68(4):1903–1929. https://doi.org/10.1007/s11423-020-09793-2
39. Ouyang F, Jiao P (2021) Artificial intelligence in education: the three paradigms. Comput Educ: Artif Intell 2:100020. https://doi.org/10.1016/j.caeai.2021.100020

40. Tapalova O, Zhiyenbayeva N (2022) Artificial intelligence in education: AIEd for personalised learning pathways. Electron J E-Learn 20(5):639–653
41. Kabudi T, Pappas I, Olsen DH (2021) AI-enabled adaptive learning systems: A systematic mapping of the literature. Comput Educ: Artif Intell 2:100017. https://doi.org/10.1016/j.caeai.2021.100017
42. Blikstein P (2013) Multimodal learning analytics. In: Proceedings of the third international conference on learning analytics and knowledge, pp 102–106. https://doi.org/10.1145/2460296.2460316
43. Chowdhary KR (2020) Natural language processing. In: Fundamentals of artificial intelligence, pp 603–649
44. Jähne B, Haußecker H (2000) Computer vision and applications
45. Prabowo R, Thelwall M (2009) Sentiment analysis: a combined approach. J Informet 3(2):143–157. https://doi.org/10.1016/j.joi.2009.01.003
46. Romero C, Ventura S (2013) Data mining in education. Wiley Interdiscip Rev: Data Min Knowl Discov 3(1):12–27. https://doi.org/10.1002/widm.1075
47. Minn S (2022) AI-assisted knowledge assessment techniques for adaptive learning environments. Comput Educ: Artif Intell 3:100050. https://doi.org/10.1016/j.caeai.2022.100050
48. Skaggs G, Wilkins JL, Hein SF (2016) Grain size and parameter recovery with TIMSS and the general diagnostic model (2015). Int J Test 16(4):310–330. https://doi.org/10.1080/15305058.2016.1145683
49. DeCarlo LT (2011) On the analysis of fraction subtraction data: the DINA model, classification, latent class sizes, and the Q-matrix. Appl Psychol Meas 35(1):8–26. https://doi.org/10.1177/0146621610377081
50. Yuan K, Qi Q (2019) KDD cup 2010: educational data mining challenge. Final Project Report
51. Stamper J, Pardos ZA (2016) The 2010 KDD cup competition dataset: engaging the machine learning community in predictive learning analytics. Learn Anal 3(2):312–316. https://doi.org/10.18608/jla.2016.32.16
52. Heffernan NT (2014) ASSISTmentsData. Retrieved 21 June 2023, from https://sites.google.com/site/assistmentsdata/datasets/
53. Feng M, Heffernan N, Koedinger K (2009) Addressing the assessment challenge with an online system that tutors as it assesses. User Model User-Adap Inter 19(3):243–266. https://doi.org/10.1007/s11257-009-9063-7
54. Henrie CR, Bodily R, Larsen R, Graham CR (2018) Exploring the potential of LMS log data as a proxy measure of student engagement. J Comput High Educ 30(2):344–362. https://doi.org/10.1007/s12528-017-9161-1
55. Okur E, Alyuz N, Aslan S, Genc U, Tanriover C, Arslan Esme A (2017) Behavioral engagement detection of students in the wild. In: Artificial intelligence in education: 18th international conference, AIED 2017, Wuhan, China, 28 June–1 July, 2017, Proceedings 18. Springer, pp 250–261. https://doi.org/10.1007/978-3-319-61425-0_21
56. Su Y-S, Ding T-J, Lai C-F (2017) Analysis of students' engagement and learning performance in a social community supported computer programming course. Eurasia J Math Sci Technol Educ 13(9):6189–6201. https://doi.org/10.12973/eurasia.2017.01058a
57. Suero Montero C, Suhonen J (2014) Emotion analysis meets learning analytics: online learner profiling beyond numerical data. In: Proceedings of the 14th Koli calling international conference on computing education research, pp 165–169. https://doi.org/10.1145/2674683.2674699
58. Oviatt S, Hang K, Zhou J, Yu K, Chen F (2018) Dynamic handwriting signal features predict domain expertise. ACM Trans Interact Intell Syst (TiiS) 8(3):1–21. https://doi.org/10.1145/3213309
59. Hsiao I-H, Huang P-K, Murphy H (2017) Integrating programming learning analytics across physical and digital space. IEEE Trans Emerg Top Comput 8(1):206–217. https://doi.org/10.1109/TETC.2017.2701201
60. Sharma K, Dillenbourg P, Giannakos M (2019) Stimuli-based gaze analytics to enhance motivation and learning in MOOCs. In: 2019 IEEE 19th International conference on advanced learning technologies (ICALT). IEEE, pp 199–203. https://doi.org/10.1109/ICALT.2019.00052
61. Schneider B, Sharma K, Cuendet S, Zufferey G, Dillenbourg P, Pea R (2018) Leveraging mobile eye-trackers to capture joint visual attention in co-located collaborative learning groups. Int J Comput-Support Collab Learn 13:241–261. https://doi.org/10.1007/s11412-018-9281-2
62. Noel R et al (2018) Exploring collaborative writing of user stories with multimodal learning analytics: a case study on a software engineering course. IEEE Access 6:67783–67798. https://doi.org/10.1109/ACCESS.2018.2876801
63. Martin K, Wang EQ, Bain C, Worsley M (2019) Computationally augmented ethnography: emotion tracking and learning in museum games. In: Advances in quantitative ethnography: first international conference, ICQE 2019, Madison, WI, USA, 20–22 Oct 2019, Proceedings. Springer, pp 141–153. https://doi.org/10.1007/978-3-030-33232-7_12
64. Cukurova M, Zhou Q, Spikol D, Landolfi L (2020) Modelling collaborative problem-solving competence with transparent learning analytics: is video data enough? In: Proceedings of the tenth international conference on learning analytics & knowledge, pp 270–275. https://doi.org/10.1145/3375462.3375484
65. Asadipour A, Debattista K, Chalmers A (2017) Visuohaptic augmented feedback for enhancing motor skills acquisition. Vis Comput 33:401–411. https://doi.org/10.1007/s00371-016-1275-3
66. Sriramulu J, Lin J, Oviatt S (2019) Dynamic adaptive gesturing predicts domain expertise in mathematics. In: 2019 International conference on multimodal interaction, pp. 105–113. https://doi.org/10.1145/3340555.3353726
67. Junokas MJ, Lindgren R, Kang J, Morphew JW (2018) Enhancing multimodal learning through personalized gesture recognition. J Comput Assist Learn 34(4):350–357. https://doi.org/10.1111/jcal.12262
68. Ou L, Andrade A, Alberto R, Van Helden G, Bakker A (2020) Using a cluster-based regime-switching dynamic model to understand embodied mathematical learning. In: Proceedings of the tenth international conference on learning analytics & knowledge, pp 496–501. https://doi.org/10.1145/3375462.3375513
69. Paans C, Molenaar I, Segers E, Verhoeven L (2019) Temporal variation in children's self-regulated hypermedia learning. Comput Hum Behav 96:246–258. https://doi.org/10.1016/j.chb.2018.04.002
70. Monkaresi H, Bosch N, Calvo RA, D'Mello SK (2017) Automated detection of engagement using video-based estimation of facial expressions and heart rate. IEEE Trans Affect Comput 8(1):15–28. https://doi.org/10.1109/TAFFC.2016.2515084
71. Nakagawa H, Iwasawa Y, Matsuo Y (2018) End-to-end deep knowledge tracing by learning binary question-embedding. In: 2018 IEEE international conference on data mining workshops (ICDMW). IEEE, pp 334–342. https://doi.org/10.1109/ICDMW.2018.00055
72. Di Mitri D, Scheffel M, Drachsler H, Börner D, Ternier S, Specht M (2017) Learning pulse: a machine learning approach for predicting performance in self-regulated learning using multimodal data. In: Proceedings of the seventh international learning analytics & knowledge conference, pp 188–197. https://doi.org/10.1145/3027385.3027447
73. Mills C, Fridman I, Soussou W, Waghray D, Olney AM, D'Mello SK (2017) Put your thinking cap on: detecting cognitive load using EEG during learning. In: Proceedings of the seventh international learning analytics & knowledge conference, pp 80–89. https://doi.org/10.1145/3027385.3027431

74. Yin Z, Zhao M, Wang Y, Yang J, Zhang J (2017) Recognition of emotions using multimodal physiological signals and an ensemble deep learning model. Comput Methods Programs Biomed 140:93–110. https://doi.org/10.1016/j.cmpb.2016.12.005
75. Nodine TR (2016) How did we get here? a brief history of competency-based higher education in the United States. J Competency-Based Educ 1(1):5–11. https://doi.org/10.1002/cbe2.1004
76. Bailey A, Vaduganathan N, Henry T, Laverdiere R, Pugliese L (2018) Making digital learning work: success strategies from six leading universities and community colleges. Boston Consulting Group, Boston
77. Johnson C, Zone E (2018) Achieving a scaled implementation of adaptive learning through faculty engagement: a case study. Curr Issues Emerg eLearning 5(1):7
78. Žliobaitė I, Bifet A, Gaber M, Gabrys B, Gama J, Minku L, Musial K (2012) Next challenges for adaptive learning systems. ACM SIGKDD Explorations Newsl 14(1):48–55. https://doi.org/10.1145/2408736.2408746
79. Joy J, Pillai RVG (2022) Review and classification of content recommenders in E-learning environment. J King Saud Univ—Comput Inf Sci 34(9):7670–7685. https://doi.org/10.1016/j.jksuci.2021.06.009
80. Khusro S, Ali Z, Ullah I (2016) Recommender systems: issues, challenges, and research opportunities. In: Information science and applications (ICISA) 2016. Springer, pp 1179–1189. https://doi.org/10.1007/978-981-10-0557-2_112
81. Poje T, Zaman Groff M (2022) Mapping ethics education in accounting research: a bibliometric analysis. J Bus Ethics 179(2):451–472. https://doi.org/10.1007/s10551-021-04846-9
82. Sallam M (2023) The utility of ChatGPT as an example of large language models in healthcare education, research and practice: systematic review on the future perspectives and potential limitations. medRxiv. https://doi.org/10.1101/2023.02.19.23286155
83. Topsakal O, Topsakal E (2022) Framework for a foreign language teaching software for children utilizing AR, voicebots and ChatGPT (large language models). J Cogn Syst 7(2):33–38. https://doi.org/10.52876/jcs.1227392
84. Farsi I (2023) Navigating the post-pandemic world: challenges and new horizons in reshaping higher education in Morocco. Moroc J Quant Qual Res 5(2). https://doi.org/10.48379/IMIST.PRSM/mjqr-v5i2.46996

Evaluation of Research Progress and Trends on Renewable Energy and Sustainable Development: A Bibliometric Analysis

Yousra Benyetho and Abdelilah El Attar

Abstract

This paper uses the SCOPUS database for a bibliometric analysis of the renewable energy/sustainable development (RE/SD) field. The study focuses on research trends, performance analysis, and scientific mapping analysis. VOSviewer and Microsoft Excel 365 were used for this bibliometric analysis. The results reveal a rapid growth in publication numbers over the last 15 years, with emerging research areas such as green finance and environmental sustainability. The study identifies influential authors, productive journals, and significant publications from countries and academic institutions. It indicates that China and the USA are the leading countries in terms of publications, with moderate international collaboration. Furthermore, the study observes that "Adebayo, Tomiwa S." is the most influential author, while "Resources Policy" is the most productive journal about RE/SD. The findings provide valuable insights for future research in this field, emphasizing the critical role of technological innovation in promoting renewable energy and sustainable development. The study recommends comparing the outputs of multiple databases to gain a more comprehensive understanding of research trends in RE/SD.

Keywords

Renewable energy · Sustainable development · Bibliometric analysis

Y. Benyetho (✉) · A. El Attar
Research Laboratory in Instrumentation and Management of Organizations LURIGOR, University Mohammed the First, Oujda, Morocco
e-mail: yousra.benyetho@ump.ac.ma

1 Introduction

Modern society relies heavily on energy consumption as a critical driver for economic development and daily activities. However, contemporary energy systems face several significant challenges, including the depletion of fossil fuels, the abundance of waste products, the impact of climate change, and the exponential growth of the human population [1]. These factors have prompted the world community to look for alternative solutions to satisfy the increasing energy demand and mitigate the associated sustainability issues, namely greenhouse gas emissions, air pollution, water use, and poverty.

Renewable energy sources have emerged as a promising approach to address these challenges and pave the way for sustainable energy development. Renewable energy technologies offer a viable solution to reduce dependence on fossil fuels and mitigate their associated negative environmental impacts [2]. Consequently, renewable energy has become a focal point of various sustainability-related policies and initiatives worldwide.

Significant changes in human attitudes toward the environment have facilitated the transition to a renewable energy system. In the past, the focus was on maximizing the acquisition of all natural resources, with little regard for the environment. However, the second half of the twentieth century saw a shift toward a more responsible attitude and tremendous respect for nature, prompted by documents and publications from the 1960s onward [3].

Despite the growing interest in renewable energy and sustainable development research, there is a lack of research focusing on examining and evaluating scientific publications from a general perspective. Therefore, this paper aims to explore the temporal distribution patterns of renewable energy and sustainable development journal articles and identify the contributions of productive authors, leading countries,

A. Moumen et al. (eds.), *Artificial Intelligence and Data Analytics for Innovative Applications in Engineering, Sustainability and Technology*, Sustainable Artificial Intelligence-Powered Applications, https://doi.org/10.1007/978-3-031-90318-2_16

and the most prolific academic institutions. Additionally, this paper aims to identify the commonly used terminologies and research subjects, assess the leading countries based on their primary applications, and offer valuable suggestions on future collaborations and research paths.

This study is expected to benefit researchers, policymakers, and individuals seeking to understand the research trends in renewable energy and sustainable development and discover potential opportunities for future research. In conclusion, the transition to renewable energy is a necessary step toward achieving sustainable development, and this study aims to provide insights into the current state of research in this field.

2 Methods

The study of bibliometric analysis is a systematic approach that allows for a comprehensive understanding of general research patterns in a specific field, utilizing the information derived from scholarly literature databases. This approach is beneficial in differentiating bibliometric analysis papers from review papers, primarily intended to discuss a particular topic's latest progress, challenges, and future directions [1].

This paper uses the SCOPUS database to perform a bibliometric analysis of renewable energy/sustainable development (RE/SD). The retrieval settings are as follows:

(TITLE-ABS-KEY (("renewable energy" OR "renewable energies" OR "renewable-energy" OR "renewables" OR "renewable resource" OR "renewable resources" OR "alternative energy" OR "sustainable energy" OR "clean energy" OR "green energy" OR "non-fossil energy") AND ("sustainable development" OR "green development" OR "sustainable growth" OR "green growth" OR "green economic growth" OR "economic growth" OR "economic sustainability" OR "sustainability")) AND PUBYEAR > 1989 AND PUBYEAR < 2024) AND ("economic growth" OR "economic development") AND (LIMIT-TO (SRCTYPE, "j")) AND (LIMIT-TO (DOCTYPE, "ar")) AND (LIMIT-TO (SUBJAREA, "ECON")) AND (LIMIT-TO (LANGUAGE, "English")).

Our final dataset contains 1727 articles from 1990 to 2023 for bibliometric analysis. It is essential to mention that we limited our search to articles published in journals from the subject area of "Economics, Econometrics and Finance" and searched within the results for "economic growth" or "economic development" to emphasize the importance of search in the economic area, as well as articles published in the English language, to obtain more precise results in our field of research.

This paper focuses on two primary bibliometric analyses: performance analysis and scientific mapping analysis. As described by [4, 5], the former involves analyzing various factors such as the country or region of publication, the institution, and the author. Several well-established bibliometric indicators are used to evaluate the essential characteristics of publications, including the number of publications (NP) and citations (NC).

The latter, scientific mapping analysis, can visualize the knowledge structure and organization of a specific study topic or journal [4]. In our study, we used the VOSviewer software for visualization. This paper mainly includes the following analyses: countries' co-authorship analysis, authors' co-authorship analysis, and keyword co-occurrence analysis.

We have comprehensively analyzed the RE/SD field research trends through bibliometric analysis, focusing on performance and scientific mapping analyses. Our findings provide valuable insights that can be utilized by researchers, policymakers, and practitioners in the field to advance research and development in renewable energy and sustainable development.

3 Results and Discussion

3.1 Evolution of Publication Output and Research Interest in RE/SD

As depicted in Fig. 1, the number of publications on renewable energy and sustainable development (1727 articles) has increased over time. To comprehend the change in publication amount, this study divides the period into two phases: the development phase and the rapid growth phase. Before 2007, the number of published articles on renewable energy and sustainable development was below ten and increased slowly. While some researchers presented an extensive analysis of renewable energy's economic and scientific aspects, most articles did not delve into the environmental and social aspects.

However, after 2007, a series of energy accidents occurred worldwide. For instance, the subprime mortgage crisis that began when the US housing bubble burst and sparked a global financial crisis in 2007 and 2008, as highlighted by [5], caused a weaker US dollar exchange rate and pushed oil prices upward. This trend profoundly affected the global economy, imposing a longer recovery time as oil is one of the most important production inputs. At this point, the number of publications began to increase, but still at a slow pace.

The formation of the International Renewable Energy Agency (IRENA) in 2009 marked the beginning of the rapid development phase of renewable energy and sustainable development. One of the pivotal events in this period was the adoption of the sustainable development goals (SDGs) by the United Nations in 2015, which aimed to achieve sustainable development worldwide by 2030 [6]. Since their implemen-

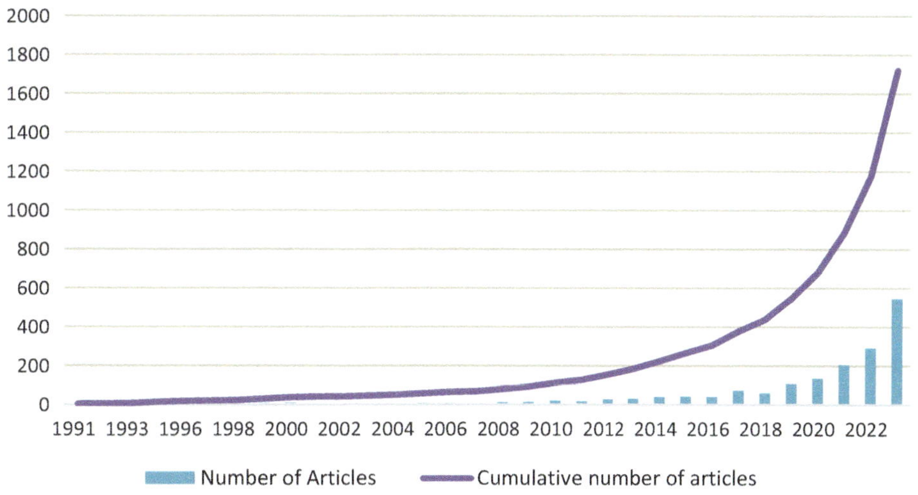

Fig. 1 Annual and cumulative growth of Scopus-indexed articles on RE/SD (1990–2023)

tation, the SDGs have been a topic of academic debate. Many scholars have attempted to provide an overview of the scholarly discussion through literature reviews, specifically on the 7th SDG goal that defines the importance of worldwide clean, affordable, and modern energy systems [3].

Another significant factor that contributed to the rapid increase in scientific articles is the outbreak of the COVID-19 pandemic in 2019. The pandemic profoundly impacted electricity demand, especially during the implementation of lockdown measures. It resulted in a decline in electricity demand, which gradually improved as the restrictions were lifted. Additionally, there was a noticeable shift in the electricity mix during this period, with a more significant proportion of renewable energy sources being utilized. Conversely, there was a substantial reduction in energy production from nuclear, fossil coal, and oil during the same period [7].

Finally, the energy crisis triggered by Russia's invasion of Ukraine has attracted the interest of many researchers. Russia is one of the top exporters of oil, gas, and coal, and the war has had a significantly adverse impact on the energy sector. As of September 2022, a third of the wealthy world's inflation rate of 9% is attributable to energy due to Russia's invasion of Ukraine. Therefore, politicians are discussing ways to be largely independent of fossil fuels, at least in electricity supply, by 2035 [8].

In conclusion, the number of publications on renewable energy and sustainable development has increased over time, with a rapid increase in the last decade. The SDGs, the Coronavirus pandemic, and the incursion of Russia into Ukraine have significantly determined the research interest in the field.

3.2 Preferred Journals in RE/SD

Our study involved a comparative analysis of journals that have published articles in the field under consideration. Our analysis revealed that the top 10 journals collectively published 999 articles, accounting for 63.4% of the articles published in this field. Of these top 10 journals, the three most productive journals were published by Elsevier, EconJournals, and Springer Nature, respectively. This suggests a higher concentration of these journals' productivity throughout the considered period. Additionally, we found that Elsevier Publishing Group published five of the top ten journals with the highest citation scores in 2022, followed by Springer Nature, with two journals in the top 10 list.

The most productive journal in our study was Resources Policy, covering 15.9% of the total publications with 250 articles, followed closely by the International Journal of Energy Economics and Policy, with 249 articles (15.8%). Other notable journals with high productivity include Environment, Development and Sustainability, with 149 articles (9.5%), and Energy Economics, with 106 articles (6.7%) (Table 1).

3.3 Distribution of Top Countries, Leading Institutions, and International Collaboration in RE/SD

Figure 2 analyzes the top 15 countries contributing to the RE/SD research field. China has emerged as a critical player, accounting for approximately 20% of global publications in this field, with 516 publications across various journals. Pakistan and the USA were ranked second and third, respectively, with 180 and 141 publications. Interestingly, only three countries, namely the USA, Italy, and Indonesia, had more than half of their publications authored by researchers from the same country, suggesting moderate intra-country collaboration. The remaining countries showed a balance between intra-country and international collaboration, with more than 35% of publications featuring multi-country authorship.

Table 1 Ten most prolific journals in the RE/SD field of study, along with their most cited paper

	Journal	TP (%)	TC	CiteScore 2022	The most cited paper (reference)	Times cited	Publisher
1	Resources Policy	250 (15.9)	19,194	11.3	The linkages between natural resources, human capital, globalization, economic growth, financial development, and ecological footprint: the moderating role of technological innovations	103	Elsevier
2	International Journal of Energy Economics and Policy	249 (15.8)	5556	3.9	Mitigating emissions in India: accounting for the role of real income, renewable energy consumption and investment in energy	44	EconJournals
3	Environment, Development and Sustainability	149 (9.5)	13,787	7.2	The nexus between urbanization, renewable energy consumption, financial development, and CO_2 emissions: evidence from selected Asian countries	48	Springer Nature
4	Energy Economics	106 (6.7)	27,184	14.7	How does green finance affect green total factor productivity? Evidence from China	124	Elsevier
5	Economic Research-Ekonomska Istrazivanja	68 (4.3)	5639	6.2	The nexus between COVID-19 fear and stock market volatility	56	Taylor & Francis
6	Frontiers in Energy Research	47 (3.0)	9057	2.9	Sustainable energy transition for renewable and low carbon grid electricity generation and supply	22	Frontiers Media S.A
7	Ecological Economics	42 (2.7)	12,453	11	The impact of fintech innovation on green growth in China: mediating effect of green finance	35	Elsevier
8	Resources, Conservation and Recycling	37 (2.3)	42,404	20.3	Challenges toward carbon neutrality in China: strategies and countermeasures	187	Elsevier
9	Journal of the Knowledge Economy	28 (1.8)	1734	4.2	The futures of Europe: society 5.0 and industry 5.0 as driving forces of future Universities	18	Springer Nature
10	Economic Analysis and Policy	23 (1.5)	4032	6.9	Enhancing green economic recovery through green bonds financing and energy efficiency investments	6	Elsevier

TP: total publications; TC: total citations

Rank	Country	TPc	SCP (%)	The most productive academic institution	TPi
1	China	516	46.8	Beijing Institute of Technology	28
2	Pakistan	180	35.2	Ilma University	29
3	USA	141	56.9	Indiana University Bloomington	5
4	Malaysia	117	37.5	Universiti Utara Malaysia	16
5	Turkey	116	37.7	Uluslararası Kıbrıs Üniversitesi	20
6	UK	105	40.1	University of Portsmouth	8
7	India	90	42.7	Lebanese American University	8
8	Saudi Arabia	86	31.4	King Saud University	21
9	Australia	85	39.9	The Australian National University	12
10	Russian. F	69	44.8	Ural'skiĭ Federal'nyĭ Universitet	13
11	Indonesia	65	51.6	Universitas Padjadjaran	7
12	Nigeria	65	43.9	Covenant University	19
13	France	61	37.7	Excelia Business School	6
14	Viet Nam	59	40.1	University of Economics Ho Chi Minh City	18
15	Italy	49	53.3	Azerbaijan State University of Economics UNEC	5

TPc is the total publications of a given country; TPi is the total publications of a given academic institution; SCP is single-country publications

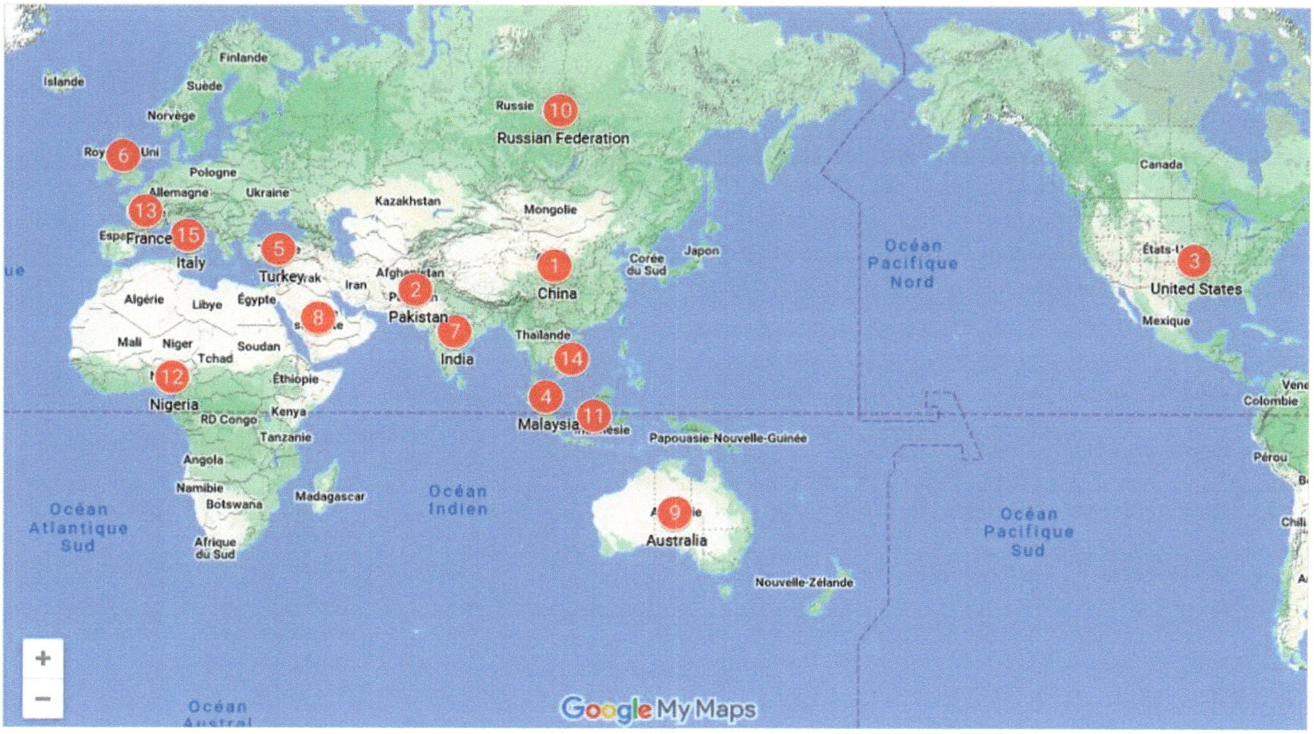

Fig. 2 Top 15 leading countries and academic institutions in RE/SD

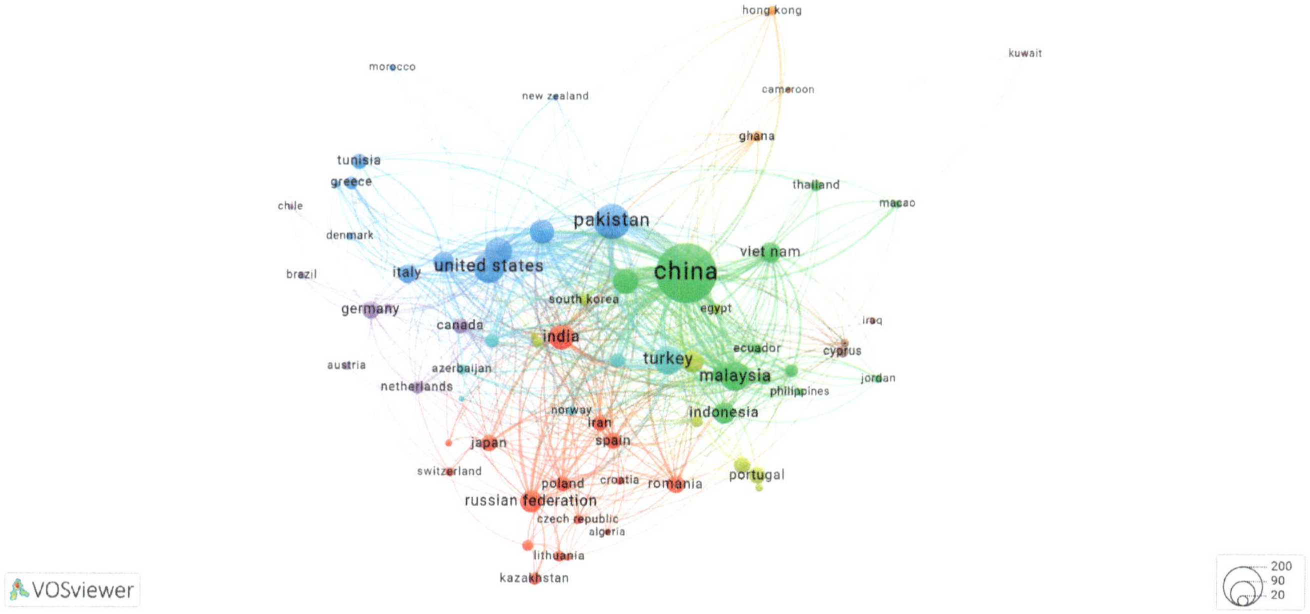

Fig. 3 Visualization of bibliometric map through co-authorships network. The URL for accessing the figure in VOSviewer is http://tinyurl.com/yt3a8ors

Figure 3 also displays the distribution of countries/territories across different regions. The strength of the relationship between two countries is indicated by the proximity of their location in VOSviewer. At the same time, the thickness of the line represents the strength of the link between them. Clusters 2 and 3 mainly comprise countries from Asia and Europe, respectively, while the other clusters comprise countries from various regions worldwide.

Among the international collaborative projects, China has the most robust collaboration with other countries, with 56 links and 516 co-authorships. Pakistan emerges as its closest collaborator with a link strength of 110. This shows that Chinese researchers have been actively collaborating with their peers from other countries to research renewable energy and sustainable development. Pakistan ranks second in terms of link strength, with 47 links and 180 co-authorships, followed by the USA with 39 links and 141 co-authorships, the UK with 39 links and 105 co-authorships, and Malaysia with 36 links and 117 co-authorships.

3.4 Leading Authors in RE/SD

Table 2 lists the top 15 authors in the RE/SD field and their respective countries. Notably, only Portugal, China, and India have two authors each on the list. The authors' first publications range from 1992 to 2020, with 11 first authors, four co-authors, and one last author. While the authors' position is not strictly regulated, the last position typically signifies seniority and the supervisory role.

Table 2 List of the 15 most productive authors in renewable energy and sustainable development research area

	Author	Scopus author ID	Year of 1st publication*	TP	*h*-index	TC	Current affiliation	Country
1	Adebayo, Tomiwa S	57218099170	2020[a]	15	57	8511	Uluslararası Kıbrıs Üniversitesi	Cyprus
2	Bekun, Festus V	57193455217	2016[b]	13	56	9848	İstanbul Gelişim Üniversitesi	Turkey
3	Fuinhas, José A	36168979700	2010[b]	11	33	3634	Universidade de Coimbra, Faculdade de Economia	Portugal
4	Marques, António C	36169680100	2010[a]	11	33	3437	Universidade da Beira Interior	Portugal
5	Shahbaz, Muhammad	57218886081	2007[a]	11	107	38,132	Beijing Institute of Technology	China
6	Taghizadeh-Hesary, Farhad	56291956400	2013[a]	11	50	8136	Tokai University	Japan
7	Abbas, Shujaat	56438900000	2014[a]	10	15	835	Ural′skiĭ Federal′nyĭ Universitet	Russian Federation
8	Apergis, Nicholas	6701803017	1992[a]	10	61	16,222	University of Piraeus	Greece
9	Ridzuan, A. R	57201919567	2012[b]	10	12	438	Universiti Teknologi MAR	Malaysia
10	Dong, Kangyin	57189246704	2016[b]	9	47	7927	University of International Business and Economics	China
11	Gyamfi, Bright A	57216591351	2020[a]	9	24	1781	Sir Padampat Singhania University	India
12	Shahzad, Umer	57206773899	2019[a]	9	40	4948	Adnan Kassar School of Business	Lebanon
13	Tiwari, Aviral K	57204698496	2010[a]	9	56	12,398	Indian Institute of Management Bodh Gaya	India
14	Khobai, Hlalefang	57190732292	2016[a]	8	8	212	University of Johannesburg	South Africa
15	Murshed, Muntasir	57204031604	2018[a]	8	52	7327	North South University	Bangladesh

* Role in co–authorship, superscripts
[a] First author
[b] Co-author
[c] Last author

Tomiwa S. Adebayo, hailing from Cyprus, tops the list with 15 publications since 2020, a 57 *h*-index, and 8511 citations. Festus V. Bekun, from Turkey, is the second author, while José A. Fuinhas and António C. Marques, both from Portugal, are the third and fourth top authors, respectively. Notably, Nicholas Apergis is the most cited author (16,222) and has the oldest publication date (1992).

3.5 Keyword Occurrences in RE/SD

The analysis of 3629 author keywords reveals 46 keywords that have been used at least 20 times in the mapping in VOSviewer. A keyword overlay visualization map, presented in Fig. 4, illustrates the publication trends over time, with the node color representing the publication period. The lighter colors highlight more recent research focused on a specific area. The figure describes the keyword overlay visualization map of RE/SD publications. Among the frequently appearing keywords, "renewable energy" and "economic growth" stand out with the same number of occurrences (376). "Sustainable development" follows with 164 occurrences, while "sustainability," "energy consumption," "natural resources," "renewable energy consumption," "CO_2 emissions," "financial development," "carbon emissions," "energy efficiency," "energy," "green finance," "environmental sustainability," and "energy transition" also appear frequently. The figure shows the keywords with high research interest between 2018 and 2022, with "energy transition," "financial development," "green finance", "technological innovation," and "environmental sustainability" being some of the recent focus areas for researchers.

All identified keywords are interrelated and crucial in transitioning to a more sustainable energy system. For example, "renewable energy" is often associated with reducing carbon emissions and mitigating the impacts of climate change. At the same time, "sustainable development" involves balancing economic, environmental, and social considerations to ensure long-term societal well-being. Moreover, "energy consumption" and "energy efficiency" are essential for reducing the environmental impact of energy production and consumption. Reduced energy consumption and increased energy efficiency can lead to lower greenhouse gas emissions, reduced dependence on fossil fuels, and increased affordability of energy services for households and businesses.

Additionally, "financial development" and "green finance" relate to financing renewable energy and sustainable development projects, which is critical to their deployment. Financial institutions increasingly recognize the importance of incorporating environmental, social, and governance (ESG) factors into their decision-making processes.

As for "Technological innovation," technological advancements have led to the development of more efficient renewable energy sources such as solar, wind, and geothermal power. Additionally, it has led to the creation of more sustainable transportation options, such as electric vehicles. Overall, technological innovation plays a critical role in advancing the field of renewable energy and sustainable development.

Lastly, "environmental sustainability" encompasses many issues, such as biodiversity conservation, natural resource management, and pollution control. Achieving environmental sustainability requires a holistic approach that involves protecting and restoring the natural environment while ensuring the well-being of human societies.

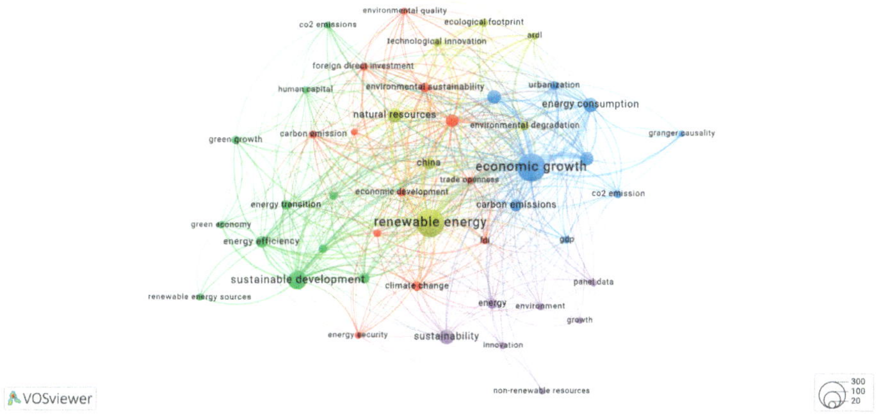

Fig. 4 Bibliometric map with overlay visualization mode based on author keywords co-occurrence. Node colors represent the publication year (purple and blue for older, green and yellow for newer), and node sizes indicate keyword frequency. Minimum occurrences of a keyword set to twenty. Access the map through the following URL: http://tinyurl.com/yrmvmswv

4 Conclusion

This study provides an in-depth analysis of the research trends in renewable energy and sustainable development. It is based on a review of 1727 publications indexed in the Scopus database. The analysis indicates that there has been a rapid growth in publication numbers over the last 15 years, and this upward trend is expected to continue. The study identifies countries and academic institutions, such as China and the USA, with significant publications and moderate international collaboration. This presents an opportunity for academic researchers from countries like Pakistan and Saudi Arabia to widen their studies' collaborations. The study observed that "Adebayo, Tomiwa S." is the most influential author, and "Resources Policy" is the leading journal about RE/SD.

The review highlights that research in areas such as energy efficiency and consumption has been well-explored and has contributed significantly to the environmental dimension of sustainable development by reducing CO_2 emissions. Additionally, the analysis identifies emerging research areas, such as green finance and environmental sustainability, which hold great potential for future studies in renewable energy and sustainable development. Furthermore, the study emphasizes the critical role of technological innovation in promoting renewable energy and sustainable development.

In conclusion, we recommend that upcoming research compare the information derived from several databases, including Scopus and Web of Science, to gain a more comprehensive understanding of research trends in renewable energy and sustainable development. The study did not incorporate databases such as gray literature (Google Scholar) to improve the findings' dependability. Overall, the findings of this study contribute to the growing body of knowledge in renewable energy and sustainable development and provide valuable insights for future research in this field.

Future research should explore how financial innovations, like green bonds, drive renewable energy investments. Effective policy frameworks could include tax incentives, subsidies for clean technologies, green infrastructure funding, and stricter emissions regulations. By examining these frameworks and sustainable business models, researchers can provide insights to enhance economic growth and environmental outcomes, supporting global sustainability goals and accelerating clean energy transitions.

References

1. Md Khudzari J, Kurian J, Tartakovsky B, Raghavan GSV (2018) Biochem Eng J 136:51
2. Hepbasli A (2008) Renew Sustain Energy Rev 12:593
3. Geraldo Schwengber J, Grünfelder T, Wieland J (eds) (2023) Sustainable development goals: perspectives from Vietnam. Metropolis-Verlag, Marburg
4. Iwami S, Ojala A, Watanabe C, Neittaanmäki P (2020) Scientometrics 122:3
5. Yoshino N, Taghizadeh-Hesary F (2014) Int J Monet Econ Finance 7:157
6. Akpan J, Olanrewaju O (2023) Energies 16:7049
7. Chong CT, Fan YV, Lee CT, Klemeš JJ (2022) Energy 241:122801
8. Umar M, Riaz Y, Yousaf I (2022) Resour Policy 79:102966

The Nexus Between Energy Consumption and Economic Growth in Morocco

Yousra Benyetho and Abdelilah El Attar

Abstract

The use of clean energy globally has certainly gained momentum since the 1970s with its many advantages over fossil fuels. Morocco, in turn, started exploiting this type of resource (except hydroelectric power) only in the 2000s. In addition, several authors have tried to analyze and verify the presence of a relationship between energy consumption and economic growth. In our study, we focused on the correlation between energy consumption by its two categories (fossil and renewable) and the economic growth of our country, Morocco. Using Eviews software, two tests were conducted to validate cointegration and causality, throughout the period 1990–2016. The ARDL test demonstrates a cointegration between GDP per labor engaged (GDP/L) as the economic growth indicator on one hand, and the three variables: Economic complexity indicator (ECI), total non-renewable energy consumption (TNRE), and total renewable energy consumption (TRE) on the other hand. The Toda-Yamamoto test proves unidirectional causalities from TRE to GDP/L, and from TNRE to GDP/L. Thus, we were able to confirm the growth hypothesis: Energy consumption (both fossil and renewable) has a positive impact on economic growth in the Kingdom of Morocco.

Keywords

Energy consumption · Economic growth · Morocco · ARDL model · Toda-Yamamoto causality

Y. Benyetho (✉) · A. El Attar
Research Laboratory in Instrumentation and Management of Organizations LURIGOR, University Mohammed the First, Oujda, Morocco
e-mail: yousra.benyetho@ump.ac.ma

1 Introduction

Global warming, oil costs, renewable energies, and so on are all contemporary issues that all countries are concerned about. First, the 1973 oil crisis disrupted the economic status of countries that imported this type of fuel. Second, global warming, which is mostly the result of human activity, has raised alarms about a scenario that threatens not just our survival but also the environmental health of our world. The blows continue, on the one hand, with increased unemployment and inflation caused by the aforementioned economic crisis and on the other hand, with the escalation of natural calamities driven mostly by global warming. These aspects have laid the groundwork for understanding the significance of examining the effects of energy on both the economy and the environment on a global scale. Governments have been interested in the topic of energy saving, taking into account the implications of this approach on the previously stated economic and environmental problems.

Renewable energy, without a doubt, represents an unavoidable answer to several risks that threaten the economic and governmental stability of the world's multiple nations. At the environmental level, so-called clean resources allow for lower carbon dioxide emissions, thereby mitigating the negative consequences of climate change. At the economic level similarly, they present several advantages; they reduce the burden of countries importing fossil fuels by reducing their reliance on this type of energy, as well as they ensure an adequate climate that attracts foreign investors, resulting in the creation of a significant number of jobs and the prosperity of the country in question. For countries that already have fossil resources, renewable energy includes endless resources, allowing them to avoid the possibility of fossil energy exhaustion. However, all of these foregoing assets have not prevented the world's economy from using fossil resources and even exploring new deposits to extract them. This attachment

A. Moumen et al. (eds.), *Artificial Intelligence and Data Analytics for Innovative Applications in Engineering, Sustainability and Technology*, Sustainable Artificial Intelligence-Powered Applications, https://doi.org/10.1007/978-3-031-90318-2_17

is mostly due to the political vision of countries seeking to assert their supremacy through the use of conventional energy, particularly oil while awaiting the global dominance of renewable energy.

The connection between economic growth and energy consumption has certainly been the subject of statistical analysis by several researchers, most notably in the form of time series. The aim of these analysts is not only to prove the existence of this relationship but also to analyze the direction of causality; in other words, does economic growth cause changes in energy consumption, or is it the other way around?

According to various research, there are four main hypotheses. The first one states that using renewable energy helps the economy grow (the growth hypothesis) [1]. The second one affirms that the economy and renewable energy help each other (the feedback hypothesis) [2]. The third one asserts that the economy affects renewable energy, but not the other way around (the conservation hypothesis) [3]. The fourth one explains that the economy and renewable energy don't affect each other (the neutrality hypothesis) [4].

In addition, it is also necessary to specify the variables that can assist in the composition of the function dealing with this link. In this context, we intend to explore the connection between energy consumption in its two forms (fossil and renewable) and Morocco's economic growth.

2 Methods

While most scientific papers studying the link between economic growth and energy consumption include a wide number of nations [1, 5], this does not rule out the existence of research that focuses on a single country [6, 7]

In light of this, we have decided to investigate the growth/energy link only at the level of the Kingdom of Morocco. The period under consideration ranges from 1990 to 2016. Our study's selection of variables was inspired by the empirical analysis of Gozgor et al. (2018), which uses the GDP/labor as an endogenous variable to express economic growth and the economic complexity indicator (ECI), total non-renewable energy consumption (TNRE), and total renewable energy consumption as exogenous variables (TRE), respectively.

The equation of our model is:

$$\log\left(\frac{Y}{L}\right)_t = \alpha_0 + \alpha_1 \mathrm{ECI}_t + \alpha_2 \log \mathrm{TNRE}_t + \alpha_3 \log \mathrm{TRE}_t + \varepsilon_t$$

t expresses the time-series character of our model, α_0 is the constant term, α_1, α_2 et α_3 are the respective coefficients of each variable in the model and ε_t is the error term which expresses the deviation between reality and the model estimate.

Data on real GDP and total labor force are obtained from the World Bank website (the World Development Indicators (WDI)). Real GDP is expressed in constant 2010 USD Billions. Thus, $\left(\frac{Y}{L}\right)_t$ indicates the Gross Domestic Product per Labor (GDP/Labor). The economic complexity indicator is obtained from the official website of the Observatory of Economic Complexity (OEC). The data on energy consumption in its two forms were collected from the statistical review of World Energy of the British (BP), expressed in Millions of Tons of Oil Equivalent (MTOE).

In the Eviews software, we initially performed the Augmented Dickey-Fuller (ADF) [9] test to ensure that the variables were stationary. We then used the ARDL test to try to specify the optimal model for our work. Next, we applied the Pesaran et al. [10] bounds cointegration test to determine whether or not a cointegration existed between the variables in our study. And finally, we adopted the Toda-Yamamoto causality analysis [11] to determine the direction of causality among those variables.

3 Results and Discussion

First, we notice that all the variables in our study are stationary at level I (1) (1st difference). Therefore, the Johansen cointegration test is the most compatible with our model, since the order of differentiation is the same for all the variables studied (I (1)). However, we considered it more appropriate to apply the ARDL cointegration test [10] given the many advantages it offers. This approach can be utilized even when numerous variables of various orders are present, as long as the order does not exceed I (1). To that purpose, this test enabled us to find the model that best complies with our equation among a large number of ARDL models. In this study, the optimal ARDL model detected under the Schwarz criterion (SIC) is ARDL (3, 2, 2, 4).

Second, the ARDL model was used as a basis for the bounds test. This test proved the existence of cointegration (long-term relationship) between the variables of our model (Fisher's F > upper bound; 14.27 > 5.61 at 1% significance).

Third, the Toda-Yamamoto causality test (Table 1) supports this finding by emphasizing the causal links observed specifically in the total consumption of fossil and renewable energy. Indeed, both of the energy consumption types have a positive impact on economic growth (growth hypothesis).

Lastly, the existence of a long-term relationship often hints at the presence of a short-run one. We ran the short and long-run ARDL tests to confirm this. These tests confirmed that total fossil energy consumption has a favorable long and short-run effect on economic growth. Renewable energy consumption, in turn, exhibits a positive, albeit minor, short-run effect,

Table 1 Results of the Toda-Yamamoto causality test

Dependent variable	Independent variable			
	GDP/labor	ECI	TNRE	TRE
GDP/labor	–	4.56 (0.2067)	47.19 (0.0000)***	20.23 (0.0002)***
ECI	1.97 (0.5759)	–	3.28 (0.3498)	0.35 (0.9500)
TNRE	1.66 (0.6463)	1.16 (0.7628)	–	2.20 (0.5311)
TRE	4.82 (0.1854)	1.69 (0.6399)	3.47 (0.3252)	–

***, **, and * indicate statistical significance at the 1%, 5%, and 10% levels, respectively

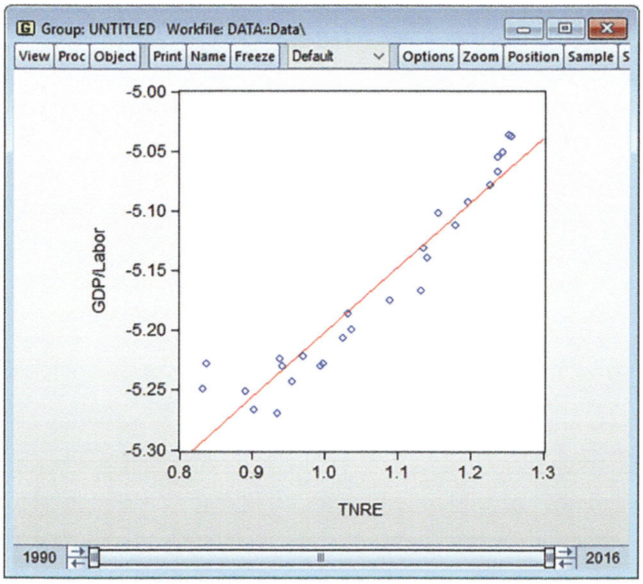

Fig. 1 Correlation between total fossil energy consumption (TNRE) and economic growth (GDP/labor)

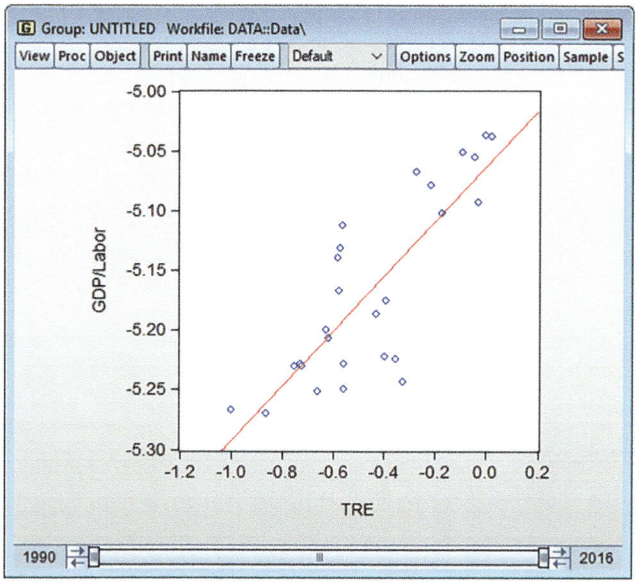

Fig. 2 Correlation between total renewable energy consumption (TRE) and economic growth (GDP/labor)

and no effect in the long-run. Regarding the ECI, neither a long-run nor short-run effect is seen. The graphical evolution of these variables in relation to economic growth reinforces the initial findings (Figs. 1 and 2).

4 Conclusion

Undoubtedly, Morocco has made considerable efforts in promoting the implementation of its renewable energy strategy, whether through the enactment of laws in this context, the implementation of large and small projects, or its strong involvement in the energy sector worldwide.

This study makes a valuable theoretical contribution by offering compelling evidence that supports the growth hypothesis, which emphasizes the crucial role of energy consumption in promoting economic development specifically in Morocco. Additionally, the study's findings have important managerial implications as they underscore the significance of implementing sustainable energy policies to foster continuous economic growth within the country.

Nevertheless, we proved through this analysis that nonrenewable energy still has a stronger impact on the economic growth of our Kingdom. These findings could be explained by the use of fossil fuels which is still dominant in terms of total energy consumption. The consumption of renewable resources has increased since the 2000s, but their share remains considerably small compared to that of fossil fuels.

Consequently, Morocco, which has begun its energy transition, must learn from the experiences of countries that have been pioneers in renewable energy practices. Among the lessons to be gained is that diversifying energy resources is unavoidable to meet local demand. Aside from that, addressing finance issues based on luring investors requires liberalization of the power sector.

References

1. Bhattacharya M, Paramati SR, Ozturk I, Bhattacharya S (2016) Appl Energy 162:733
2. Kahia M, Aïssa MSB, Lanouar C (2017) Renew Sustain Energy Rev 71:127
3. Menyah K, Wolde-Rufael Y (2010) Energy Econ 32:1374
4. Menegaki AN (2011) Energy Econ 33:257
5. Çakmak EE, Acar S (2022) J Clean Prod 352:131548
6. Alam MJ, Begum IA, Buysse J, Rahman S, Van Huylenbroeck G (2011) Renew Sustain Energy Rev 15:3243
7. Bui Minh T, Bui Van H (2023) Energy Rep 9:609
8. Gozgor G, Lau CKM, Lu Z (2018) Energy 153:27
9. Dickey DA, Fuller WA (1979) J Am Stat Assoc 74:427
10. Pesaran MH, Shin Y, Smith RJ (2001) J Appl Econom 16:289
11. Toda HY, Yamamoto T (1995) J Econom 66:225

Geospatial Data-Driven Cadaster for Moroccan Land-Use Planning: Solar Cadaster Case Study

Youssef Rissouni, Elhassan Jamal, Hicham Jamil, Rachid El Ansari, Bouabid El Mansouri, Jamal Chao, and Aniss Moumen

Abstract

Morocco's ambitious renewable energy goals necessitate an advanced, data-driven approach to urban planning, wherein a big data-enabled and AI-supported solar cadaster can play a pivotal role (Elhassan et al. in Proceedings of the 4th edition of international conference on Geo-IT and water resources 2020, Geo-IT and water resources 2020. ACM, Al-Hoceima Morocco, pp 1–5, 2020). By integrating large-scale geospatial datasets and applying AI-driven analytics, this cadaster can optimize land-use planning, enhance regulatory transparency, and foster investment readiness. However, its potential is hindered by fragmented data systems, outdated technological infrastructure, and limited cross-agency collaboration. This study examines these challenges through stakeholder insights from land tenure specialists, policymakers, spatial planners, and investors. Employing big data and AI tools, the research highlights critical inefficiencies in data governance and regulatory alignment. It proposes a comprehensive strategy combining centralized data repositories, predictive analytics, and AI-driven site assessments to unlock the solar cadaster's transformative capabilities.

Keywords

Big data analytics · Fit-for-purpose cadaster · Land information management · Data governance · Sustainable development

Y. Rissouni (✉) · E. Jamal · H. Jamil · R. E. Ansari · A. Moumen
Laboratory of Engineering Sciences, National School of Applied Sciences, Ibn Tofaïl University, Kenitra, Morocco
e-mail: youssef.rissouni@gmail.com

B. E. Mansouri · J. Chao
Laboratory of Natural Resources and Sustainable Development, Ibn Tofaïl University, Kenitra, Morocco

1 Introduction

The last few years has seen growing interest in integrating renewable energy into urban planning, particularly as Morocco's cities continue their rapid expansion. Successfully merging these two domains demands a robust, data-centric approach to land information management. A solar cadaster—which is a specialized tool to assess urban rooftop solar potential [2]—shows particular promise for enhancing regulatory transparency, optimizing land-use choices, and draw investment toward photovoltaic infrastructure development. Through the application of big data analytics and AI technologies, this cadaster enables precise site selection, predictive modeling, and automated evaluation of rooftop suitability, marking a significant step forward in renewable energy planning.

Previous research has emphasized how data-driven land information systems benefit urban development, especially when creating fit-for-purpose cadasters that provide transparency and decision-making efficiency in land management [3]. The work of Polo and García [4] revealed the crucial role of digital surface models (DSM) in urban PV planning, showing that DSM accuracy affects solar potential estimates on rooftops. Building on this foundation, El Assal and Rochdane's [5] study highlights geospatial data integration's vital role in optimizing energy site selection within smart city initiatives, while emphasizing the necessity of collaborative governance structures for successful renewable energy projects [5].

Yet despite these advancements, Morocco faces several obstacles in implementing a comprehensive solar cadaster, including disconnected data systems, aging infrastructure, and insufficient cross-agency collaboration [6].Our research expands upon these findings through an in-depth examination of various stakeholder perspectives—including land

A. Moumen et al. (eds.), *Artificial Intelligence and Data Analytics for Innovative Applications in Engineering, Sustainability and Technology*, Sustainable Artificial Intelligence-Powered Applications, https://doi.org/10.1007/978-3-031-90318-2_18

tenure experts, policymakers, urban planners, and investors—on Morocco's current geospatial infrastructure and data requirements. This paper identifies critical challenges while proposing an integrated data framework that leverages big data analytics, AI-driven site suitability assessments, and predictive modeling capabilities.

2 Methodology

The methodological approach for this study integrates both qualitative stakeholder insights and a comprehensive spatial data inventory, creating a robust framework for advancing urban planning and renewable energy integration.

- **Stakeholder Interviews**: The research initiated with a series of structured, in-depth interviews involving a broad spectrum of stakeholders, including land tenure specialists, policy advisors, urban planners, and investment representatives. These interactions provided crucial insights into the multifaceted needs and challenges faced by each group, particularly concerning data accessibility, land tenure security, and regulatory alignment. This phase of engagement was pivotal in identifying systemic constraints and setting precise objectives, ensuring that the cadaster's design aligns closely with stakeholder requirements.
- **Geospatial Data Inventory**: Following the stakeholder engagement, the study conducted an exhaustive spatial data inventory, gathering and analyzing urban and environmental datasets from a variety of secondary sources, including cadastral records, land-use maps, socioeconomic data, and environmental assessments (see Table 2 for data inventory). This inventory phase facilitated a detailed assessment of existing land-use configurations, cadastral delineations, and socioenvironmental parameters critical to spatial decision-making. By cataloging and evaluating these data layers, the inventory establishes a comprehensive spatial foundation, ensuring that urban planning and renewable energy site selection are grounded in accurate, context-specific data.

This dual approach, merging qualitative stakeholder insights with an extensive spatial data inventory, provides a well-rounded framework for designing a responsive, data-driven solar cadaster tailored to Morocco's evolving urban and energy landscapes.

3 Results and Discussion

3.1 Stakeholder Analysis

The study employed a series of structured, in-depth interviews to pinpoint specific needs and challenges faced by stakeholders in Morocco's land tenure, policy, and investment sectors, thereby highlighting essential objectives for advancing data-driven land-use planning (Table 1).

Land tenure experts underscored the necessity for clear and secure ownership rights, adaptable legal frameworks, and the recognition of customary practices, which are often hindered by fragmented and outdated cadastral data. Integrating big data analytics and interoperable geospatial data systems was identified as a key priority to enhance tenure security and enable the flexible recording of various tenure types. Public policy and governance authorities emphasized the critical importance of accessible and reliable land data in effective policy-making, particularly supporting sustainable development initiatives. Current regulatory challenges, fragmented data repositories, and outdated infrastructure obstruct transparency and impede effective data sharing. Spatial planners and urban development professionals require high-quality, interoperable geospatial data and advanced analytics to facilitate precise urban expansion planning and optimal site identification for renewable energy installations, particularly solar projects. However, existing limitations in data sources and analytical capabilities currently hinder effective land-use forecasting. For the investment community, the transparency and accuracy of land data are vital for mitigating tenure risks and making informed site selections for renewable energy projects. Ongoing issues with opaque data-sharing mechanisms and inconsistencies highlight the urgent need for a reliable and cost-effective cadaster that leverages big data analytics and interoperable systems to secure and share tenure information.

3.2 Urban Data Analysis

The urban environment is a space formed by adding various urban forms, each reflecting the conception of the city and communal life at a particular time. Each major ideology has produced a unique urban form. This form incorporates the associated urban functions: housing, economic, cultural, and leisure activities, among others. Indeed, an urban area

Table 1 Summary of the stakeholder analysis, focusing on their needs, challenges, and objectives related to a geospatial data-driven cadaster for land-use planning in Morocco

Stakeholder group	Needs	Challenges	Objectives
Land tenure specialists	Clear and secure land ownership rights Adaptable legal frameworks Recognition of customary practices Efficient tenure systems	Data fragmentation and lack of updated cadastral info hinder tenure security	Ensure inclusive cadaster that supports all tenure forms (customary and formal) Flexible legal approach for ownership recording
Public policy and governance authorities	Reliable land data for policy-making Support for sustainable development policies Centralized data sharing	Regulatory barriers Lack of centralized data sharing Outdated infrastructure affects transparency and execution	Provide a unified data repository for better policy integration Enable transparent data exchange and support for SDGs
Spatial planners and urban development experts	High-quality geospatial data for urban expansion Optimal locations for solar installations Improved land-use forecasting	Outdated data sources Insufficient geospatial integration Limited analytical capacity for future planning	Enable precise mapping of renewable energy sites, especially solar Improve spatial planning with integrated data
Investment community (private and public sectors)	Transparent, accurate land data Reduced risks in land tenure Efficient site selection for renewable energy projects	Opaque data-sharing mechanisms Potential data inconsistencies Concerns over tenure risks	Establish a trustworthy, cost-effective cadaster Facilitate investment in renewable energy with clear land data

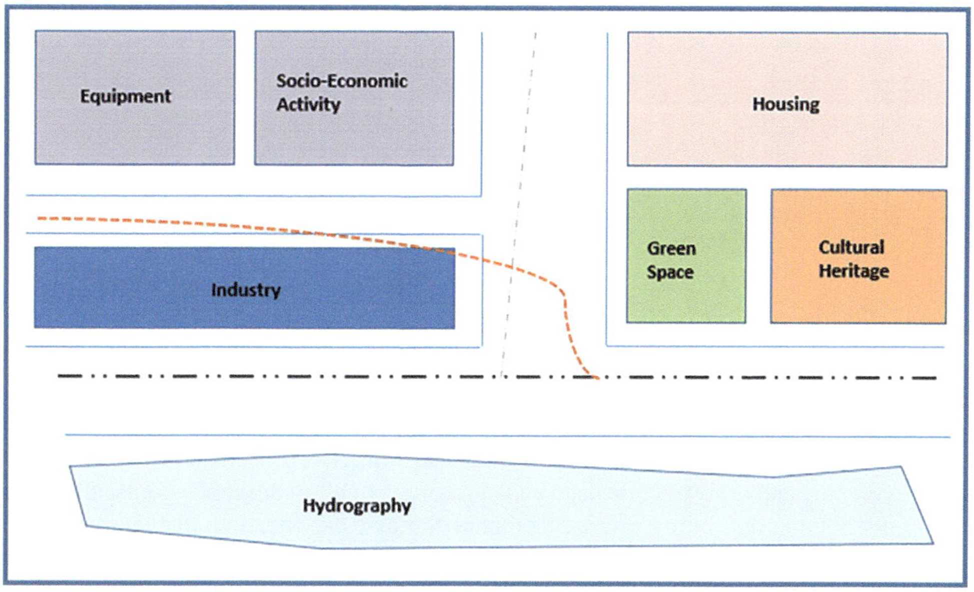

Fig. 1 Schematic representation of the urban environment

comprises a whole series of environments, more or less interconnected and overlapping. Functional diversity is thus discussed when at least two of these functions are represented within a neighborhood (Fig. 1).

Urban data is fundamental to developing a geospatially driven solar cadaster, underpinning sustainable urban energy planning and integration. Key data layers include foundational spatial references, cadastral information, urban planning frameworks, environmental assessments, and infrastructure networks. Foundational spatial data, such as maps and satellite imagery, provides accurate spatial positioning, essential for identifying optimal sites for urban solar installations. Legal and cadastral data clarify land tenure and ownership, minimizing investment risks by ensuring secure property rights. Urban planning data, including zoning regulations and development frameworks, aligns solar projects with existing land-use strategies, promoting cohesive urban development. Environmental data enables the assessment of solar irradiance and

Table 2 Data inventory

Category	Description	Key data types	Primary use	Relevance to solar cadaster
Foundational spatial reference data	Essential spatial layers that provide a base for mapping and spatial analysis	Cartographic data, geodetic information, satellite imagery, orthophotos, administrative boundaries	Foundational reference for all spatial data	Provides base maps and precise spatial coordinates for potential solar site analysis
Legal and ownership data	Data related to land ownership, registration status, and governance	Registered property data, non-registered land information, ownership records	Supports land tenure security and legal governance	Ensures clear ownership information, reducing tenure risks and facilitating investment in solar sites
Territorial and urban planning data	Data used for national and regional planning, including urban zoning and land-use development strategies	Zoning plans, urban and regional development frameworks, sectoral strategies (industry, agriculture, tourism)	Guides land-use planning and infrastructure development	Identifies areas suitable for solar projects by aligning with existing zoning and development plans
Environmental and natural resource Data	Data describing environmental attributes and natural resources for sustainable land management	Geological, hydrological, soil, and ecological data, environmental indicators	Assesses sustainability and ecological impact	Evaluates environmental suitability, including solar irradiance, to ensure optimal, low-impact site selection
Socioeconomic and infrastructure data	Data on population demographics, economic indicators, and infrastructure networks	Population distribution, development metrics, economic activity, transportation, utilities, public facilities	Supports strategic planning, investment, and public services	Assesses access to infrastructure (roads, electricity) and potential impact on communities near solar installations

ecological impact, guiding site selection to maximize solar potential while minimizing adverse environmental effects. Socioeconomic and infrastructure data, encompassing population distribution and utility access, facilitates the assessment of infrastructure readiness and community impact, supporting equitable and socially responsible energy deployment. Together, these integrated urban data layers enable a robust, data-driven approach, advancing Morocco's solar cadaster and contributing to its broader renewable energy objectives (Table 2).

The current state of geospatial data for land-use planning and renewable energy initiatives in Morocco reveals critical deficiencies that impede effective, integrated decision-making across multiple sectors. While foundational datasets, including cadastral maps, topographic outlines, and partial utility network records, provide a starting point, they frequently lack the accuracy, detail, and interoperability essential for comprehensive spatial analysis. For example, cadastral records often remain outdated, lacking secure classifications that integrate both formal and customary land rights, which are vital for enhancing tenure security among land tenure specialists. Policymakers encounter further limitations due to fragmented socioeconomic datasets and the absence of a centralized repository, hampering efforts to integrate sustainable development goals into urban planning frameworks effectively. Spatial planners, reliant on precise topographic and infrastructure data to forecast urban growth and identify optimal sites for solar installations, face additional constraints due to incomplete or insufficient datasets. This data scarcity extends to utility and infrastructure managers, who require detailed and accurate maps of utility networks to streamline maintenance and support service expansions.

Table 3 A matrix outlining actors, data needs, and specific requirements for a geospatial data-driven cadaster system, focusing on urban planning and solar energy initiatives in Morocco

Actors	Existing data	Data needs	Specific requirements
Land tenure specialists	Cadastral maps, ownership records	Updated cadastral boundaries, clear land tenure classification	Secure, clear property boundaries; integration of both customary and formal land rights
Public policy and governance	Socioeconomic data, zoning regulations	Centralized, integrated cadastral and urban planning data	Reliable, up-to-date land data for policy-making; support for sustainable urban development
Spatial planners	Basic topography, partial building footprints, limited infrastructure data	High-resolution topography, complete building footprints, transportation networks	Accurate, high-quality spatial data; optimal solar site selection
Utility and infrastructure managers	Utility network layouts, partial maps of public facilities	Full infrastructure layers (water, electricity, gas), facility details	Comprehensive infrastructure mapping for maintenance and expansion; data security and controlled access
Environmental agencies	Green space inventory, natural resource maps	Environmental impact data, land availability and usage restrictions	Detailed resource constraints; ecological sensitivity for site selection
Investment community	Property maps, limited ownership data	Transparent cadastral data, tenure risk assessment	Access to reliable tenure information; accurate site data to mitigate investment risks
General public/citizens	Limited public facility information, general maps	Renewable project locations, community impact data	Transparent, accessible information on renewable energy sites and their community effects

Table 4 Proposed system features

Proposed system features	Description and functionality
Centralized big data repository	Implementing a centralized data repository enables the consolidation of geospatial and cadastral information from multiple sources, ensuring consistent, accessible data for stakeholder reference and promoting cross-agency data interoperability.
Blockchain technology for secure data exchange	Integrating blockchain protocols within the cadaster system enhances data security and transaction transparency, ensuring auditable and secure data exchanges across stakeholders. This approach is intended to address concerns of data fragmentation and inconsistencies.
Analytics for solar cadaster	Analytics tools support solar energy site selection by identifying land parcels with optimal solar irradiance while also considering environmental and spatial constraints. This feature aligns the cadaster with Morocco's renewable energy targets and advances informed investment decisions.

Furthermore, the investment community, dependent on transparent and reliable land tenure information, confronts inconsistencies that heighten tenure risks and complicate site selection for renewable energy projects. Addressing these gaps necessitates the development of a centralized, interoperable geospatial data repository that can provide comprehensive, accessible, and up-to-date data (Table 3).

3.3 Proposed System Features

The proposed system (Table 4 and Fig. 2) introduces a big data and geospatially driven cadaster designed to address essential needs for data consistency, transparency, and analytical support across land tenure, policy, spatial planning, and investment sectors in Morocco. We recommend establishing

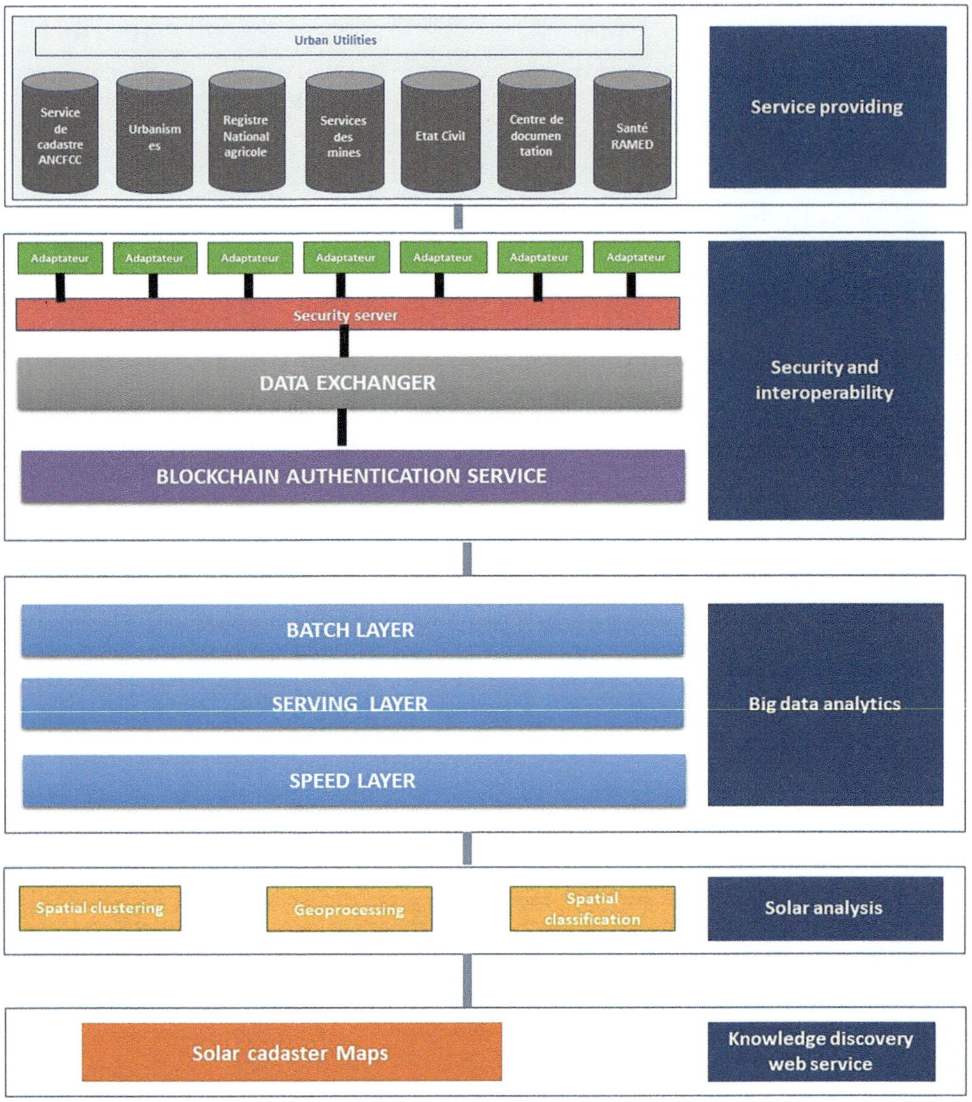

Fig. 2 Conceptual framework of the solar cadaster [3]

a centralized Big Data Repository to consolidate geospatial and cadastral information from various sources. This initiative will enhance interoperability and ensure that all stakeholders have access to accurate and up-to-date land data.

The repository will be pivotal in supporting policy-making, spatial planning, and secure investments by providing a unified source of truth for land information. To further enhance security and transparency, we propose integrating blockchain technology for data security. This technology will facilitate secure, auditable data transactions among agencies and stakeholders, offering significant benefits to investors by mitigating tenure risks. In addition, we suggest implementing predictive analytics for solar cadaster tools to assist in identifying optimal sites for solar energy projects. By harnessing big data analytics, this feature will enable spatial planners and investors to assess solar irradiance and environmental constraints, aligning site selection with Morocco's renewable energy targets.

4 Conclusion

Our findings highlight how Morocco can transform its urban planning by integrating sustainable urban planning and renewable energy through a big data and AI-driven solar cadaster. The current challenges we uncovered—scattered data sources, outdated systems, and poor communication between agencies—could be effectively tackled by creating one unified, transparent platform. By talking with stakeholders on the ground, we learned that combining smart analytics, location-based insights, and automated site analysis could help Morocco achieve its clean energy goals more efficiently. Beyond just serving Morocco's sustainability goals, this new approach could inspire other developing nations to modernize how they handle urban growth and land-use decisions by giving decision-makers and different government departments the tools to work together with reliable, up-to-date information.

Reference

1. Elhassan J, Aniss M, Jamal C (2020) Big data analytic architecture for water resources management: a systematic review. In: Proceedings of the 4th edition of international conference on Geo-IT and water resources 2020, Geo-IT and water resources 2020. ACM, Al-Hoceima Morocco, p 1–5. https://doi.org/10.1145/3399205.3399225
2. Alqahtani N, Balta-Ozkan N (2021) Assessment of rooftop solar power generation to meet residential loads in the city of Neom, Saudi Arabia. Energies 14(13), Article no 13. https://doi.org/10.3390/en14133805
3. Rissouni Y et al (2024) Fit-for-purpose cadaster architecture for Moroccan land-use planner: proposal and perception. In: E3S web conference, vol 489, p 04017. https://doi.org/10.1051/e3sconf/202448904017
4. Polo J, García RJ (2023) Solar potential uncertainty in building rooftops as a function of digital surface model accuracy. Remote Sens 15(3):567. https://doi.org/10.3390/rs15030567
5. El Assal Z, Rochdane H (2023) Citizens motivation towards solar energy in the context of the smart city, the case of Casablanca Morocco. E3S web conference, vol 412, p 01049. https://doi.org/10.1051/e3sconf/202341201049
6. Sejati AW, Buchori I, Rudiarto I, Silver C, Sulistyo K (2020) Open-source web GIS framework in monitoring urban land use planning: participatory solutions for developing countries. J Urban Reg Anal 12(1). https://doi.org/10.37043/JURA.2020.12.1.2

A Comprehensive Assessment of the Most Effective Neural Network Models for Arabic Sentiment Analysis

Youssra Zahidi and Yassine Al-Amrani

Abstract

Sentiment Analysis (SA) involves classifying text as positive, neutral, or negative. Arabic Sentiment Analysis (ASA) faces unique challenges due to the complexity of Arabic language features. To address these challenges, deep learning (DL), a branch of machine learning (ML), employs various neural network (NN) models. This research evaluates the effectiveness of key NN models in ASA, including Artificial Neural Networks (ANNs), Recurrent Neural Networks (RNNs), Convolutional Neural Networks (CNNs), Long Short-Term Memory (LSTM), Gated Recurrent Units (GRU), Attention Mechanisms, and Transformer-based models. CNNs excel in ASA by automatically extracting relevant features from text with minimal preprocessing, making them suitable for tasks like educational content analysis. RNNs, particularly LSTM and GRU, handle sequential data effectively, capturing the context and nuances in long Arabic texts. Attention mechanisms enhance RNNs by focusing on relevant text parts, improving accuracy. By leveraging these mechanisms, transformer-based models set a new standard in ASA, processing text in parallel and capturing complex word relationships with exceptional performance. Our analysis shows that CNNs and LSTM models are highly effective for ASA. Transformer-based models are emerging as the leading choice for superior accuracy in ASA.

Keywords

Deep learning · Arabic sentiment analysis · Artificial neural networks · Recurrent neural networks · Convolutional neural networks · Long short-term memory · Gated recurrent units · Attention mechanisms · Transformer-based models

1 Introduction

Across various social media platforms, users regularly share their emotions and opinions on a multitude of topics. However, much of this content is unstructured and lacks a standardized format, making it challenging to classify the data effectively. SA has emerged as a powerful tool to address this challenge, offering the capability to predict the sentiment expressed in text. In our exploration of ASA, we thoroughly evaluate a range of existing NN models to determine the most effective options for ASA tasks.

DL, a pivotal branch of ML, has dramatically reshaped the landscape of modern problem-solving, especially in SA. Key advancements in this field have been driven by various NN models, including ANNs, CNNs, RNNs, LSTM networks, GRU, Attention Mechanisms, and Transformer-based models.

Due to its many dialects, intricate morphology, and scarcity of comprehensive linguistic materials, Arabic poses particular difficulties that are more severe than those faced by languages like English. Because of these aspects, ASA research is especially complex, requiring a thorough evaluation of NN models to determine which ones are most suited to meet the particular requirements of ASA.

Although ANNs are the fundamental building blocks of many DL models, their more straightforward architecture may make it more difficult for them to handle the many subtleties of the Arabic language. CNNs, on the other hand, are particularly

Y. Zahidi (✉) · Y. Al-Amrani
Information Technology and Modeling Systems Research Team, Abdelmalek Essaadi University, Tetuan, Morocco
e-mail: youssra.zahidi-etu@uae.ac.ma

Y. Al-Amrani
e-mail: yassine.alamrani@uae.ac.ma

A. Moumen et al. (eds.), *Artificial Intelligence and Data Analytics for Innovative Applications in Engineering, Sustainability and Technology*, Sustainable Artificial Intelligence-Powered Applications, https://doi.org/10.1007/978-3-031-90318-2_19

effective in situations when little preprocessing is required because of their ability to automatically identify and recognize local patterns in text, such as phrases and word connections. When evaluating lengthy Arabic texts, RNNs are ideal for capturing the temporal characteristics of language because they are built to handle sequential data. The capacity of LSTM networks and GRU to maintain long-term relationships and effectively manage the computational demands involved in processing sequential data makes them stand out among RNNs versions.

By allowing the model to focus on the most pertinent portions of the input sequence, attention mechanisms significantly improve the performance of RNNs and increase their interpretability and accuracy in sentiment classification tasks. Meanwhile, Transformer-based models, which capitalize on these attention mechanisms, have set new standards in SA across multiple languages, including Arabic. Their ability to process text in parallel rather than sequentially, combined with their effectiveness in capturing complex word relationships, makes them particularly powerful for ASA, where a deep understanding of language subtleties is essential.

Our extensive evaluation aims to equip ASA researchers with the insights necessary to choose the most effective NN models for their projects, allowing them to make informed decisions that consider the specific challenges posed by the Arabic language. The remainder of this research is organized as follows: Sect. 2 explores the differences between simple NNs, ANNs, and DL models. Section 3 offers an overview of the NN models employed in ASA, along with a comparative analysis of their effectiveness. Section 4 presents a detailed discussion of the comparison results, and Sect. 5 concludes with a summary of findings and recommendations for future research in ASA.

Table 1 Comparison between Simple ANNs, ANNs, and DL models based on various characteristics

Aspect	Simple ANNs	ANNs	DL models
Definition	Basic neural network models for simple tasks	A broad category of neural network models with various architectures	A subset of ANNs focused on complex, multi-layered networks
Network depth	Shallow (1–2 hidden layers)	Can be shallow or deep	Deep (many hidden layers, often more than 5 layers)
Feature learning	Basic feature extraction	Basic to advanced feature extraction	Hierarchical feature learning with multiple levels
Training techniques	Basic algorithms like gradient descent	Standard algorithms plus advanced techniques	Advanced techniques like Adam optimizer, dropout, etc
Architectural examples	Single-layer perceptron, shallow MLP	Feedforward NNs, basic CNNs for image recognition, simple RNN for sequences	ResNet for deep image classification, BERT for NLP tasks, deep CNNs, LSTM

Table 2 Comparison between DL and simple NNs Models at the level of complexity

Criterion	Simple NNs	DL models
Parameters	Fewer parameters, with weights and biases for each connection	Significantly more parameters
Complexity	Less complex, less computationally demanding	More complex, requires more resources and computation due to layers and parameters
Model examples	Perceptron, simple feedforward network	Autoencoders, CNN, RNN with multiple layers
Autoencoders	–	Detect anomalies, compress data, and facilitate generative modeling

2 The Difference Between Simple ANNs, ANNs, and DL Models

In the following section, we examine the distinctions between simple ANN, ANN, and DL models. Tables 1, 2, 3, and 4 offer a comparative analysis of simple ANN versus DL models across various criteria.

3 The Most Effective Neural Network Models for Arabic Sentiment Analysis

In ASA, advanced NN algorithms like CNNs and RNNs have shown notable success. CNNs are effective at identifying local text patterns, while RNNs excel at processing sequential data, making them essential for capturing context in SA. Variants like LSTM networks and GRU enhance RNNs by addressing issues like long-term dependencies, improving the analysis of lengthy texts. By concentrating on important input segments, attention mechanisms improve the accuracy and interpretability of sentiment models. By using attention to process text in parallel, transformer-based models like BERT capture intricate word relationships and establish new benchmarks in SA. This research compares these advanced techniques (CNNs, RNNs (including LSTM and GRU), Attention

Table 3 Comparison between DL and simple NNs Models at the level of model training

Criterion	Simple NNs	DL models
Training speed	Faster training	Longer and more expensive training
Learning capability	Limited to simple tasks	Capable of learning complex models and handling sophisticated tasks
Resources and data	Less resource-intensive, smaller datasets	Requires more resources and larger datasets for training

Table 4 Comparison between DL and simple NNs Models at the level of model performance

Criterion	Simple NNs	DL models
Problems	Solves basic problems like simple pattern recognition	Effective for complex tasks like NLP, speech recognition, and processing large data volumes
Task types	Basic pattern recognition, and information classification	Complex tasks such as NLP, speech recognition, etc

Mechanisms, and Transformer-based models) with traditional ANNs to determine their effectiveness in ASA, focusing on accuracy, computational efficiency, and generalization across datasets.

3.1 Artificial Neural Network

The ANN, the most basic type of NN, is distinguished by its simple feedforward architecture, which makes it easy to understand and simple. Information flows from input nodes through hidden layers to the output node in a single direction.

3.2 Convolutional Neural Network

Renowned for its efficacy, the CNN is a potent model with one or more convolutional layers that are essential for identifying spatial hierarchies in data and have shown particular utility in ASA.

3.3 Recurrent Neural Network

The RNN, being more complex, processes data in both directions and is designed to handle sequences of varying lengths by using its internal memory cells to learn from past information and make predictions. Among RNNs.

3.4 Long Short-Term Memory

The LSTM network stands out for its ability to learn long-term dependencies and manage lengthy sequences, which has made it a prominent choice for advanced SA tasks in contemporary research.

3.5 Gated Recurrent Units

The GRU similar to LSTMs but with a simplified architecture. Balances performance and computational efficiency.

3.6 Attention Mechanisms

Allows models to focus on specific parts of the input sequence, improving the representation of important words or phrases, enhancing sentiment classification.

3.7 Transformer-Based Models

Uses an attention-based architecture revolutionizing natural language processing. Models like BERT, GPT, and RoBERTa pretrain on large text corpora and fine-tune for specific tasks. BERT captures the context from both left and right, making it powerful for SA.

Table 5 illustrates the various criteria for evaluating these neural network models. From this table, we can conclude that every neural network model has its distinct features.

4 Results and Discussion

The effectiveness of neural network techniques in SA is a subject of significant interest, particularly given the advancements in DL architectures. In our comprehensive assessment, we explored several key models: ANN, CNN, RNN, LSTM networks, GRU, Attention Mechanisms, and Transformer-based models. Each of these models has unique benefits, and choosing the best strategy for SA tasks requires an awareness of both their advantages and disadvantages.

Because of their simpler architecture, which typically consists of a few fully connected layers, ANNs are the foundation of DL and have been widely applied across various domains, including SA. However, while ANNs can perform adequately on simpler tasks, they often struggle to capture the

Table 5 Comparison of the neural network models using various criteria

	ANN	CNN	RNN	LSTM	Gated recurrent units (GRU)	Attention mechanisms	Transformer-based models
Data type	Tabular data, text data	Image data	Sequence data	Sequence data	Sequence data	Sequence data/text data	Text data
Parameter sharing	No	Yes	Yes	Yes	Yes	No	No
Fixed length input	Yes	Yes	No	No	No	No	No
Exploding and vanishing gradient	Yes	Yes	Yes	Yes	No (partially resolved)	No	No
Recurrent connections	No	No	Yes	Yes	Yes	No	No
Spatial relationship	No	No	No	No	No	No	No
Strengths	• Possesses a distributed memory structure • Operates effectively with incomplete knowledge • Capable of storing information across the entire network Exhibits fault tolerance	• Automatically identifies significant features without human guidance Incorporates weight sharing	• Can be used to expand the effective neighborhood of pixels • Retains information over time Due to its ability to remember previous inputs, this model is particularly effective for time series prediction; this feature is known as LSTM	• They overcome short-term memory limitations more effectively than 'Vanilla' RNNs They have the ability to model long-term dependencies in sequences	Balances performance and computational efficiency	Improves representation of important words or phrases, increasing sentiment classification accuracy	Effectively captures the context of words from both left and right
Limitations	• Unexplained network behavior • Dependence on hardware Determining the optimal network structure	• Cannot maintain spatial invariance for input data • Does not capture object orientation or position Requires a large amount of training data	• Training an RNN can be challenging • Struggles with gradient exploding and vanishing issues Ineffective for handling very long sequences when using hyperbolic tangent or rectified linear unit activation functions	• The memory requirements are higher than those of 'Vanilla' RNNs due to the presence of multiple memory cells They increase computational complexity compared to RNN models by introducing more parameters to learn	Less powerful than LSTMs for some specific tasks	• Higher computational complexity compared to models without attention	Requires large amounts of data for pretraining and significant computational resources
ASA works	[1, 2]	[3–17]	[9, 18–21]	[3–9, 13, 15, 16, 20–28]	[13–16, 20, 21, 24, 25, 29]	[19, 21, 26]	[16, 17, 25, 27, 28, 30–35]

complexities of sequential or spatial data, which are common in SA. For example, ANNs may not be able to accurately model the dependencies between words in a sentence, which results in less accurate sentiment classification. For SA tasks, more sophisticated models, such as CNNs and RNNs, are usually preferred.

Given their capacity to automatically extract significant features from textual input, CNNs have shown themselves to be especially successful in SA. Their architecture is highly suited for applications where the spatial structure of data is crucial because it is made to detect local patterns, including phrases and word relationships. With little to no preprocessing, CNNs are excellent at identifying sentiment-indicating patterns, such as particular word pairings or n-grams, in SA. CNNs are very useful in fields like education, where they can accurately classify feelings in reviews or feedback due to their capacity to handle massive volumes of data and detect important elements.

RNNs are perfect for SA tasks that require comprehending the temporal context of text because they are made to handle sequential data, especially their variants like LSTM and GRU. The ability of RNNs to retain information over time, in contrast to CNNs, is essential for analyzing lengthy texts or sequences in which the meaning of one word in a sentence affects that of subsequent words. Long-term dependencies are particularly well-captured by LSTM networks, which guarantees that crucial information is preserved as the sequence develops. Because of this, LSTMs are especially helpful for tasks like evaluating lengthy reviews or feedback in educational settings. Even though GRUs are easier to use and require less computing power than LSTMs, they are still very effective at processing sequential data, which makes them a good substitute for many SA tasks.

RNNs have been greatly improved by attention mechanisms, which enable models to selectively focus on the most pertinent portions of the input sequence, improving the interpretability and accuracy of sentiment classification, especially in tasks that demand a thorough comprehension of context. For instance, in SA of educational texts, where the sentiment may depend heavily on specific parts of the input, attention mechanisms can help the model prioritize these key segments, leading to more accurate sentiment predictions.

Transformer-based models, which leverage attention mechanisms, have revolutionized SA by setting new benchmarks across various languages, including Arabic, English, German, French, and Turkish. Unlike RNNs, which process text sequentially, transformers process text in parallel, allowing them to capture complex relationships between words more effectively. This parallel processing capability, combined with their ability to understand context at a granular level, makes transformers the state-of-the-art model for SA. Their exceptional performance across various tasks, including SA, has made them the preferred choice for researchers aiming to achieve the highest accuracy. Transformers' ability to generalize well across different languages and tasks further cements their status as the cutting-edge approach in the field.

Overall, our analysis shows that CNNs and LSTM networks have gained popularity in the field of ASA because of their superior performance, even though each of these NN models has advantages and uses in SA. Transformer-based models are the go-to option for sentiment classification tasks in the modern era, offering unparalleled accuracy and efficiency, and have recently emerged as the most advanced approach, demonstrating remarkable success and setting a new standard for SA across multiple languages. SA researchers and practitioners should evaluate these models according to the particular requirements of their tasks, striking a balance between the need for accuracy, computational efficiency, and the ability to handle complex and diverse data.

5 Conclusion

To sum up, this study carefully examined and assessed the best neural network models for ASA, offering a thorough analysis of the unique traits and advantages of each model. Based on the particular needs and difficulties of ASA, our analysis provides researchers and practitioners with insightful information that helps them choose the best model for their SA task.

References

1. Moraes R, Valiati JF, Gavião Neto WP (2013) Document-level sentiment classification: an empirical comparison between SVM and ANN. Expert Syst Appl 40:621–633. https://doi.org/10.1016/j.eswa.2012.07.059
2. Zahidi Y, El Younoussi Y, Al-Amrani Y (2021) A powerful comparison of deep learning frameworks for Arabic sentiment analysis. Int J Electr Comput Eng 11:745–752. https://doi.org/10.11591/ijece.v11i1.pp745-752
3. Ombabi AH, Ouarda W, Alimi AM (2020) Deep learning CNN–LSTM framework for Arabic sentiment analysis using textual information shared in social networks. Soc Netw Anal Min 10:1–13. https://doi.org/10.1007/s13278-020-00668-1
4. Heikal M, Torki M, El-Makky N (2018) Sentiment analysis of Arabic tweets using deep learning. Procedia Comput Sci 114–122
5. Abu Kwaik K, Saad M, Chatzikyriakidis S, Dobnik S (2019) LSTM-CNN deep learning model for sentiment analysis of dialectal Arabic. In: Communications in computer and information science. Springer, pp 108–121
6. Alayba AM, Palade V, England M, Iqbal R (2018) A combined CNN and LSTM model for Arabic sentiment analysis. Lecture notes in computer science (including subseries lecture notes in artificial intelligence and lecture notes in bioinformatics). 11015 LNCS, pp 179–191. https://doi.org/10.1007/978-3-319-99740-7_12/FIGURES/5

7. Abdullah M, Hadzikadicy M, Shaikhz S (2019) SEDAT: sentiment and emotion detection in Arabic text using CNN-LSTM deep learning. In: Proceedings—17th IEEE international conference on machine learning and applications, ICMLA 2018. Institute of Electrical and Electronics Engineers Inc., pp 835–840
8. Al Omari M, Al-Hajj M, Sabra A, Hammami N (2019) Hybrid CNNs-LSTM deep analyzer for Arabic opinion mining. In: 2019 6th International conference on social networks analysis, management and security, SNAMS 2019. Institute of Electrical and Electronics Engineers Inc., pp 364–368
9. Wahdan A, Hantoobi SAL, Salloum SA, Shaalan K (2020) A systematic review of text classification research based on deep learning models in Arabic language. Int J Electr Comput Eng (IJECE) 10. https://doi.org/10.11591/IJECE.V10I6.PP%P
10. Al-Azani, S., El-Alfy, ESM (2017) Hybrid deep learning for sentiment polarity determination of Arabic microblogs. In: Lecture notes in computer science (including subseries lecture notes in artificial intelligence and lecture notes in bioinformatics). Springer, pp 491–500
11. Omara E, Mosa M, Ismail N (2019) Deep convolutional network for Arabic sentiment analysis. In: 2018 Proceedings of the Japan-Africa conference on electronics, communications, and computations, JAC-ECC 2018. Institute of Electrical and Electronics Engineers Inc., pp 155–159
12. Dahou A, Elaziz MA, Zhou J, Xiong S (2019) Arabic sentiment classification using convolutional neural network and differential evolution algorithm. Comput Intell Neurosci 2019. https://doi.org/10.1155/2019/2537689
13. Sharukh Bhathena J, Chaudhry A, Sai GS, Mantri H (2023) Enhanced multilingual sentiment analysis using ensemble learning and tree structured Parzen estimator hyperparameter optimization. Harbin Gongcheng Daxue Xuebao/J Harbin Eng Univ 44:2025–2045
14. Abdelgwad MM, Soliman THA, Taloba AI, Farghaly MF (2021) Arabic aspect based sentiment analysis using bidirectional GRU based models. J King Saud Univ—Comput Inf Sci 34:6652–6662. https://doi.org/10.1016/j.jksuci.2021.08.030
15. Saleh H, Mostafa S, Gabralla LA, Aseeri AO, El-Sappagh S (2022) Enhanced Arabic sentiment analysis using a novel stacking ensemble of hybrid and deep learning models. Appl Sci 12:8967. https://doi.org/10.3390/APP12188967
16. Hameed RA, Abed WJ, Sadiq AT (2023) Evaluation of hotel performance with sentiment analysis by deep learning techniques. Int J Interact Mob Technol (iJIM). 17:70–87. https://doi.org/10.3991/IJIM.V17I09.38755
17. Bourahouat G, Abourezq M, Daoudi N (2023) Improvement of Moroccan dialect sentiment analysis using Arabic BERT-based models. J Comput Sci 20:157–167. https://doi.org/10.3844/JCSSP.2024.157.167
18. Jerbi MA, Achour H, Souissi E (2019) Sentiment analysis of code-switched Tunisian dialect: exploring RNN-based techniques. In: Communications in computer and information science. Springer, pp 122–131
19. Al-Dabet S, Tedmori S (2019) Sentiment analysis for Arabic language using attention-based simple recurrent unit. In: 2019 2nd International conference on new trends in computing sciences, ICTCS 2019—Proceedings. https://doi.org/10.1109/ICTCS.2019.8923072
20. Alhuri LA, Aljohani HR, Almutairi RM, Haron F (2020) Sentiment analysis of COVID-19 on Saudi trending hashtags using recurrent neural network. In: International conference on developments in eSystems engineering. Dec 2020, pp 299–304. https://doi.org/10.1109/DESE51703.2020.9450746
21. Berrimi M, Oussalah M, Moussaoui A, Saidi M (2023) Attention mechanism architecture for Arabic sentiment analysis. ACM Trans Asian Low-Resour Lang Inf Process 22:107. https://doi.org/10.1145/3578265/ASSET/088DB44B-7680-4603-AB93-79BE9C9325FC/ASSETS/GRAPHIC/TALLIP-21-0241-F06.JPG
22. Elfaik H, Nfaoui EH (2021) Deep bidirectional LSTM network learning-based sentiment analysis for Arabic text. J Intell Syst 30:395–412. https://doi.org/10.1515/jisys-2020-0021
23. Albayati AQ, Al-Araji AS, Ameen SH (2020) Arabic sentiment analysis (ASA) using deep learning approach. J Eng 26
24. Omara E, Mosa M, Ismail N (2022) Applying recurrent networks for Arabic sentiment analysis. Menoufia J Electron Eng Res. 31:21–28. https://doi.org/10.21608/MJEER.2022.218776
25. Al Wazrah A, Alhumoud S (2021) Sentiment analysis using stacked gated recurrent unit for Arabic tweets. IEEE Access 9:137176–137187. https://doi.org/10.1109/ACCESS.2021.3114313
26. Ombabi AH, Ouarda W, Alimi AM (2024) Improving Arabic sentiment analysis across context-aware attention deep model based on natural language processing. Lang Resour Eval. https://doi.org/10.1007/S10579-024-09741-Z
27. Alosaimi W, Saleh H, Hamzah AA, El-Rashidy N, Alharb A, Elaraby A, Mostafa S (2024) ArabBert-LSTM: improving Arabic sentiment analysis based on transformer model and long short-term memory. Front Artif Intell 7:1408845. https://doi.org/10.3389/FRAI.2024.1408845/BIBTEX
28. Karfi IE, Fkihi SE (2023) A combined Bi-LSTM-GPT model for Arabic sentiment analysis. Int J Intell Syst Appl Eng 11:77–84
29. AL-Smadi M, Hammad MM, Al-Zboon SA, AL-Tawalbeh S, Cambria E (2023) Gated recurrent unit with multilingual universal sentence encoder for Arabic aspect-based sentiment analysis. Knowl Based Syst 261:107540. https://doi.org/10.1016/J.KNOSYS.2021.107540
30. El Karfi I, El Fkihi S (2022) An ensemble of Arabic transformer-based models for Arabic sentiment analysis. Int J Adv Comput Sci Appl. 13:561–567. https://doi.org/10.14569/IJACSA.2022.0130865
31. Mohamed O, Kassem AM, Ashraf A, Jamal S, Mohamed EH (2023) An ensemble transformer-based model for Arabic sentiment analysis. Soc Netw Anal Min 13:1–14. https://doi.org/10.1007/S13278-022-01009-0/METRICS
32. Antoun W, Baly F, Hajj HM (2020) AraBERT: transformer-based model for Arabic language understanding. arxiv
33. Chouikhi H, Chniter H, Jarray F (2021) Arabic sentiment analysis using BERT model. Commun Comput Inf Sci 1463:621–632. https://doi.org/10.1007/978-3-030-88113-9_50
34. Ashraf M, Marzouk M, Kora R, Mohammed A (2023) An enhanced transformer-based approach with meta-ensemble learning for Arabic sentiment analysis. In: 3rd International mobile, intelligent, and ubiquitous computing conference, MIUCC 2023. pp 67–73. https://doi.org/10.1109/MIUCC58832.2023.10278312
35. Baniata LH, Kang S (2024) Switch-transformer sentiment analysis model for Arabic dialects that utilizes a mixture of experts mechanism. Mathematics 12:242. https://doi.org/10.3390/MATH12020242

In-Depth Evaluation of Leading Neural Network Models with Word Embedding Approach for Arabic Sentiment Analysis

Youssra Zahidi and Yassine Al-Amrani

Abstract

The expansion of social media platforms has given users a powerful platform to express their opinions on a wide range of topics, significantly influencing the field of sentiment analysis (SA) within natural language processing (NLP). While SA is essential for deriving insights and informing decision-making based on public sentiment, it encounters specific challenges with the Arabic language due to its diverse dialects and complex morphology. Deep learning (DL) techniques, particularly convolutional neural networks (CNNs) and long short-term memory (LSTM) networks, have greatly improved SA by capturing critical features more effectively than traditional machine learning (ML) approaches. This paper explores two key architectures: the CNN approach, which utilizes the FastText word embedding model, and the LSTM approach, also incorporating FastText, for Arabic sentiment analysis (ASA). The study assesses the performance of these models across two distinct datasets and validates the findings of prior comparative research. The results show that both models perform robustly, with higher classification accuracy observed on the first dataset compared to the second, reinforcing the conclusions drawn from earlier research in aspect-based sentiment analysis.

Keywords

Arabic sentiment analysis · Deep learning · Convolutional neural networks · Long short-term memory · FastText

1 Introduction

Recently, popular social networks like Twitter, Instagram, and Facebook have attracted a large user base, enabling interactive communication and collaboration. These platforms allow people to express their sentiments using various forms of social data. The daily production of substantial data on social networks reflects audience sentiment on diverse topics such as social issues, business, and politics. Social data is often "unstructured," "informal," and "rapidly evolving," making traditional analysis methods resource-intensive and time-consuming to apply.

SA, also known as opinion mining, aims to discern the attitudes of users within a social network towards specific subjects or events [1]. It can be conducted at various levels, including word, sentence, document, and topic levels.

Our research primarily focuses on analyzing sentiment at the sentence level for Arabic text [2], aiming to determine whether users' sentences convey a positive or negative sentiment. However, the complex structure, rich morphological system, ambiguity, insufficient resources, and the presence of varying dialects in the Arabic language present numerous challenges and obstacles to advancements in ASA [3].

DL [4] techniques, particularly CNNs LSTM networks, have revolutionized SA by adeptly capturing and interpreting complex features with exceptional precision and depth. These sophisticated models exceed the capabilities of traditional machine learning (ML) methods by revealing subtle patterns and relationships within the data, resulting in notably more accurate and insightful sentiment classification. This research

Y. Zahidi (✉) · Y. Al-Amrani
Information Technology and Modeling Systems Research Team, Abdelmalek Essaadi University, Tetuan, Morocco
e-mail: youssra.zahidi-etu@uae.ac.ma

Y. Al-Amrani
e-mail: yassine.alamrani@uae.ac.ma

A. Moumen et al. (eds.), *Artificial Intelligence and Data Analytics for Innovative Applications in Engineering, Sustainability and Technology*, Sustainable Artificial Intelligence-Powered Applications, https://doi.org/10.1007/978-3-031-90318-2_20

aims to validate and showcase the effectiveness of two DL architectures using the FastText [5] word embedding model: the CNN approach and the LSTM approach for ASA [6–8]. Our study builds on previous research, categorized into comparative and practical studies. This paper, as a practical study, complements a series of comparative evaluations conducted at various level.

We focused on our previous deep studies, which can be categorized into two types: a "comparative study" and a "practical study".

Our research paper, serving as a practical study, complements a series of comparative assessments conducted at various levels:

The comparative study includes the following evaluations:

- Comparison of neural network models used in SA in Arabic: In the study [9], we performed a thorough comparison and analysis of different neural network models. This evaluation revealed that, among other algorithms, "CNN" and LSTM demonstrate several advantages in the ASA domain, yielding noteworthy performance outcomes.
- Evaluation of word embedding models for SA in Arabic: In the work [10], we thoroughly explored the subject of word embedding models, especially their relevance in the field of ASA. Our evaluation study revealed that "FastText" emerges as the most effective word embedding model, producing highly favorable results, thus underlining its significance in ASA.

This study explores two DL architectures for ASA: a CNN model combined with FastText and an LSTM model also utilizing FastText. The primary objective is to validate the effectiveness of these architectures. Our experimental results confirm that both the CNN and LSTM approaches, when used with FastText embedding, significantly enhance the accuracy of Arabic text classification across two separate datasets. This confirmation supports the conclusions of our earlier research, highlighting the efficacy of FastText. By incorporating these advanced features, we aim to further improve ASA by leveraging the strengths of CNN and LSTM models alongside the efficient FastText embedding technique.

The remaining sections of this paper are structured as follows: Sect. 2 outlines the proposed architectures. Section 3 details the experimental results, including an in-depth evaluation of the two architectures: CNN with FastText and LSTM with FastText. Section 4 delves into a thorough discussion of the obtained experimental results, and Sect. 5 concludes our research work.

2 Proposed Architectures

To classify Arabic text effectively, the recommended architectures leverage both the CNN and LSTM approaches, utilizing the FastText word embedding model. Figures 1 and 2 illustrate the various steps involved in these proposed architectures, showcasing the processes that enable accurate sentiment classification.

2.1 Data Collection Phase

The datasets used in our approach are:

- The First Dataset [11]: We used an open-access dataset specifically gathered for SA of Arabic dialect social media posts. This dataset contains approximately "52,155" tweets, which are divided into three primary sentiment categories: "Positive", "Neutral", and "Negative". It is noteworthy that the majority of these tweets are labeled as Neutral (Table 1 and Fig. 3).

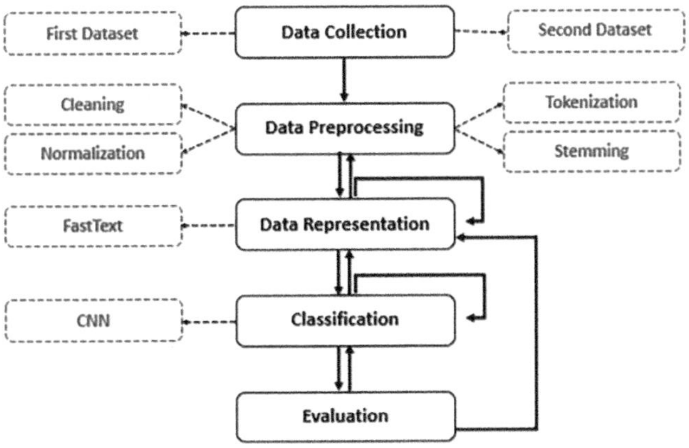

Fig. 1 The first architecture: CNN approach with FastText

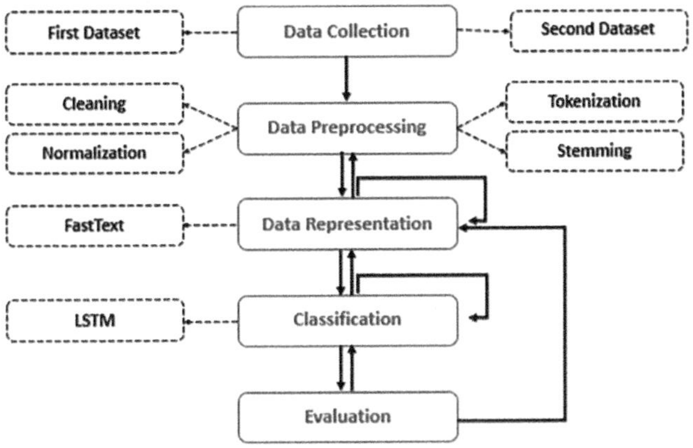

Fig. 2 The second architecture: LSTM approach with FastText

Table 1 Number of tweets per label for sentiment analysis of the first dataset

Sentiment analysis	Positive	Negative	Neutral
Number of tweets	6777	15,362	30,016

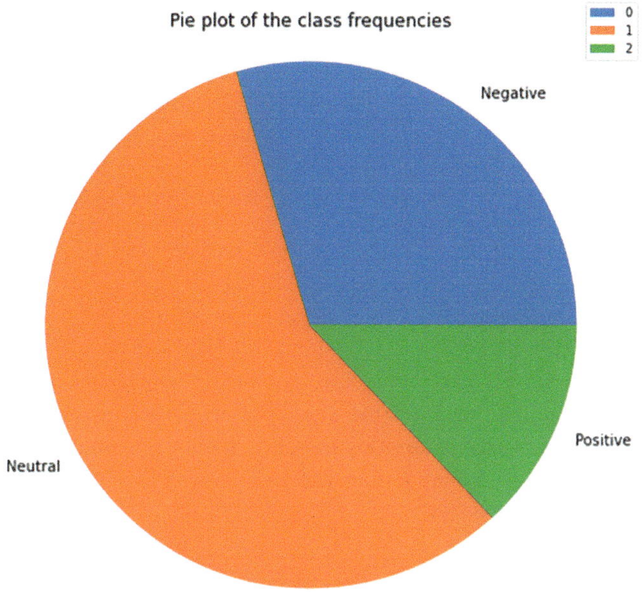

Fig. 3 The class frequencies pie chart of the first dataset

Table 2 Number of comments per label for sentiment analysis of the second dataset

Sentiment analysis	Positive	Negative
Number of comments	3673	6581

- The Second Dataset [12]: The dataset used in our ASA assessment comprises 10,254 comments extracted from Arabic Facebook discussions related to the 2016 Moroccan elections. These comments are written in both Standard Arabic and the Moroccan dialect (Table 2 and Fig. 4).

2.2 Data Preprocessing and Cleaning

This phase focuses on filtering the extracted and selected data before analysis, identifying, and removing "non-textual" and "irrelevant" content from the research domain. To prepare the data for classifier treatment, several steps are followed based on a suggested architecture and methodology. The pretreatment process includes "Cleaning," "Normalization," "Tokenization," and "Stemming." During cleaning, incorrect or missing records are corrected, and "extra white spaces," "null values," and "foreign characters" like non-Arabic digits and punctuation marks are removed. Normalization converts Arabic text to its basic form by replacing "hamza" in Arabic letters (آ ,أ ,إ) with (ا), "shorten alif" with (ya: ي), the letter (ta: ة) with (ha: ه), and removing the "tatweel character" and "diacritics." Tokenization splits sentences into "tokens" by identifying word boundaries, primarily using white spaces. Stemming reduces words to their root form by removing "prefixes" and "suffixes." The NLTK library, a free toolkit for NLP designed for Arabic and other languages, is utilized in this process.

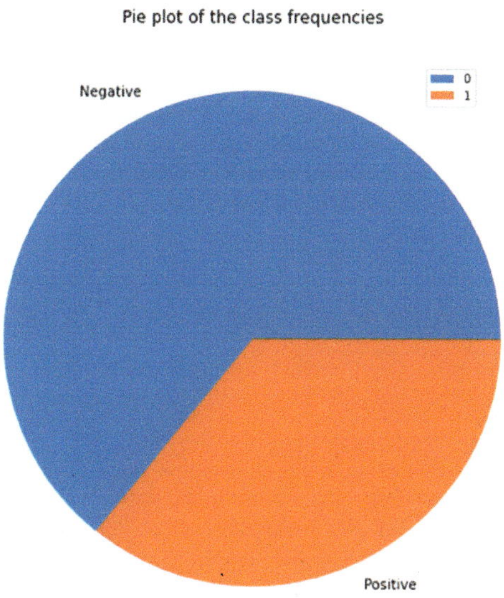

Fig. 4 The class frequencies pie chart of the second dataset

2.3 Data Representation

In this study, data representation is adeptly managed using the FastText [13, 14] word embedding model, which plays a crucial role in structuring the data for advanced analysis and processing. Unlike traditional models, FastText interprets words as compositions of character-level "n-grams," breaking them down into smaller "sub-word" units and associating each "n-gram" with a distinct vector. The comprehensive word representation is achieved by aggregating these vectors. This method allows FastText to capture intricate semantic relationships between words that share common character sequences, providing a richer understanding of their meanings. Additionally, FastText excels in generating embedding for rare or unseen words by synthesizing the vectors of their constituent "n-grams" derived from known character sequences. The specific steps and procedures for implementing FastText are detailed in Table 3.

Table 3 Our main steps to use FastText

<table>
<tr><th>Steps</th><th colspan="3">Description</th></tr>
<tr><td rowspan="5">First step</td><td rowspan="5">For facilitating the work with FastText, download a pre-trained model of pre-formed Arabic word vectors called "cc.ar.300.vec." This model exhibits the following characteristics:</td><td>Download and installation</td><td>https://dl.fbaipublicfiles.com/fasttext/vectors-crawl/cc.ar.300.wave.gz and unpack it</td></tr>
<tr><td>Number of queryable words</td><td>2,000,000</td></tr>
<tr><td>Character encoding</td><td>UTF-8 (Unicode)</td></tr>
<tr><td>Word space size</td><td>1.2 Go</td></tr>
<tr><td>Dimension embedding</td><td>300</td></tr>
<tr><td>Second step</td><td colspan="3">The pre-trained model is loaded to extract word vectors, leveraging the FastText model's predefined format for this purpose</td></tr>
<tr><td rowspan="2">Third step</td><td rowspan="2">The embedding layer utilizes the embedding matrix, which contains the weights corresponding to each word in the training data. This process involves:</td><td colspan="2">Listing each word individually that was present in the tokenized word index of the training dataset</td></tr>
<tr><td colspan="2">Finding the embedding weight for each word by fetching the corresponding weight from the FastText model</td></tr>
</table>

2.4 Classification

Text classification involves categorizing Arabic text based on its content, typically using DL methods. Our study focuses on CNNs and LSTM networks combined with the FastText word embedding model. We aim to optimize this classification model to demonstrate the effectiveness of these architectures on two different datasets to confirm the results of our comparative studies previously conducted.

CNNs [15] are renowned for their effectiveness and are widely used across various domains, including automatic speech recognition (ASR). This computational framework includes one or more convolutional layers, which may be fully connected, and employs multiple layers of perceptrons. CNNs are particularly adept at handling data structured in a grid format, such as images. Unlike traditional ML methods that depend on predefined descriptors for classification, CNNs can learn and extract task-specific descriptors during training. This capability eliminates the need for manual feature engineering and enhances the efficiency and broad adoption of CNNs.

LSTM [16], a neural network algorithm used in DL and artificial intelligence, features a cell and "input," "output," and "forget" gates. These gates control the flow of information into and out of the cell, enabling it to maintain values over varying periods. Unlike conventional feedforward NNs, LSTMs include feedback connections, making them highly suitable for classifying, processing, and forecasting time series data. They can effectively manage lags of different durations between significant events in a time series.

LSTMs were developed to address the "vanishing gradient" problem encountered in training traditional RNNs. They offer several advantages, including a relatively low sensitivity to gap length, which exceeds the capabilities of Markov models and other sequence learning methods in many applications.

2.5 Evaluation

The classifier's performance is evaluated using the accuracy metric, which measures the quality of its predictions. Accuracy reflects how accurately the classifier assigns the correct class labels to the input data, offering a clear indication of its overall effectiveness.

- Accuracy: It is a metric that measures the percentage of correctly classified entities in a dataset, represented as the ratio of correctly classified entities to the total number of entities.

$$\text{Accuracy} = \frac{\text{Correct Predictions}}{\text{Total Number}}$$

3 Experimental Results

This experiment aims to showcase the classification outcomes achieved by configuring the CNN parameters using two different datasets. To establish the parameters, we employ the training set (80%, comprising 41,724 comments) and the testing set (20%, comprising 10,431 comments) for the first dataset. Besides, the training set (80%, comprising 8,203 comments) and the testing set (20%, comprising 2,051 comments) in the second dataset.

The results can significantly differ depending on the various combinations of CNN parameters employed during the classifier run. The key parameters used are outlined in Table 4.

To determine the optimal parameter values for achieving significant accuracy, Table 5 displays the results as validation accuracy percentages (%) for two datasets, evaluated across various epochs and batch sizes.

Table 4 The major parameters employed in our study

The major parameters	Signification	Values
Epochs	This is the number of iterations	10, 20, 30, 40
Batch size	This is the number of samples per gradient update	100, 250, 500
Classification accuracy	This is a method to be applied to evaluate the results	Val accuracy
Optimizer	This is a method to adjust the weights according to the input values	Adam

3.1 Analysis of CNN with FastText Classification Results

Experiment Results Analysis of the Classification of CNN with FastText in the First Dataset:

- The first dataset consistently achieves high validation accuracy, generally stabilizing around 84%.
- The highest validation accuracy (84.4%) is achieved with a batch size of 500 after 20 epochs.
- Smaller batch sizes (100 and 250) also perform consistently well.
- Performance shows good consistency across various batch sizes and epochs.

Experiment Results Analysis of the Classification of CNN with FastText in the Second Dataset:

- The second dataset shows more variation in validation accuracy compared to the first.
- The highest validation accuracy (78.3%) is achieved with a batch size of 500 after 40 epochs.
- Performance is generally better with larger batch sizes, though it fluctuates more.

Comparative Summary of the Results in the First and Second Datasets:

- First Dataset: It achieves higher and more stable accuracy, peaking at 84.4% with a batch size of 500 after 20 epochs.
- Second Dataset: It shows more variability with a peak accuracy of 78.3% with a batch size of 500 after 40 epochs.

In general, the first dataset performs consistently well across different configurations, while the second dataset benefits more from larger batch sizes and longer training. This analysis helps in understanding the optimal training strategies for each dataset, suggesting that the first dataset might not need as many epochs or as large batch sizes to achieve high accuracy, whereas the second dataset requires more epochs and larger batch sizes for optimal performance.

3.2 Analysis of LSTM with FastText Classification Results

Experiment Results Analysis of the Classification of LSTM with FastText in the First Dataset in the First Dataset:

- The first dataset consistently achieves high validation accuracy, with all values above 83%, stabilizing around 84%.
- The highest validation accuracy (84.2%) is achieved with a batch size of 250 after 20 epochs.
- Smaller batch sizes (100 and 250) tend to perform better than the largest batch size (500).

Table 5 Classification results of CNN and LSTM approaches with FastText using Adam optimizer based on two various datasets

Epochs		10			20			30			40		
Batch Size		100	250	500	100	250	500	100	250	500	100	250	500
Val_ Accuracy of CNN classification (%)	First dataset	84.1	84.1	83.8	83.5	83.7	84.4	84	84.1	83.9	84.2	83.9	84
	Second dataset	77.8	77.1	77.5	77.6	77.5	78	77.8	77.8	77.8	77.3	77.7	78.3
Val_ Accuracy of LSTM classification (%)	First dataset	83.9	84.1	83.4	84	84.2	84	84.1	83.9	84.1	84.1	84.1	84.0
	Second dataset	80	77.5	77	78.8	76.2	76.8	78.1	76.4	76.9	78.4	76.9	77.1

Experiment Results Analysis of the Classification of LSTM with FastText in the First Dataset in the Second Dataset:

- The second dataset shows more variation in validation accuracy compared to the first dataset.
- The highest validation accuracy (80%) is achieved with a batch size of 100 after 10 epochs.
- Performance tends to decrease with larger batch sizes and more epochs, indicating potential overfitting or suboptimal learning rates.
- The optimal performance is observed with a batch size of 100 at earlier epochs, suggesting smaller batch sizes might be more effective for this dataset.

Comparative Summary of the Results in the First and Second Datasets:

- First Dataset: It shows higher and more stable accuracy across different configurations, with an optimal batch size of 250 and 20 epochs.
- Second Dataset: It exhibits more variability with optimal performance at a smaller batch size (100) and fewer epochs (10).

The analysis suggests that while both datasets can achieve high accuracy, their performance depends significantly on batch size and epochs. The first dataset performs best with a moderate batch size and epochs, whereas the second dataset performs better with a smaller batch size and fewer epochs, indicating different optimal training strategies for each dataset.

4 Discussion

This comparative study emphasizes the choice of CNN and LSTM for their notable advantages and high performance in ASA field. It is important to note that other neural network models might also offer benefits in this domain. The evaluation employed the FastText word embedding model, with experiments designed to describe and assess two specific architectures: CNN with FastText and LSTM with FastText. These architectures were tested using the renowned Adam optimizer, exploring various parameters such as epochs (10, 20, 30, 40), batch sizes (100, 250, 500), and classification accuracy across two different datasets to validate their effectiveness and relevance in ASA.

The following section delves into the results from the CNN with FastText experiments, uncovering several key trends. For the first dataset, the highest validation accuracy was 84.4%, achieved with a batch size of 500 after 20 epochs. In contrast, the second dataset reached its peak validation accuracy of 78.3% with a batch size of 500 after 40 epochs. Generally, validation accuracy increases with the number of epochs for both datasets, though this trend is inconsistent and varies with batch size. The first dataset shows stable performance with minor fluctuations around its maximum value, while the second dataset experiences more significant variations.

In the subsequent section, the results from the LSTM with FastText experiments are analyzed, leading to the following conclusions: For the first dataset, the highest accuracy of 84.20% was achieved with a batch size of 250 after 20 epochs. Accuracy generally stayed above 83%, with slight fluctuations and batch sizes of 100 and 500 also produced relatively high accuracies. In the second dataset, the highest accuracy of 80% was achieved with a batch size of 100 after 10 epochs. Accuracy varied across epochs and batch sizes, with smaller batch sizes yielding the highest accuracy.

The first dataset consistently showed slightly higher accuracies across various batch sizes and epochs when comparing the two datasets. However, the differences in accuracy between the two datasets were minimal, indicating comparable performance.

5 Conclusion

In this evaluation, we analyzed the performance of CNN and LSTM networks with the FastText embedding model for ASA. Our experiments showed that both CNN and LSTM models improve text classification accuracy across two datasets. Notably, the first dataset achieved higher accuracy than the second with both architectures.

References

1. Al-Amrani Y, Lazaar M, Elkadiri KE (2017) Sentiment analysis using supervised classification algorithms. In: ACM International conference proceeding series. Association for Computing Machinery
2. Zahidi Y, Younoussi YEL, Al-Amrani Y (2020) Arabic sentiment analysis approaches: an overview. In: Proceedings—10th international conference on virtual campus, JICV 2020. https://doi.org/10.1109/JICV51605.2020.9375763
3. Zahidi Y, Younoussi YEL, Al-Amrani Y (2020) Arabic sentiment analysis problems and challenges. In: Proceedings—10th international conference on virtual campus, JICV 2020. https://doi.org/10.1109/JICV51605.2020.9375650
4. Zahidi Y, El Younoussi Y, Azroumahli C (2019) Comparative study of the most useful Arabic-supporting natural language processing and deep learning libraries. In: 2019 International conference on optimization and applications, ICOA 2019. Institute of Electrical and Electronics Engineers Inc.
5. Umer M, Imtiaz Z, Ahmad M, Nappi M, Medaglia C, Choi GS, Mehmood A (2022) Impact of convolutional neural network and

FastText embedding on text classification. Multimed Tools Appl. https://doi.org/10.1007/s11042-022-13459-x
6. Ombabi AH, Ouarda W, Alimi AM (2020) Deep learning CNN–LSTM framework for Arabic sentiment analysis using textual information shared in social networks. Soc Netw Anal Min 10:1–13. https://doi.org/10.1007/s13278-020-00668-1
7. Alotaibi F, Gupta V (2022) Sentiment analysis system using hybrid word embeddings with convolutional recurrent neural network. Int Arab J Inf Technol. 19:330–335. https://doi.org/10.34028/iajit/19/3/6
8. Alayba AM, Palade V (2021) Leveraging Arabic sentiment classification using an enhanced CNN-LSTM approach and effective Arabic text preparation. J King Saud Univ—Comput Inf Sci. https://doi.org/10.1016/j.jksuci.2021.12.004
9. Zahidi Y, El Younoussi Y, Al-Amrani Y (2022) Arabic sentiment analysis based on neural network models: overview and comparison, pp 77–80. https://doi.org/10.5220/0010728700003101
10. Zahidi Y, El Younoussi Y, Al-Amrani Y (2022) An overview of word embedding models evaluation for Arabic sentiment analysis. In: Lecture notes in networks and systems. 489 LNNS, pp 411–427. https://doi.org/10.1007/978-3-031-07969-6_31
11. Boujou E, Chataoui H, El Mekki A, Benjelloun S, Chairi I, Berrada I (2021) An open access NLP dataset for Arabic dialects : data collection, labeling, and model construction
12. Elouardighi A, Maghfour M, Hammia H (2017) Collecting and processing Arabic Facebook comments for sentiment analysis. In: Lecture notes in computer science (including subseries lecture notes in artificial intelligence and lecture notes in bioinformatics). 10563 LNCS, pp 262–274. https://doi.org/10.1007/978-3-319-66854-3_20/COVER
13. Dharma EM, Gaol FL, Warnars HLHS, Soewito B (2022) The accuracy comparison among Word2Vec, Glove, and Fasttext towards convolution neural network (CNN) text classification. J Theor Appl Inf Technol 100:349–359
14. Aziz Altowayan A, Elnagar A (2017) Improving Arabic sentiment analysis with sentiment-specific embeddings. In: Proceedings—2017 IEEE international conference on big data, big data 2017. Jan 2018, pp 4314–4320. https://doi.org/10.1109/BIGDATA.2017.8258460
15. Fesseha A, Xiong S, Emiru ED, Diallo M, Dahou A (2021) Text classification based on convolutional neural networks and word embedding for low-resource languages: Tigrinya. Information (Switzerland) 12:1–17. https://doi.org/10.3390/info12020052
16. Abu Kwaik K, Saad M, Chatzikyriakidis S, Dobnik S (2019) LSTM-CNN deep learning model for sentiment analysis of dialectal Arabic. Commun Comput Inf Sci 1108:108–121. https://doi.org/10.1007/978-3-030-32959-4_8/COVER

Advanced Multi-touch Attribution for Improved Marketing Analytics

Zine El Abidine El Mekkaoui and Hatim Benyoussef

Abstract

Multi-touch attribution (MTA) has emerged as an imperative analytical framework in marketing analytics, driving the strategic optimization process through accurate attribution insights in the context of the widening complexities of digital consumer behavior. The following bibliometric analysis explores research relevant to MTA, including patents, theses, conference proceedings, and journal articles published from 2010 to 2024, to present an outlook on significant development, critical methodologies, and shifting trends. Conclusions point to a radical shift from the rule-based traditional approaches to more AI-enhanced data-driven methods by machine learning and sophisticated statistical techniques that will continue to enhance predictive accuracy. Further, blockchain technologies have also been underlined among factors enhancing data privacy in MTA systems. Prominent contributors include Kannan, Li, and Pauwels. Publicly available datasets, such as those published by Criteo, have catalyzed immense methodological innovation and provided empirical evidence for model development. Despite these many works, several challenges remain in model interpretability, data quality, and privacy considerations within MTA applications. Future research will likely concentrate on federated learning, privacy-preserving analytics, and real-time data processing, enabling further improvements to the MTA framework. This research combines updated information about MTA, serving as primary material for scholars in delimiting potential ways to improve the methodology of MTA in digital marketing.

Z. E. A. El Mekkaoui (✉) · H. Benyoussef
Ibn Tofail University, Kenitra, Morocco
e-mail: zineelabidine.elmekkaoui@uit.ac.ma

Keywords

Multi-touch attribution · Attribution model · Marketing analytics · Digital marketing

1 Introduction

Measuring marketing campaign effectiveness and determining optimal marketing spending have always been critical topics for marketing academicians and practitioners [1]. Indeed, digital marketing has made customer purchase cycles more complex and generated vast amounts of information at the level of individual users [2]. This shift has led to increased studies related to multi-touch attribution (MTA) modeling during the 2010s.

Multi-touch attribution modeling is defined by [3] as "an advertising measuring technique that scores the value of each touch point (viewing an advertisement) leading to conversion (sale of the product)." The attribution modeling development has moved from simple single-touch models, such as the last-touch model, which attributes credit to the very last touch before conversion, to rule-based heuristic models, an example being the time decay model that favors touches that occur closer to the conversion time. Further developments led to more complex data-driven approaches incorporating statistical techniques and machine learning to handle the complexities of marketing communications and customer journeys, thus facilitating the prediction of future interactions or sales [4]. As this model helps form marketing decisions, the importance of using accurate data to support and improve decisions is again similar to more prominent theories of decision-making [5]. Initial investigations into MTA predominantly focused on forecasting outcomes through diverse methodological approaches, such as logistic regression [6, 7], Markov chain models [8, 9], survival analysis [10, 11], and the Shapley value [12]. Since 2018, a significant trend has emerged toward

A. Moumen et al. (eds.), *Artificial Intelligence and Data Analytics for Innovative Applications in Engineering, Sustainability and Technology*, Sustainable Artificial Intelligence-Powered Applications, https://doi.org/10.1007/978-3-031-90318-2_21

integrating artificial intelligence (AI) and machine learning in MTA research [13–16].

This research provides an overarching review of the current state of scholarship within this emerging field. We define critical terms for a shared understanding and outline the core database used to conduct our analysis. From there, we explore current trends in research, identify the leading authors in the field and their critical publications, and examine the evolution of attribution modeling methods, highlighting the most influential approaches today. This approach will enable us to address the following four basic questions: What are the contemporary research directions and established methodologies in multi-touch attribution? What are the leading academic and scientific contributors in the MTA area? Which are the most common methodologies used to derive MTA models? How have the methods of attribution modeling in the discipline developed over time?

2 Definitions

With all the different ways one might define key terms inherent in multi-touch attribution, which can lead to confusion, especially for those just entering the field, we attempt to establish a set of coherent and constant definitions.

Attribution: Based on varied definitions [4, 17, 18], attribution in marketing is an analytics-driven process that involves identifying, assessing, and placing value on the various touchpoints and interactions along the customer journey responsible for desired outcomes such as conversions. It applies advanced analytical methods to correctly attribute credit to each marketer's touch. It thus helps marketers optimize their strategies and better allocate budgets, considering the impact of various marketing touches on changing consumer behavior and conversion rates.

Attribution modeling is an analytical process wherein the credit for customer conversions is attributed to the touchpoints in the customer journey. It deploys advanced analytics that measures the effectiveness of every online or offline interaction in driving a customer's desired actions. Identifying and quantifying various channels and touchpoints with their respective contributions facilitates strategic marketing decisions for resource optimization across the conversion path [19–22].

Multi-touch attribution is an analytics approach responsible for measuring the effectiveness of different advertising touchpoints in driving conversion and tracking users' cross-device and cross-format interaction. This approach assigns the value of a conversion to multiple touchpoints, thereby enabling thorough ROI evaluation at the individual touchpoint level. The importance of MTA has grown because, through it, granular insights about advertising effectiveness are made possible, which helps refine marketing strategies. [3, 23–25].

Touchpoint and credit allocation: Touchpoints refer to engagements between a consumer and a brand across different mediums, such as advertisements or social media content; every touchpoint is assigned a contribution probability. These probabilities may be compiled based on specific attributes of the touchpoint (for instance, channel, positioning, or creative element) to establish a cumulative weight for that attribute. Giving these touchpoints "credit" requires assigning a value or weight to each component, such as a channel, creative element, or even an interaction with a search ad, depending on the dimension used by the individual or organization conducting the attribution [23].

3 Methodology

We developed an extensive database of scholarly articles related to multi-touch attribution by combining the functionalities of Zotero and Publish or Perish (PoP) software programs. Zotero is a free, open-source application that allows users to gather, organize, and cite research sources while capturing detailed, nuanced bibliographic information such as publication types, author(s), title, abstract, and publisher. PoP, developed by Harzing in 2007, is a software program for citation analysis. It retrieves data from sources such as Google Scholar and Scopus and then calculates metrics, including the H-index and total citations.

In this review, 203 MTA publications and patents were collated from academic databases: Google Scholar, Scopus, Web of Science, ProQuest, and Google Patents using the Zotero connector. Additional citation data was pulled from Google Scholar using Publish or Perish via the query below, limited to 200 entries: ("Multi-Touch Attribution" OR "Multitouch Attribution" OR "Attribution Modeling" AND ("Marketing" OR "Advertising" OR "Digital Marketing" OR "Online Marketing") OR "Digital Marketing Attribution" OR "Online Marketing Attribution"). This query allowed coverage by focusing on spelling variants, narrowing the context to relevant areas for marketing, and narrowing the applications down to digital marketing. Comparing Zotero and PoP datasets using Python's "pandas" and "re" libraries yielded 93 unique PoP publications added manually to Zotero. So, the total expanded to 296 records. Refining that dataset to include only items with the word 'attribution' in the title or abstract reduced the count to 191. These entries were then merged using "FuzzyWuzzy" fuzzy matching with a similarity threshold above 80 percent. This process produced the final dataset used in this review: a CSV file containing 191 publications and patents dating from 2010 to 2024. Of

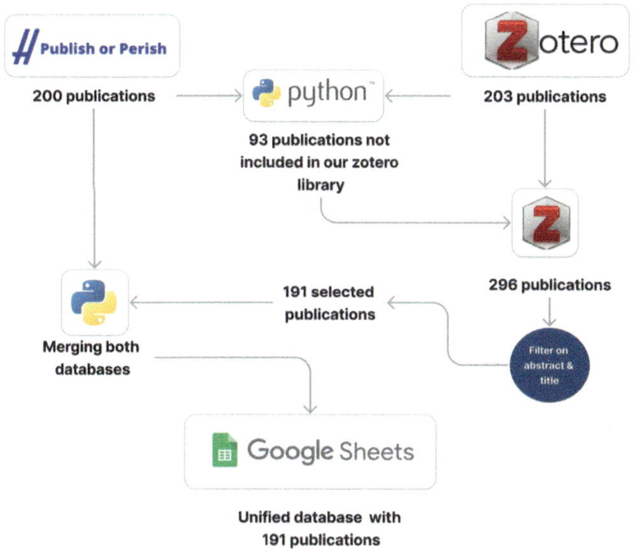

Fig. 1 Methodological framework for constructing the research database

these, only 73 publications are Scopus-indexed journal articles and conference papers, nearly 38 percent of all publications included. Critical data fields include, among others, publication type, year, author, title, abstract, citation count, and further metadata (Fig. 1).

4 Results

4.1 Trends in Publication Types and Counts Over Time

We selected our database's "year" and "publication Type" columns. We inserted a pivot table to count the occurrences of each type of publication from 2010 to 2024. We sorted the data by the publication year in ascending order to generate the stacked column chart below.

Figure 2 illustrates the growth and diversification of publications in MTA research from 2010 to 2024, with a marked increase beginning in 2014 and a peak in 2022 at 32 publications. Initially, contributions were minimal, with only one or two works published annually between 2010 and 2013. Over the years, publication types have diversified, as shown by the cumulative representation of categories, including journal articles, conference papers, patents, theses, preprints, and book sections. Between 2010 and 2024, 64 journal articles were published, of which 39 were indexed in Scopus, amassing 2697 citations—evidencing substantial academic impact. Conference papers reached 31, with 30 being indexed in Scopus and 721 citations, thus underlining their relevance to the field. The 38 patents produced during this period, which generated 570 citations, indicate commercial potential, with significant contributors such as Adobe Inc. (12 patents) and Google Inc. (5 patents) [26, 27]. Preprints, totaling 14, received 292 citations, suggesting the field's interest in prompt dissemination of findings. Five book sections were published, and four were indexed in Scopus, collectively gathering 31 citations. The growth in the volume of patents indicates the trend toward commercialization and industrial applications in MTA research. At the same time, the steady increase in journal articles and conference proceedings points to continued theoretical and academic interest. As of 2024, two preprints and four journal articles have already been published, one of which is indexed in Scopus [28]. Thus, the upward momentum in MTA research continues, signaling a growing field.

4.2 Public Datasets Utilized in MTA Studies

To identify research articles using Criteo's databases, we used the "Keyword" feature in Publish or Perish (PoP). We entered a query with the keywords "Criteo," "Criteo dataset," "multi-touch attribution," and "marketing attribution." We exported the results into a CSV format and then imported them into Google Sheets for processing. We cleaned up the data set by removing those publications that mentioned Criteo without using the company's data. The dataset was sorted in ascending order by the publication year (Pub Year). We also included three other columns, which are publication type (Pub Type), "Author (s)," and "dataset name."

Table 1 shows a chronological overview of the critical publications on MTA research using Criteo public datasets for 2017–2024. In 2017, [29] introduced the "Criteo Attribution Modeling for Bidding Dataset," demonstrating that integrating attribution modeling into bidding improves display ad efficiency. Further studies in 2018 and 2020 using this dataset were conducted by [30], where a model using sequence-to-sequence prediction combined with attention mechanisms to capture user behavior yielded significant performance gains, while [31] proposed CAMTA, which is a deep recurrent model enhances the accuracy of per-channel attribution while reducing the bias. In 2020 and 2021, respectively, [32, 33] presented an optimization methodology and reinforcement learning frameworks using a standard data basis to enhance attribution tasks in a budget constraint. In 2022, [34] worked on improving marketing returns using a machine learning-based model for B2B and B2C lead scoring, and [35] proposed causal MTA due to user confounding bias. The "Criteo Uplift Prediction Dataset" was proposed in 2023 [36] to estimate the causal effects of ads from a non-traditional attribution approach. In 2024, [37] proposed DCRMTA, which introduces causal features in large-scale behavior data using the "Criteo 1TB Click Logs dataset." Last but not least, [38] proposed a contextual bandit model for optimizing ad bids

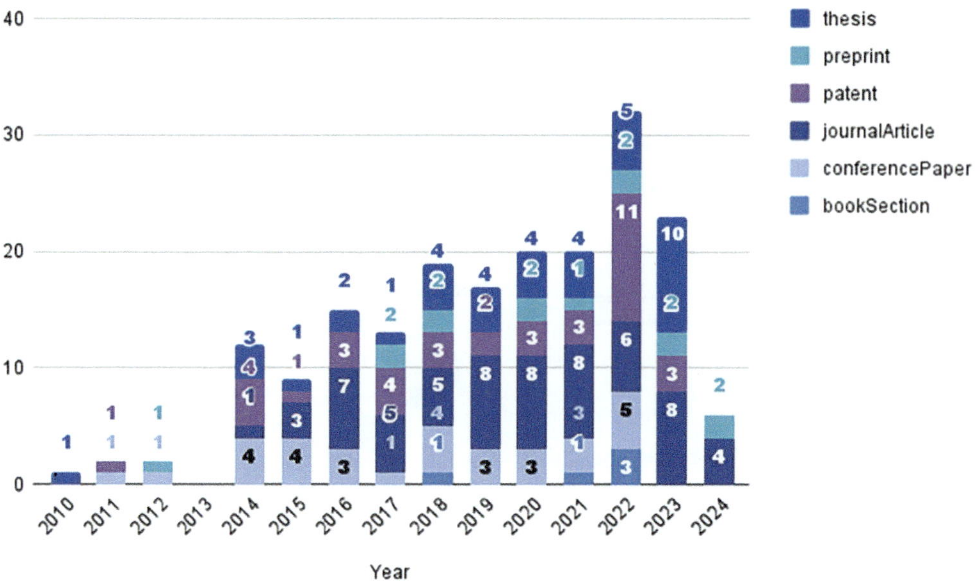

Fig. 2 Publication trends and types in MTA Research (2010–2024)

Table 1 Chronological overview of publications utilizing Criteo datasets (2017–2024)

Pub type	Author(s)	Year	Dataset name
Conference paper	Diemert, E. et al.	2017	Criteo Attribution Modeling for Bidding Dataset
Conference paper	Ren, K. et al.	2018	Criteo Attribution Modeling for Bidding Dataset
Conference paper	Kumar, S. et al.	2020	Criteo Attribution Modeling for Bidding Dataset
Conference paper	Bompaire, M. et al.	2020	Criteo Attribution Modeling for Bidding Dataset
Conference paper	Malik, M. et al.	2021	Criteo Attribution Modeling for Bidding Dataset
Conference paper	Yao, D. et al.	2022	Criteo Attribution Modeling for Bidding Dataset
Thesis (Ph.D.)	Bhatta, I	2022	Criteo Attribution Modeling for Bidding Dataset
Preprint	Betlei, A. et al.	2023	Criteo Uplift Prediction Dataset
Preprint	Bompaire, M. et al.	2023	Criteo Attribution Modeling for Bidding Dataset
Preprint	Tang, J	2024	Criteo 1 TB Click Logs Dataset
Journal article	Gigli, M.; Stella, F	2024	Criteo Attribution Modeling for Bidding Dataset

under budget constraints and further advanced the MTA research using Criteo datasets.

4.3 Influential Authors and Their Contributions

For the construction of Table 2, fractional counting was applied to every publication, allocating citation credit across co-authors to derivate proper impact. The total number of citations per year for each author was calculated by first normalizing citations of each publication by the number of years since publication, then applying that fractional rate to the authors involved. The citations per year for each paper were then summed for all of his papers to determine each author's cumulative influence. Authors were ranked by their cumulative citations per year (CpY), a more time-sensitive indication of the author's impact. We also added value by including total citations for each author (Cite), total number of publications (Nb Pub), types of publication (Pub Types), the most recent year associated with MTA research (Last MTA), and, finally, underlined each author's most cited work (Key Pub) for emphasis.

P.K. Kannan ranks first with an average of 36.78 citations per year, totaling 350.25 citations across four publications (three articles, one preprint). His most relevant work, co-authored with H. Alice Li [39], introduced a methodology for improving conversion credit attribution across marketing channels using individual-level data, emphasizing carryover and spillover effects. H. Alice Li, with an annual citation of 30.63 and total citations amounting to 301.25 from two articles and one dissertation, collaborates greatly with Kannan; the 2014 and 2016 studies are cornerstones in marketing attribution [40]. Koen Pauwels ranks third, with 27.67 citations per year, and is best known for his work with [41] on omnichannel marketing attribution. He proposes using advanced techniques like machine learning and blockchain to tackle nonlinear consumer journeys. Ron Berman comes in fourth place, with 27.37 citations annually for one article and one patent on online advertising inefficiencies. His 2018 work evidences the value of sophisticated attribution models, such as the Shapley value, over more straightforward methods like last-touch attribution. Sharing fifth place, Dimitrios Buhalis and Katerina Volchek averaged 22.17 citations per year for their collaborative 2021 study, categorizing attribution methods based on big data to assign value to customer touchpoints across digital and offline channels [23]. Anindya Ghose ranks sixth, at 20.60 citations per year. His 2021 work with Cui et al. pushed forward marketing attribution models using analytics and underlined how this could apply to practical industrial applications.

4.4 Journals Engaged in MTA Research

We have narrowed our datasets to journal articles and conference papers only from the "Publication Title" column. After that, we generated a pivot table of the following four columns: "Journal Title," "Number of Publications (Nb Pub)," citations or "cites" (sum of cites received by all published works which belong to MTA), and "Cites Per Year (CpY)." We ordered the pivot table by descending under the CpY column. Ultimately, we conducted a hardcopy search for the journals, H-index, Impact factor in Scopus "IF Sco" and Impact Factor of Web of Science "IF WoS."

Table 3 provides just a selection of examples from a more comprehensive dataset. However, it focuses exclusively on those journals that have recorded the highest number of citations annually for published work related to MTA. The "International Journal of Research in Marketing" had four contributions, which together commanded 867 citations, including two of the ten most highly cited pieces in the study area under discussion [22, 42]. The "Journal of Marketing" has published the number one ranked article in MTA with 184 citations, and the average citation was 93.2 per year. First-ranked Journal of Marketing holds an average of 93.2 citations per year per article in MTA; it holds 184 total citations and has published the number one ranked article in MTA. However, in this case, despite only two attribution-related publications, the "Journal of Marketing Research" ranking is third, given its high citation count-one of these is a highly cited article [39]. On the other hand, the same status was achieved by the works of Buhalis and Volchek, which ranked as the third most cited in the relevant domain, published in the "International Journal of Information Management." The fifth place is occupied by the journal "Marketing Science," in which an article by Li and Kannan in 2014 was the second most cited in the respective field; another article published in 2018 was written by Berman. The titles of the journals show that marketing is dominant and takes precedence over information science. This is because multi-touch attribution is more of a marketing concern by nature, as it deals with data and analytics to deal with issues arising from marketing.

4.5 Methodological Advances in MTA Research

MTA research emphasizes the need for practical means of enhancing model functionality and reliability to predict conversions and budget optimization. The quality and complexity of the data feature determine model functionalities. Evaluation criteria include simplicity, interpretability, robustness, and accuracy. Simpler models, like linear regression, may lack precision [43], while complex models, such as LSTM networks, offer greater accuracy but reduced interpretability. Balancing interpretability and accuracy remains a crucial challenge.

We have gathered 69 studies that paid close attention to practical development regarding the MTA models. We manually added a column titled "Method" to our dataset for methodological tendency analysis, listing the methods adopted. Using Google Sheets' pivot table function, we retrieved the frequency of each technique and classified RNN and LSTM as "Machine Learning." At the same time, higher-order Markov chains were categorized as "Markov chains." The result can be seen in the frequency table. Table 4 provides an overview of methodological approaches adopted by

Table 2 Leading authors in MTA research and their prominent works

Rank	Author	Nb Pub	Cite	CpY	Pub types	Last MTA	Key pub
1	Kannan, P.K	4	350.25	36.78	3 Articles, 1 preprint	2024	Li and Kannan (2014)
2	H. Alice Li	3	301.25	30.63	2 Articles, 1 Ph.D. thesis	2016	Li and Kannan (2014)
3	Pauwels, Koen	3	183	27.67	3 Articles	2021	Cui et al. (2021)
4	Berman, Ron	2	165	27.37	1 Article, 1 patent	2018	Berman (2018)
5	Buhalis, Dimitrios	1	66.5	22.17	1 Article	2021	Buhalis and Volchek (2021)
5	Volchek, Ka-terina	1	66.5	22.17	1 Article	2021	Buhalis and Volchek (2021)
6	Ghose, Anindya	2	126.5	20.60	2 Articles	2021	Cui et al. (2021)

Table 3 Leading journals in MTA research

Journal title	H-index	IF Sco	IF WoS	Nb Pub	Cite	CpY
International Journal of Research in Marketing	121	3.352	1.31	4	867	108.39
Journal of Marketing	284	11.79	2.33	1	184	93.2
Journal of Marketing Research	202	5.984	1.08	2	622	66.33
International Journal of Informa tion Management	177	5.775	5.94	1	133	44.33
Marketing Science	153	5.643	0.88	3	293	43.43

MTA-published studies between 2011 and 2024. A review of 69 articles identified these approaches, which reported the adoption of significant methodologies 75 times. This difference is explained by some publications reporting more than one approach in a single study.

Table 4 Trends in methodological approaches for MTA (2011–2024)

Method	11	12	13	14	15	16	17	18	19	20	21	22	23	24	Total
Machine learning								2	1	3	2	7	4	2	21
Markov chain		1		2	1			1	3	4		2	4		18
Shapley value							1	4	3	2		2	3		15
Logistic regression	1			2			1	1	2	1		2	1		11
Bayesian inference				2	1	1	1							1	6
Survival analysis				1		1	1						1		4
Grand total	1	1	0	7	2	2	4	8	9	10	2	13	13	3	75

Early years (2011–2013): Initial research on MTA focused on overcoming the deficiencies inherent in traditional models, where more straightforward approaches were often dominant. Logistic regression, first applied by [7], proposed a basic architecture for predicting conversions based on marketing touchpoints. Abhishek et al. [8] introduced Markov chains, which represent transition probabilities between touchpoints quite well, offering valuable insights into the sequences leading up to conversions.

Middle period (2014–2016): Between 2014 and 2016, ever more sophisticated methods were developed: Bayesian Inference [44, 45] allowed for updating attribution estimates over time as new information kept coming in, adjusting for changes in customer behavior. [11] applied survival analysis to examine whether a conversion occurred and when it did, adding a time dimension to the investigation. Although logistic regression remained prevalent due to its interpretability, Bayesian inference and survival analysis faced complex data configurations in MTA.

Recent years (2017–2024): Since 2017, sophisticated computational techniques have become predominant. Shapley Value, for instance, is used by [46], emanating from cooperative game theory, and allows fair attribution by allocating credit to each touchpoint in direct proportion to their marginal contribution to conversions. Machine learning techniques have become the most popular method, especially since 2018 [14], with models such as DNNs and RNNs uncovering complex nonlinear relationships in large datasets. The persistence of Markov chains, Bayesian inference, and survival analysis in this period underscores their adaptability to evolving data needs.

Additional methods and ensemble modeling: Other attempts have been made, such as econometric models [47], sequence mining [48], and vector autoregressive models, though less frequently. Ensemble models leverage the strengths of multiple approaches by combining techniques like Markov chains or Shapley value with machine learning to achieve both interpretability and predictive accuracy. Most recently, [49] suggested a zero-knowledge blockchain approach to improve the transparency-privacy trade-off when linking ad exposures to conversions without releasing personally identifiable information.

5 Conclusion

This review details the evolution and increased importance of MTA modeling within digital marketing based on journal articles, conference proceedings, patents, theses, and book sections published between 2010 and 2024. Indeed, since 2014, there has been rapid growth in scholarly and industry interest as the behavior of digital consumers started to rise in complexity and data began piling up through numerous channels. A synthesis of 69 publications, whereby authors developed and tested various models, points to advancing MTA methods through Bayesian inference, Markov chains, Shapley value, machine learning, and other advanced model applications. The main findings reflect a move from the traditional rule-based systems toward more progressive data-driven approaches, whereby machine learning significantly improves the precision and flexibility of methods related to MTA. Recently, blockchain technology has surfaced as a viable solution for tackling MTA's privacy and transparency issues. Public datasets, particularly those provided by organizations like Criteo, have facilitated empirical research and innovation; however, difficulties remain concerning interpretability, data quality, and privacy. Although our review encompasses a broad dataset, it may neglect contributions from non-English or less frequently cited sources, which could result in an underrepresentation of new and regionally focused developments in MTA. Moreover, an emphasis on citation metrics can distort the visibility of research, favoring well-established methodologies at the expense of innovative techniques. Subsequent investigations in MTA ought to prioritize incorporating privacy-preserving technologies, including blockchain and federated learning, enhance frameworks for real-time data processing, and explore alternative methodologies, such as transformers, which may yield improved accuracy in more complex consumer journeys. In cooperation with ad platforms, an increased variety of data sets will consolidate empirical knowledge, and an emphasis on model interpretability will make such progress more practical for professionals.

Acknowledgements The National Center for Scientific and Technical Research (CNRST) of Morocco supported this work through the Ph.D. Associate Scholarship—PASS program.

References

1. Abhishek V, Despotakis S, Ravi R (2017) Multi-channel attribution: the blind spot of online advertising. https://papers.ssrn.com/abstract=2959778
2. Malthouse E, Copulsky J (2023) Artificial intelligence ecosystems for marketing communications. Int J Advert 42:128–140. https://doi.org/10.1080/02650487.2022.2122249
3. Pattanayak S, Pati PB, Singh T (2022) performance analysis of machine learning algorithms on multi-touch attribution model. In: 2022 3rd International conference for emerging technology (INCET), pp 1–7
4. Jayawardane CHW, Halgamuge SK, Kayande U (2015) Attributing conversion credit in an online environment: an analysis and classification. In: 2015 3rd International symposium on computational and business intelligence (ISCBI), pp 68–73

5. El Mountassir YE (2020) Economic intelligence system and decision making: proposal of a theoretical model. Moroc J Quant Qual Res 2:137–152. https://doi.org/10.48379/IMIST.PRSM/mjqr-v2i3.22005
6. Shao X (2014) Method and apparatus for data-driven multi-touch at-tribution determination in multichannel advertising campaigns. https://patents.google.com/patent/US20140236705A1/en
7. Shao X, Li L (2011) Data-driven multi-touch attribution models. In: Proceedings of the 17th ACM SIGKDD international conference on knowledge discovery and data mining. Association for Computing Machinery, New York, NY, USA, pp 258–264
8. Abhishek V, Fader P, Hosanagar K (2012) Media exposure through the funnel: a model of multi-stage attribution. https://papers.ssrn.com/abstract=2158421
9. Brodersen KH, Gallusser F, Koehler J, Remy N, Scott SL (2015) Inferring causal impact using Bayesian structural time-series models. Ann Appl Stat 9. https://doi.org/10.1214/14-AOAS788
10. Ji W, Wang X, Zhang D (2016) A probabilistic multi-touch attribution model for online advertising. In: Proceedings of the 25th ACM international conference on information and knowledge management. Association for Computing Machinery, New York, NY, USA, pp 1373–1382
11. Zhang Y, Wei Y, Ren J (2014) Multi-touch attribution in online advertising with survival theory. In: 2014 IEEE international conference on data mining. IEEE, Shenzhen, China, pp 687–696
12. Mahboobi SH, Usta M, Bagheri SR (2018) Coalition game theory in attribution modeling: measuring what matters at scale. J Advert Res 58:414–422. https://doi.org/10.2501/JAR-2018-014
13. Kroon MH (2023) Explaining clickstreams by Layerwise relevance propagation in a 2D-convolutional neural network. https://thesis.eur.nl/pub/70250
14. Li N, Arava SK, Dong C, Yan Z, Pani A (2018) Deep neural net with attention for multi-channel multi-touch attribution. https://doi.org/10.48550/arXiv.1809.02230
15. Lu Z, Kannan PK (2024) Measuring the synergy across customer touchpoints using transformers. https://papers.ssrn.com/abstract=4684617
16. Yan Z, Kumar FAVKS, Dong C, Pani A, Li N (2022) Utilizing a touchpoint attribution attention neural network to identify significant touchpoints and measure touchpoint contribution in multichannel, multi-touch digital content campaigns. https://patents.google.com/patent/US11287894B2/en
17. Berman R (2018) Beyond the last touch: attribution in online advertising. Mark Sci 37:771–792. https://doi.org/10.1287/mksc.2018.1104
18. Thornton T, Thorne S, Calderon A (2021) Modelling user behaviour in market attribution: finding novel data features using machine learning. UK Academy for information systems conference proceedings 2021
19. Danaher PJ, Van Heerde HJ (2018) Delusion in attribution: caveats in using attribution for multimedia budget allocation. J Mark Res 55:667–685. https://doi.org/10.1177/0022243718802845
20. Fernández L (2020) Applications of multi-touch attribution modelling. https://repositorio.utdt.edu/handle/20.500.13098/11573
21. Gaur J, Bharti K (2020) Attribution modelling in marketing: literature review and research agenda. Acad Mark Stud J 24:1–21
22. Kannan PK, Reinartz W, Verhoef PC (2016) The path to purchase and attribution modeling: introduction to special section. Int J Res Mark 33:449–456. https://doi.org/10.1016/j.ijresmar.2016.07.001
23. Buhalis D, Volchek K (2021) Bridging marketing theory and big data analytics: the taxonomy of marketing attribution. Int J Inf Manag 56:102253. https://doi.org/10.1016/j.ijinfomgt.2020.102253
24. Dalvie S, Siddiqui S, Ratra N (2023) Identifying the base sales contribution for MTA model. Int J Sci Res Eng Manag 7. https://doi.org/10.55041/IJSREM24431
25. Ji W, Wang X (2017) Additional multi-touch attribution for online advertising. Proc AAAI Conf Artif Intell 31. https://doi.org/10.1609/aaai.v31i1.10737
26. Paulsen T, Andrus I, Purser N (2022) Performing attribution modeling for arbitrary analytics parameters. https://patents.google.com/patent/US11347809B2/en
27. Sapp S, Schnabl SF, Vaver J, Fan R (2020) Attribution modeling using withheld or near impressions. https://patents.google.com/patent/US10607254B1/en
28. Ben Mrad A, Hnich B (2024) Intelligent attribution modeling for enhanced digital marketing performance. Intell Syst Appl 21:200337. https://doi.org/10.1016/j.iswa.2024.200337
29. Diemert E, Meynet J, Galland P, Lefortier D (2017) Attribution modeling increases efficiency of bidding in display advertising. In: Proceedings of the ADKDD'17. Association for Computing Machinery, New York, NY, USA, pp 1–6
30. Ren K, Fang Y, Zhang W, Liu S, Li J, Zhang Y, Yu Y, Wang J (2018) Learning multi-touch conversion attribution with dual-attention mechanisms for online advertising. In: Proceedings of the 27th ACM international conference on information and knowledge management, pp 1433–1442
31. Kumar S, Gupta G, Prasad R, Chatterjee A, Vig L, Shroff G (2020) CAMTA: causal attention model for multi-touch attribution. In: Di Fatta G, Sheng V, CuzzocreaA, Zaniolo C, Wu X (eds) IEEE international conference on data mining workshops, ICDMW. IEEE Computer Society, pp 79–86
32. Bompaire M, D'esir A, Heymann B (2020) Bidding through the lens of attribution: pick the right labels! arXiv
33. Malik M, Gupta G, Vig L, Shroff G (2021) BCQ4DCA: budget constrained deep Q-network for dynamic campaign allocation in computational advertising. In: 2021 International joint conference on neural networks (IJCNN), pp 1–8
34. Bhatta I (2022) Optimizing marketing channel attribution for B2B and B2C with machine learning based lead scoring model, Doctoral dissertation, Capitol Technology University. https://www.proquest.com/openview/65d82057fee3c27039e2429bc1d72b00/1?pqorig site=gscholarcbl=18750diss=y
35. Yao D, Gong C, Zhang L, Chen S, Bi J (2022) CausalMTA: eliminating the user confounding bias for causal multi-touch attribution. http://arxiv.org/abs/2201.00689
36. Betlei A, Vladimirova M, Sebbar M, Urien N, Rahier T, Heymann B (2023) Maximizing the success probability of policy allocations in online systems. http://arxiv.org/abs/2312.16267
37. Tang J (2024) DCRMTA: unbiased causal representation for multi-touch attribution, http://arxiv.org/abs/2401.08875
38. Gigli M, Stella F (2024) Multi-armed bandits for performance marketing. Int J Data Sci Anal. https://doi.org/10.1007/s41060-023-00493-7
39. Li H (Alice), Kannan PK (2014) Attributing conversions in a multichannel online marketing environment: an empirical model and a field experiment. J Mark Res 51:40–56. https://doi.org/10.1509/jmr.13.0050
40. Li H (Alice), Kannan PK, Viswanathan S, Pani A (2016) Attribution strategies and return on keyword investment in paid search advertising. Mark Sci 35:831–848. https://doi.org/10.1287/mksc.2016.0987
41. Cui TH, Ghose A, Halaburda H, Iyengar R, Pauwels K, Sriram S, Tucker C, Venkataraman S (2021) Informational challenges in omnichannel marketing: remedies and future research. J Mark. https://doi.org/10.1177/0022242920968810
42. Anderl E, Becker I, Von Wangenheim F, Schumann JH (2016) Mapping the customer journey: lessons learned from graph-based online attribution modeling. Int J Res Mark 33:457–474. https://doi.org/10.1016/j.ijresmar.2016.03.001

43. Romero Leguina J, Cuevas Rumín Á, Cuevas Rumín R (2020) Digital marketing attribution: understanding the user path. Elec 9:1822. https://doi.org/10.3390/electronics9111822
44. Nottorf F (2014) Modeling the clickstream across multiple online advertising channels using a binary logit with Bayesian mixture of normals. Electron Commer Res Appl 13:45–55. https://doi.org/10.1016/j.elerap.2013.07.004
45. Xu Y, O'Connor B, Teraoka B (2016) Credit attribution based on measured contributions of marketing activities to deals. https://patents.google.com/patent/US20160063427A1/en
46. Cano Berlanga S, Vilella C Attribution models and the cooperative game theory. Recer. Dipòs. Recer. Catalunya (2017)
47. Sinha R, Saini S, Anadhavelu N (2014) Estimating the incremental effects of interactions for marketing attribution. In: 2014 International conference on behavioral, economic, and socio-cultural computing (BESC2014), pp 1–6
48. Sinha R, Mehta S, Bohra T, Krishnan A (2015) Improving marketing interactions by mining sequences. In: Wang J, Cellary W, Wang D, Wang H, Chen S-C, Li T, Zhang Y (eds) Web information systems engineering—WISE 2015. Springer International Publishing, Cham, pp 277–292
49. Knox A (2020) Zero knowledge blockchain attribution. https://patents.google.com/patent/US20200334708A1/en

Water Data Management in Morocco—Enhancing Decision-Making Through Data

Hicham Jamil, Elhassan Jamal, Youssef Rissouni, Bouabid El Mansouri, and Aniss Moumen

Abstract

With new challenges related to climate change, such as irregular precipitation, drought, and water scarcity, water management has become a key driver of important strategies in all countries worldwide. Indeed, the effective management of water resources relies on a well-structured process based on data collection, continuous coordination between various stakeholders, and an improved management system (Hicham et al. in Processing and decisions relating to water resources data: the case of Morocco. SHS web conference, 2021 [1]). To handle the increasing volume of collected and generated data, big data (BDA) has become an essential tool, offering enhanced capabilities for data processing, real-time monitoring, and predictive assessments. However, adopting these technologies alone is not sufficient; the coordination of data flow and its relevance is equally important. A forward-looking study on user needs, data flow, and requirements is just as crucial as the implementation of an information system for water resources management based on big data.

Keywords

Water resources · Big data · Big data analytics

H. Jamil (✉) · E. Jamal · Y. Rissouni · A. Moumen
Laboratory of Engineering Sciences, National School of Applied Sciences, Ibn Tofaïl University, Kenitra, Morocco
e-mail: hicham.jamil@gmail.com

B. E. Mansouri
Laboratory of Natural Resources and Sustainable Development, Ibn Tofaïl University, Kenitra, Morocco

1 Introduction

Water, commonly termed "blue gold," constitutes an essential resource for human existence and numerous economic sectors, particularly agriculture, which consumes nearly 70% of the global freshwater resources annually. This sector is also a major contributor to water pollution through the extensive use of fertilizers and pesticides. Given the increasing challenges posed by water scarcity, optimizing the management of water resources has become imperative for both developed and developing countries.

Recent developments in big data analytics, artificial intelligence, and machine learning present robust technical solutions capable of addressing these water management issues. These technologies offer significant improvements in reducing water loss, enhancing resource utilization efficiency, and proactively mitigating pollution risks, with proven practical effectiveness.

This article explores innovative and promising practices addressing water waste and pollution. We analyze the perspectives of various stakeholders and the challenges they face in effectively managing this vital resource. To improve water data processing, we propose a significant data analytics architecture that ensures:

- Enhanced data quality and preservation through an operational structure tailored to specific needs.
- Ease of maintenance due to its centralized nature.
- It reduced software maintenance and data penetration.
- Authorization for sharing data among authorized users in an organization.
- Database protection to maintain data consistency.
- User permissions to insert, edit, and delete data in the database.

A. Moumen et al. (eds.), *Artificial Intelligence and Data Analytics for Innovative Applications in Engineering, Sustainability and Technology*, Sustainable Artificial Intelligence-Powered Applications, https://doi.org/10.1007/978-3-031-90318-2_22

We also present a detailed explanation of the interviews conducted with various actors involved in water resource management. Our interview guide focuses on identifying the processes and actions needed to achieve effective analytical results, with the analysis of interview results primarily based on the NVIVO software.

Finally, we offer a discussion section to analyze the results and provide recommendations based on our findings.

2 Context

The efficient management of water resources is heavily dependent on the systematic collection, processing, and aggregation of accurate data. In Morocco, Hydraulic Basin Agencies (HBA) play a central role in this process, as they are responsible for measuring, recording, and analyzing hydrological data. This information is methodically gathered and structured within a centralized system, ensuring that all relevant stakeholders have access to standardized and reliable datasets. Establishing a structured and well-coordinated approach to data management is essential, as it provides the foundation for evidence-based decision-making and long-term planning in the water sector [2].

Moroccan water law 36-15 establishes a legal obligation for Basin Agencies to systematically transmit collected data to the Central Department of the General Directorate of Water. This regulatory framework is designed to facilitate the seamless and timely exchange of information, empowering policymakers and water resource managers to develop well-coordinated strategies for sustainable water distribution. By centralizing data management, it enhances governance through increased transparency and accountability while also strengthening collaboration between agencies at both regional and national levels. This integrated approach ensures that decision-makers have reliable, up-to-date information, allowing for more effective and adaptive water resource management [3].

Given the strategic importance of hydrological data, the information systems within the Basin Agencies must not operate in isolation. Instead, they should function as integral components of a broader, unified national water information system, managed by the General Directorate of Water. "This integration is crucial for ensuring that data collected at local and regional levels is harmonized, accessible, and compatible with national-level policy frameworks" [4, 5].

Therefore, it is necessary to conduct a detailed examination of the processes related to data collection, consolidation, and processing [6]. Since raw data only becomes meaningful once it has been systematically transformed and analyzed, establishing a standardized methodology for data integration is essential. By structuring data within an optimized processing framework, authorities can enhance the accuracy, reliability, and usability of hydrological information, ensuring its effectiveness in decision-making and long-term resource planning [7, 8].

3 Methodology

Before starting the interviews, we presented the research framework and objectives to participants to ensure a clear understanding of the study. This step helped structure discussions and encouraged relevant, detailed responses. Establishing this context also facilitated a more precise data collection process.

To evaluate data collection processes and decision-making frameworks, we employed a qualitative research approach, directly engaging with key stakeholders involved in water resource management. This approach provided valuable insights into real-world operational challenges and enabled a thorough evaluation of current data management practices.

Each interview, lasting between 40 min and an hour depending on the topic's complexity, was recorded, transcribed, and carefully analyzed using NVIVO, a qualitative data analysis tool. This software helped categorize responses, recognize recurring themes, and extract meaningful insights.

By identifying keyword patterns and thematic trends, NVIVO supported a structured analysis, ensuring a clear and systematic interpretation of the findings. The following figure outlines our research methodology, detailing the steps from interview preparation to final data analysis (Fig. 1).

3.1 Interviewees

The interviewees were selected based on their responsibilities in addressing our issue, and they are affiliated with the following three entities:

- Water Branch—The principal body responsible for formulating and implementing national water management policies.
- Hydraulic Basin Agencies (HBA)—Authorities tasked with the collection, monitoring, and regulation of hydrological data.
- Regional Environmental Directorates—Agencies involved in analyzing, interpreting, and integrating water data into environmental policies and strategic planning.

The interviews were conducted with individuals in charge of water management and key decision-makers within their respective offices.

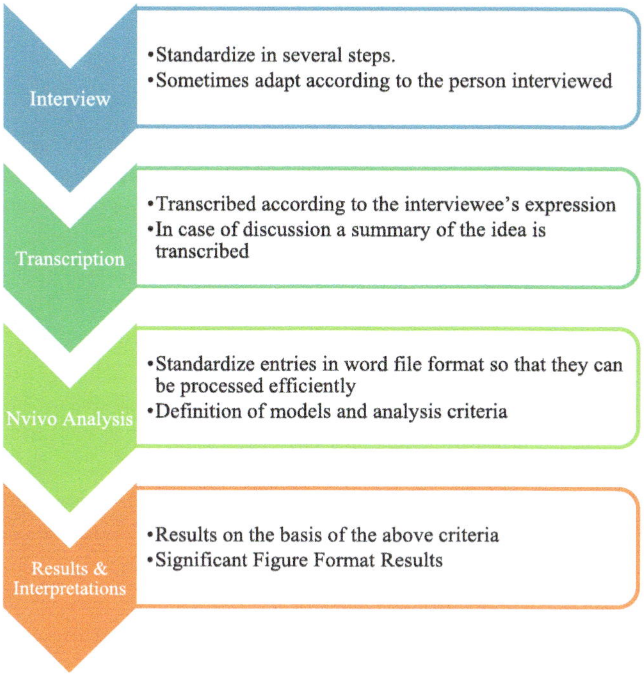

Fig. 1 Interview process

The average age of the respondents is 39, with different profiles and more than 15 years of experience in the field: water resources engineers, IT, and managers: interviews were carried out directly via a maintenance guide.

The actors involved in the interview guide are key players who are managers acting directly on the process in Table 1.

3.2 Interview Guide

The interview guide serves as a crucial tool in the research methodology, ensuring a structured and organized approach to data collection. It not only provides logistical details but also establishes a clear framework for both the interviewer and interviewee, facilitating meaningful discussions. The introduction clearly defines the study's objectives and explains the role of each participant to ensure a shared understanding of the research process.

Rather than just listing names and titles, the guide outlines each interviewee's specific expertise and contributions. It follows a well-structured logic, starting with a general introductory stage and progressing toward a more in-depth exploration, thereby facilitating a comprehensive analysis of all elements.

Figure 2 provides a visual summary of this process.

Table 1 Managers interviewed

Code	Gender	Age	Administration	Position	Role
Interview1	Man	57	Directorate General of Water	Head of the Water Resources Division	Senior Water Resources Data Manager
Interview2	Man	38	Directorate General of Water	Head of the Hydrology Service	Hydrological data operator and manager at DGW
Interview3	Man	39	HBA Errachidia	Head of the Water Resources Management Division	Personnel responsible for data from water resource measurement stations
Interview4	Man	39	Directorate General of Water	Head of the Organization and Information Systems Division	IT provider and key player in consolidating water resources data
Interview5	Woman	36	Regional Directorate of the Environment of Beni Mellal	Head of the environmental management service	Environmental data manager

Fig. 2 Interview steps

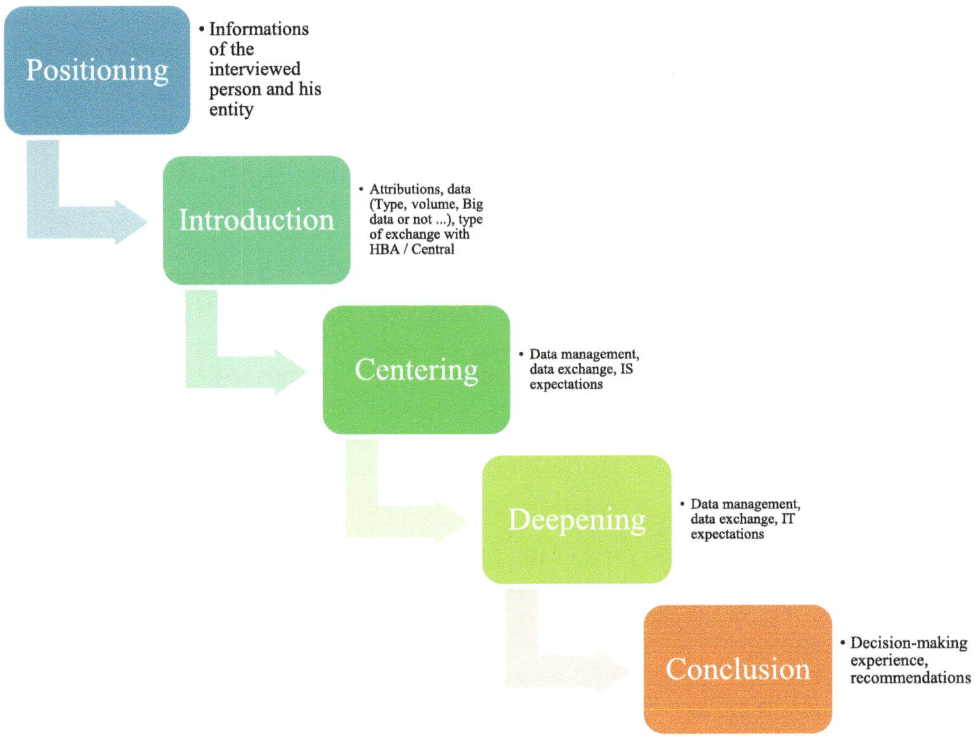

3.3 NVIVO Analysis

For the analysis of interview results, we used NVIVO, a qualitative research software that allows for the interpretation of interviews by classifying responses, identifying recurring themes, and extracting significant points discussed during the interviews.

Additionally, NVIVO's ability to process audiovisual material provided a more comprehensive analysis of non-textual data. It also enables keyword frequency analysis and thematic mapping, helping to uncover patterns within large datasets.

In this study, we leveraged NVIVO to structure the interview responses, segment discussions into relevant thematic categories, and conduct comparative analyses. The processed data was then visualized and exported in multiple formats, facilitating a systematic examination of key findings (Fig. 3).

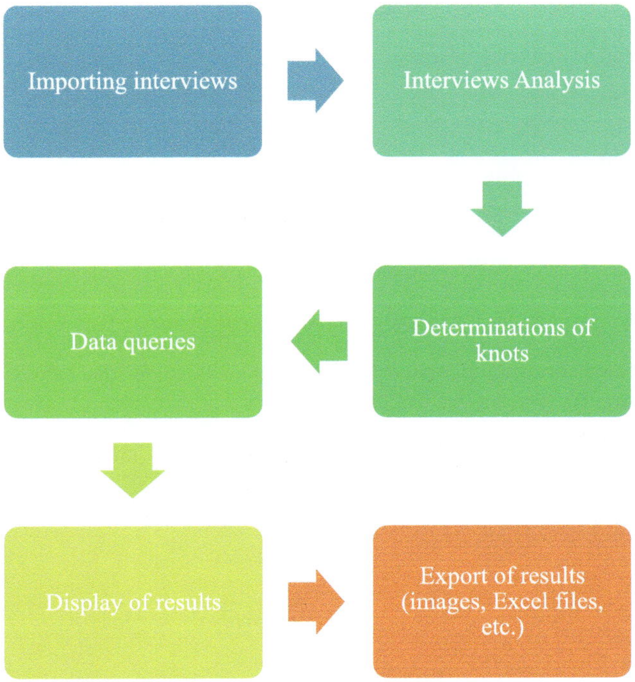

Fig. 3 NVIVO process

4 Results and Discussion

Upon importing the interviews into NVIVO, we utilized the software to generate a frequency table. This table provides insights into the percentage and frequency of each word used in the responses. The findings from this analysis will be presented in detail in Table 2, illustrating the distribution and relevance of key terms within the dataset.

The words most frequently used in our case are data, systems, processes, Excel, and partners, representing 50% of the stakeholders' problems for implementing a sharing system and processes capable of managing and analyzing the data received from partners.

Table 2 Overview of the frequency table keywords

Mot	Longueur	Nombre	Pourcentage pondéré (%)
données	7	49	3.48
système	7	20	1.42
niveau	6	18	1.28
processus	9	15	1.07
badr21	7	12	0.85
partie	6	11	0.78
pas	3	11	0.78
sous	4	11	0.78
cas	3	10	0.71
mesure	6	10	0.71
excel	5	9	0.64
qualité	7	9	0.64
abh	3	8	0.57
partenaires	11	8	0.57
service	7	8	0.57
sont	4	8	0.57
donnée	6	7	0.5
internes	8	7	0.5
phase	5	7	0.5
base	4	6	0.43
développement	13	6	0.43
externes	8	6	0.43
beaucoup	8	5	0.36
dépend	6	5	0.36
existe	7	5	0.36
gestion	7	5	0.36

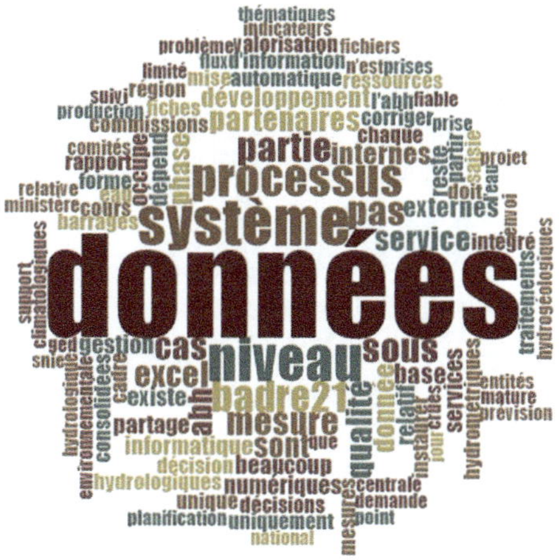

Fig. 4 General word cloud

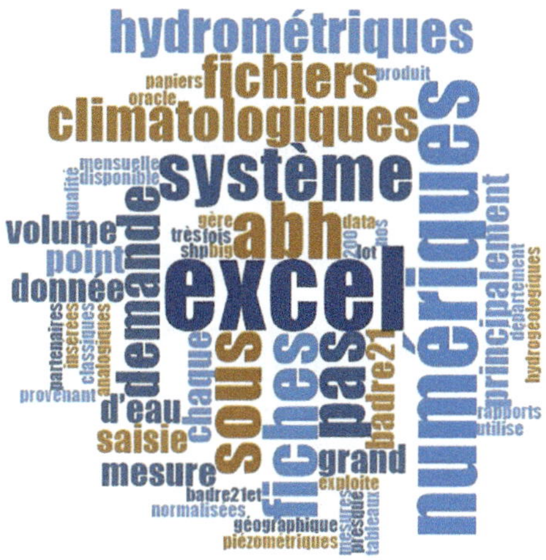

Fig. 5 Data cloud

In Fig. 4, we can say that according to the interviews and based on the previous analysis, the data is the central element mentioned in the discussions; we also find the process and the system: the two are linked because the system is based on process informatization.

4.1 Datas

According to the interviews, the managers all cited the presence of two data formats, digital and analog: digital data (more than 90%) and are generally Excel files relating to water points or Oracle data in the Badre21 system (Water resources) (Fig. 5).

The data is related to hydrology, hydrogeology, and climatology; however, data digitization and quality are problematic (Fig. 6).

We note that the NVIVO analysis results confirm the problem of formats and the variety of data that requires a database with advanced technology.

4.2 Data Sharing

According to Fig. 7 (Still under NVIVO), the sharing of information must essentially concern water but which can be impacted, according to the interviewees, by the lack of knowledge.

4.3 Decision-Making

The prominence of the term "commission" in the decision-making process underscores the typical hierarchical structure prevalent in the public sector, where decisions often stem from leadership or committee consensus. Furthermore, including terms such as "indicators" and "partners" emphasizes the importance of collaboration and the involvement of various stakeholders in developing a national water system.

Fig. 6 Data cloud (details)

Fig. 7 Data sharing

Fig. 8 Data making

This highlights the potential benefits of a comprehensive approach that effectively engages all relevant parties in addressing water-related challenges (Fig. 8).

4.4 Important Comments

The NVIVO software was utilized for interview data analysis, enabling researchers to perform structured and in-depth qualitative assessments. This tool facilitated the organization, categorization, and coding of responses, ensuring a systematic exploration of key themes and trends.

By adopting this methodological approach, the research maintained rigor, consistency, and reliability throughout the analysis process. NVIVO's capabilities allowed for objective interpretation of interviewees' perspectives, reducing bias while enhancing the clarity and depth of the findings. This structured framework ultimately strengthened data-driven decision-making and provided a comprehensive understanding of stakeholder insights (Table 3).

Table 3 Topics of analysis

Factor	Testimonial
A: Data flow	Partners either input data directly into the system or it is logged into a dedicated database overseen by the observatory unit. The central administration ensures the reliability of the information by conducting validation checks on both the collected data and the corresponding indicators
B: Data format and volume	Data is presented in both tabular and geographic formats, covering climatological, hydrometric, and hydrogeological data
C: Quality	Initially, information production is limited, with subsequent mass production allowing for quality assessment. Overall, the produced data quality tends to be satisfactory
D: Current system and model used	Resources are currently constrained, indicating a need for enhanced technical expertise among business service personnel
E: Data usage	Piezometric data is monitored via web platforms, alongside Excel-based flow data
F: Data sharing	The development of the SNIE (National Water Information System) aimed to facilitate water data sharing across different entities within the water sector
G: Decision-making	Implementation outcomes vary depending on specific circumstances

4.5 Discussion

Almost all stakeholders expressed concerns regarding the management of water-related data from collection to storage in the database. They highlighted the challenges and difficulties associated with this process, particularly in terms of data quality, accuracy, and standardization.

The interviews revealed that water data necessarily consists of multiple types, including climatic, hydrological, and hydrogeological data, among others. Although a large volume of data is collected, it is only partially utilized due to the lack of advanced processing and analysis tools. This underscores the importance of big data and its crucial analytical capabilities.

The interviewed officials emphasized the need to establish a centralized system for water data management to support decision-making and enhance collaboration.

Table 4 provides a concise summary of the key points discussed and their analysis, along with proposed improvements based on the interviews.

Table 4 Results

Issue	Current situation	Proposed solution (BDA)
Data entry	Manual, non-standard	Achievable through continuously connected sensors in real-time
Standardizing data	Non-standardized data	Standardization can be achieved using the data standardization layer in big data analytics systems
Data validation	Absent	Infinite data storage capability
Data storage	Raw storage with difficulty for large volumes	Unlimited data storage
Real-time data analysis and dynamic visualization	Despite the availability of multiple database systems and IT infrastructures, research efforts remain minimal	Big data analytics facilitates analysis across multiple dimensions
Prediction	Difficult/absent	Feasible through predictive layers in big data analytics (BDA)

5 Conclusion

Interviews with various stakeholders in the water sector highlight critical issues related to data collection, processing, and dissemination. It is essential to adopt a more organized approach at the national level to optimize coordination and decision-making, primarily based on big data. The establishment of a National Water Information System (SNIE) could significantly enhance water management.

Water sector officials also emphasized the importance of coordination among different stakeholders. Big data analysis presents a promising solution by strengthening governance and enabling data-driven decision-making, based not only on collected data but also on effective data exchange.

References

1. Hicham J, Elhassan J, El Mansouri B, Aniss M, Jamal C (2021) Processing and decisions relating to water resources data: the case of Morocco, SHS web conference. In: 3rd International conference on quantitative and qualitative methods for social sciences (QQR'21), vol 119. https://doi.org/10.1051/shsconf/202111903007
2. Elhassan J, Aniss M, Jamal C (2020) Big data analytic architecture for water resources management: a systematic review GEOIT4W-2020. In: Proceedings of the 4th edition of international conference on Geo-IT and water resources 2020, Geo-IT and water resources 2020, March 2020, Article no. 19, pp 1–5. https://doi.org/10.1145/3399205.3399225
3. Alter S (1998) Information systems: Addison-Wesley Longman Publishing Co., Inc.
4. Alter S (2003) Pervasive real-time IT as a disruptive technology for the IS field. In: Proceedings of the 36th annual Hawaii international conference on system sciences, 2003, p 10-pp
5. Loi 36-15, Chapitre X (October 2016)
6. Moumen A, El Mansouri B, Oulidi HJ, Khazaz L (2015) Système d'Information sur l'Eau au Maroc: Etat d'art, Problématique, Approche et Prototype. Conference paper, Nov 2015
7. Stair R, Reynolds G (2013) Fundamentals of information systems: Cengage learning
8. Jamil H, Jamal E, Rissouni Y, El Mansouri B, Moumen A, Chao J (2024) Data process and water resources management in Morocco: issues and challenges. In: E3S web of conferences, Feb 2024. https://doi.org/10.1051/e3sconf/202448904018

GPSR Compliance

The European Union's (EU) General Product Safety Regulation (GPSR) is a set of rules that requires consumer products to be safe and our obligations to ensure this.

If you have any concerns about our products, you can contact us on ProductSafety@springernature.com

In case Publisher is established outside the EU, the EU authorized representative is:

Springer Nature Customer Service Center GmbH
Europaplatz 3
69115 Heidelberg, Germany

Batch number: 10755916

Printed by Printforce, the Netherlands